Springer Series in Optical Sciences

Volume 11

Edited by David L. MacAdam

Springer Series in Optical Sciences

Edited by David L. MacAdam

Editorial Board: J. M. Enoch   D. L. MacAdam   A. L. Schawlow   T. Tamir

P. S. Theocaris   E. E. Gdoutos

Matrix Theory of Photoelasticity

With 93 Figures

Springer-Verlag Berlin Heidelberg GmbH 1979

Professor PERICLES S. THEOCARIS, D.Sc.
Professor EMMANUEL E. GDOUTOS, Ph.D.

Athens National Technical University, Athens, Greece

Library of Congress Cataloging in Publication Data. Theocaris, Pericles S 1921—. Matrix theory of photoelasticity. (Springer series in optical sciences; v. 11) Bibliography: p. Includes index. 1. Photoelasticity. I. Gdoutos, E. E., 1948—. joint author. II. Title. TA418.12.T48 620.1'1295 78-14275

ISBN 978-3-662-15807-4   ISBN 978-3-540-35789-6 (eBook)
DOI 10.1007/978-3-540-35789-6
© by Springer-Verlag Berlin Heidelberg 1979
Originally published by Springer-Verlag Berlin Heidelberg New York in 1979.
Softcover reprint of the hardcover 1st edition 1979

2153/3130-543210

Preface

Photoelasticity as an experimental method for analyzing stress fields in mechanics was developed in the early thirties by the pioneering works of Mesnager in France and Coker and Filon in England. Almost concurrently, Föppl, Mesmer, and Oppel in Germany contributed significantly to what turned out to be an amazing development. Indeed, in the fifties and sixties a tremendous number of scientific papers and monographs appeared, all over the world, dealing with various aspects of the method and its applications in experimental stress analysis. All of these contributions were based on the so-called Neumann-Maxwell stress-optical law; they were developed by means of the classical methods of vector analysis and analytic geometry, using the conventional light-vector concept. This way of treating problems of mechanics by photoelasticity indicated many shortcomings and drawbacks of this classical method, especially when three-dimensional problems of elasticity had to be treated and when complicated load and geometry situations existed.

Meanwhile, the idea of using the Poincaré sphere for representing any polarization profile in photoelastic applications was introduced by Robert in France and Aben in the USSR, in order to deal with problems of polarization of light passing through a series of optical elements (retarders and/or rotators). Although the Poincaré-sphere presentation of any polarization profile constitutes a powerful and elegant method, it exhibits the difficulty of requiring manipulations in three-dimensional space, on the surface of the unit sphere. However, other graphical methods have been developed to bypass this difficulty. By a parallel projection of the points of the sphere on its equatorial plane, Kuske developed the j-circle method; Aben and others used the stereographic projection connected with the Wulff net, which maps polarization states so as to preserve angles. Another type of mapping, based on the Lambert-Schmidt projection, which preserves areas is currently under development by Theocaris.

Concurrently with these graphical methods, some analytical methods were developed, based on the Stokes and Jones vectors that are extensively used in crystallography. These methods are the powerful Mueller and Jones calculi. Finally, the analytical method based on quaternia, which uses the Pauli spin matrices should also be mentioned. All of these modern graphical and analytical methods have been developed by use of various techniques based either on conventional light-vector concept, using vector calculus, or on matrix methods. However, no unified method exists to date that incorporates all of these modern ideas.

The purpose of this book is to present a novel and unified interpretation of all of the problems of two- and three-dimensional photoelasticity by use of the modern methods of description of polarized light, based on the concepts of the Poincaré sphere, as well as on Mueller and Jones calculi. Although these powerful methods were conceived many years ago, only in the last two decades they have been applied in the solution of current problems of photoelasticity. The simplicity and elegance of these methods enable one to solve complicated problems in straightforward and comprehensive manners and to devise simple solutions of complex problems that, because of their nature, could not be solved by use of classical methods. Furthermore, although the Poincaré-sphere representation can help enormously to visualize and formulate problems and theorems in a qualitative manner, geometric manipulations that involve spherical trigonometry and the simple and efficient procedures of matrix analysis, provide accurate results in the sequence.

This book, as conceived here, is based on matrix theory, because this powerful calculus was judged to be the most suitable for the presentation, in a simple and unified manner, of all of the theorems related to these methods. Thus, apart from the fact that this monograph is the first to exhibit all of these new methods, based on the Poincaré sphere, and on Jones and Mueller calculi, it reformulates already-known theorems of photoelasticity and derives others on the basis of the present approach.

This book is intended to be an instrument of learning, rather than a review of contributions in these modern techniques related to photoelasticity. Therefore, from the great number of papers and monographs published on photoelasticity, we have selected only those that seemed to suit the purpose of this book. Likewise, references and the bibliography at the end of the book are limited to papers that are closely related to the subjects treated in this monograph.

Particular care was taken throughout the book to give clear, straightforward, and simple presentations of the various topics. The authors hope that the monograph will be used as a basis on which the future researcher and student can develop his own new and fruitful ideas and promote research in formulation of problems of photoelasticity by matrices, which, despite its significance, has been given little attention as yet.

Finally, the authors are particularly indebted to Mr. Constantin Thireos, Assistant at the Laboratory, who prepared the groundwork of the bibliography contained at the end of the book and who also contributed in the preparation of the manuscript.

Athens, Greece   P. S. THEOCARIS · E. E. GDOUTOS
September, 1978

Contents

1. Introduction

The science of photoelasticity, under the generalized meaning of the term, including all methods of experimental stress analysis that are based on Neumann-Maxwell stress-optical law, has been found to be the most powerful and effective method for full-field analysis of two- or three-dimensional stress fields. Although the temporary or artificial birefringence of noncrystalline materials or the so-called photoelastic phenomenon was discovered in 1816 by Sir David Brewster, because of the lack of suitable model materials, the development of photoelasticity started at the beginning of the twentieth century, with the pioneering works of Coker and Filon. Classical two- or three-dimensional photoelasticity, based on the relative stress-optical retardation along the principal stress-directions, was then increasingly rapidly developed and reached its full maturity at the middle of our century.

Owing to the difficulties of separation the principal stresses by the methods of classical photoelasticity, other phenomena of optical physics, such as interferometry, holography, and light scattering, were used in photoelasticity in attempts to evaluate completely the principal stresses by optical methods. Up to now a large number of new techniques that aim to serve this purpose have been developed.

However, although photoelasticity has been applied successfully to the solution of two-dimensional problems, many difficulties are encountered in the investigation of three-dimensional states of stress. In these cases, the rotation of the principal stresses along the light path complicates the interpretation of the experimental data. The frozen-in method and the light-scattering technique are so far the most popular methods of three-dimensional photoelasticity.

Because in the science of photoelasticity only phenomena involving propagation of continuous light beams are encountered, leaving apart all other phenomena in which interaction of light and matter takes place, the electromagnetic or the wave theories of light have been adapted to explain the photoelastic effect, since the beginning of the advanced approach of photoelasticity. Both of these theories are based on the concept of the light vector, which in the electromagnetic theory is described as the electric- or the magnetic-field vector, whereas in the wave theory it is defined as the position vector of the ether particle at each instant, from some origin. All of the basic formulas upon which photoelasticity was founded, involving the optics of polariscopes, compensation methods, and interpretation of experimental data in the various holo-interferometric methods were derived by use of the simple concept of the light vector.

However, although the introduction of the light vector in physical optics proved to be adequate and was successfully used for the solution of many optical problems, there were other problems for which the resulting calculations become cumbersome, tedious, and error prone. Such problems are encountered in the propagation of light through a series of optical elements, such as polarizers and retarders. Even the simplest problems of two-dimensional photo-elasticity, such as those that involve optics either in polariscopes, or in other optical arrangements, are difficult to treat by the conventional light-vector concept. Such difficulties are greatly aggravated in three-dimensional problems. In these problems, the variation of principal stresses from point to point results in the rotation of principal directions along the light path. Thus, a three-dimensional photoelastic model behaves like a pile of birefringent plates with different principal directions and birefringences.

Optical phenomena that involve propagation of light through a series of optical elements can be more adequately treated by the powerful and elegant methods that are based on the modern description of polarized light by use of the concepts of the Poincaré sphere, the Stokes vector, and the Jones vector. Although the Stokes vector was introduced in 1852 and the Poincaré sphere in 1898, these methods have received little attention in the literature. Related to the Stokes vector, a new calculus was developed by Mueller in 1948, the so-called Mueller calculus, for treating phenomena of light propagation. The Jones vector and the corresponding Jones calculus were conceived by Jones in 1941.

The power of these methods lies in the fact that the result of an optical element, such as a polarizer or a retarder, on a beam of polarized light, can be predicted in a simple and unified manner. Thus, according to the Poincaré-sphere representation, each polarization state is represented by a point on a unit-radius sphere, and the result of an optical element on a polarized light can be found by a suitable transformation on the surface of the sphere. In the Mueller calculus, each light beam is represented by a four-element column vector (the Stokes vector) and each optical element by a four-by-four matrix; the Stokes vector at the output of an optical device is found by multiplying the matrix of the input Stokes vector by the matrix of the element. A similar procedure is adopted in the Jones method, in which each polarized light beam is represented by a two-element complex column matrix and each optical device by a two-by-two complex matrix.

Because the representation of each polarization state by a point on the Poincaré sphere or by a Stokes or Jones vector is unique, and the transformation on the Poincaré sphere and the Mueller and Jones matrices, depicting the influence of an optical element on a light beam, are well defined, it is easy to find the output polarization form from the corresponding input form. For this task, the corresponding vectors and matrices of all polarization states and optical elements have been tabulated.

The book is divided into thirteen chapters. In the second, introductory chapter, the basic principles of the electromagnetic theory of light, which are concerned with propagation of light in isotropic and anisotropic media, are presented. Chapter 3 provides the various methods of description of polarized light. Beginning with the classical vectorial presentation of polarized light, it introduces the contemporary and advanced methods, based on the Poincaré sphere, the Stokes vector, and the Jones vector. In Chapter 4 the methods introduced previously for the description of polarized light are used to predict the form of the polarized light that emerges from an optical element, from knowledge of the incident-light form and the characteristics of the optical element. Suitable manipulations on the Poincaré sphere, as well as the Mueller and Jones calculi, based on the Stokes and Jones vectors, respectively, for solution of each problem are presented. In this chapter, the power of these new methods in the solution of any problem that involves polarizers and retarders in a simple and unified manner is shown. This chapter ends with a useful "*equivalence theorem*", which characterizes the combination of any number of optical retarders. Chapter 5 gives the various methods of measuring the characteristics of an elliptically polarized light, with particular emphasis on photoelectric methods.

After the introduction of the general characteristics of contemporary methods of description of polarized light in these four chapters, these methods are used to study the basic problems encountered in photoelastic investigations of two- and three-dimensional states of stress.

Chapter 6 deals with the basic photoelastic phenomenon, upon which the science of photoelasticity is based. The generalized Maxwell-Neumann law, which connects the variations of the refractive indices of a three-dimensional model, when it is stressed, with the values of the principal stresses, is deduced from the assumption of the coaxiality of Fresnel's optical and Cauchy's stress ellipsoids.

In Chapter 7, all of the basic formulas of two-dimensional photoelasticity, including the optics of plane and circular polariscopes and the Senarmont and Tardy compensation methods, are proved, by use of the graphical method of the Poincaré sphere and the matrix methods of the Mueller and Jones calculi.

Chapter 8 studies the general problem of the photoelastic determination of the state of stress in a three-dimensional body by using the modern methods of description of polarized light. The problem is first formulated by use of the Jones calculus, and the corresponding differential equations are deduced. It is concluded that, from the integrated optical effect, which results from illuminating the three-dimensional body by a light beam of a given wavelength, only three characteristic quantities, related to the state of stress in the body, can be measured. Thus, the solution of the problem necessitates additional data. For this purpose, the idea of using various wavelengths is introduced. However, this concept cannot be applied to the solution of plate problems where the integrated optical effect is zero. For such problems, and for the general three-dimensional problem, the Faraday and Kerr effects, which result in the rotation of the plane of polarization and insertion of additional birefringence, respectively, are introduced to provide additional optical data. The Mueller-calculus and the Poincaré-sphere methods are also used for formulation and solution of the problem. The particular case when the ratio of the birefringence per unit length to the corresponding rotation of the principal stresses is constant is considered; the well-known Drucker-Mindlin formulas are immediately deduced.

In Chapter 9, the light-scattering phenomenon is introduced for the solution of the three-dimensional problem. After a brief discussion of the application of this phenomenon to three-dimensional photoelasticity by the conventional methods and the difficulties that result from the rotation of the principal axes, the problem is formulated and solved by the introduction of the matrix methods of polarized light.

Chapter 10 deals with the formulation of the interferometric methods of stress analysis by matrices. The general expressions that yield the light intensity in a Mach-Zehnder interferometer when the light rays pass through the specimen, or the formation of Fizeau fringes by the interference of the light rays reflected from the front and rear faces of the specimen, are derived by the elegant and concise method of the Jones calculus.

In Chapter 11, the basic principles of the method of holography and its applications to interferometry, the so-called holographic interferometry, are presented. The holographic phenomenon is explained, both mathematically, by deriving the basic equations that govern the recording and reconstructing processes of the method, and by use of only physical concepts that are based on the relationship between interference and diffraction, as well as on the mechanical interference (moiré) phenomenon. The same procedure is also used for the study of the method of holographic interferometry and the inter-pretation of the thus-obtained interferograms. The methods of real-time and double-exposure holographic interferometry and their applications to deter-mination of small displacements that result from application of a system of loads to deformable bodies are particularly studied. This chapter ends with a study of the application of holographic interferometry to photoelasticity, the so-called holographic photoelasticity, which leads to the direct deter-mination of the isochromatic and isopachic fringe patterns. The advantages of the method of holographic photoelasticity compared with the method of interferometric photoelasticity, studied in the previous chapter, are also discussed.

In Chapter 12, the method of birefringent coatings is developed, for applica-tion to the photoelastic determination of surface strains in two- and three-dimensional opaque bodies. After a general discussion of the principles of the method and the interpretation of the obtained photoelastic patterns, particular emphasis is given to the various factors that influence the accuracy of strain measurement by use of birefringent coatings. The factors that influence the accuracy of the method, like the reinforcing phenomenon, the edge-effect, the strain gradient and curvature effects, and the effects of different Poisson's ratios of the body and the coating are discussed and analyzed. This chapter ends with a brief description of the concepts of photoelastic strain gauges for determination of the local state of stress in a body, by use of birefringent coatings.

Finally, in Chapter 13, the graphical and numerical methods used in polarization optics, which are based on the Poincaré sphere and the Jones calculus, are discussed and analyzed. These methods include the techniques based on the Wulff net, the j circle, and quaternions. The first two are graphical methods; they are based on the properties of the stereographic and parallel projections of the Poincaré sphere, respectively, on planes parallel or coincident with the equatorial plane of the sphere. The method of quaternions is a numerical method, based on the representation of the unitary Jones matrices in terms of the four basic Pauli matrices. After the basic principles and properties of these methods are studied, they are applied to the solution of relevant problems of polarization optics and two-dimensional photoelasticity, thus exemplifying the use of the methods.

A bibliography, including a large number of the published contributions related to the various chapters of this book, is provided at the end of the book.

2. Electromagnetic Theory of Light

2.1 The Nature of Light

Since the ancient Greek era, several theories of the nature of light have been postulated. The Greek philosopher Aristotle and his contemporaries first attempted to explain how the sensation of vision is caused and how light comes to earth from the sun or the stars. They were also concerned with the nature of light and the mechanism of vision. However, their postulates on the nature of light, due to the lack of experiments, were only of philosophical character and they were not able to formulate a systematic theory of light capable of explaining the optical phenomena involved in the propagation of light or the interaction of light and matter.

Perhaps it is worthwhile to mention here one of the investigators of the middle ages, also of Greek origin, who contributed significantly and very elegantly to the development of Science—Franciscus Maurolycus (1494–1577). He was a prolific writer and published many books related to mathematics, physics, optics, and philosophy. Some of his contributions in optics are contained in his book "Theoremata de lumine et umbra ad perspectivam radiorum incidentium" (Venice 1575). This great scientist may be considered the precursor not only of the science of optics, but also of the modern scientific thought.

Newton, in 1660, formulated the first systematic theory of light, according to which all luminous bodies emit small particles or corpuscles that are propagated in space with the velocity of light. By his corpuscular theory, Newton explained the phenomenon of reflection of light. However, his explanation of refraction of light on a surface that separates two media led him to the unrealistic conclusion that the velocity of light in a dense medium is greater than the velocity of light in air.

Newton's corpuscular theory of light was replaced by the wave theory, proposed by Huygens in 1678. According to this theory a luminous body acts as a source of disturbances in a hypothetical medium, filling space, which was called *the ether*. These disturbances are propagated through the ether in the form of transverse waves that, when they reach the eye, produce the sensation of vision.

The wave theory of light easily explains the phenomena of reflection, refraction, interference, and diffraction. However, the explanation of the phenomenon of polarization necessitates that light waves must be transverse, which, because of the great velocity of propagation of light, requires unrealistic properties of the ether. On this basis, ether would have to be considered a fluid that fills space and that has a very great modulus of elasticity and a low density.

All of these difficulties that result from considering the light as a wave motion in a medium are avoided by considering light as an electromagnetic wave. The electromagnetic theory of light was introduced by Maxwell in 1864. According to this theory, light waves are of the same character as the electromagnetic waves, that are caused by a rapidly oscillating electric current. Electromagnetic waves are propagated through space by varying electric and magnetic fields, whose directions are mutually perpendicular, as well as perpendicular to the direction of propagation. One of these two directions can be considered to be that of the electric field. Great support for this theory results from the fact that the velocities of propagation of light and electromagnetic waves are equal. This theory explains all of the phenomena of geometric optics, as well as the phenomena of interference, diffraction, and polarization.

In both the wave theory of Huygens and the electromagnetic theory of Maxwell, the representative quantity of the light disturbance is the *light vector*, which in the wave theory represents the displacement of the vibrating particle, whereas in the electromagnetic theory, it represents either the electric or the magnetic field vector.

Newly discovered phenomena concerning the interaction of light and matter, such as the photoelectric and Compton effects, could not be explained by the Huygens or Maxwell theories. These phenomena were explained by a particle theory of light developed in its modern form by Planck and Einstein in 1905. This theory is based on the assumption that exchange of energy between radiation and matter, such as the emission or absorption of light by atoms, takes place in a discontinuous manner by definitely determined amounts of energy that depend only on the frequency of the radiation.

The particle theory of light, although it explains the phenomena related to the interaction of light and matter, is inadequate for the explanation of some phenomena that are related to the wave character of light, such as interference and diffraction. For a unified explanation of all of the phenomena related to the particle and wave nature of light, the quantum theory was introduced in recent years by de Broglie, Schrödinger, and Heisenberg. This theory attributes to light the characteristics of both particles and waves; it is a combination of the particle and the wave theories of light.

In this monograph, which deals with the problems of photoelasticity by use of the powerful methods of description of polarized light by the Poincaré sphere, and the Mueller and Jones calculi, only the phenomena of geometrical optics and the wave nature of light are encountered. Problems related to interaction of light and matter are excluded. Therefore, for the study of the problems in this book, the electromagnetic theory of light will be used throughout. However, the same problems can also be studied by the wave theory of light, which, like the electromagnetic theory, is based on the concept of the light vector.

In the following, the basic concepts of the electromagnetic theory of light encountered in the propagation of light in optically isotropic and anisotropic bodies will be studied.

2.2 Maxwell's Equations

The propagation of an electromagnetic wave is governed by Maxwell's equations, which are the mathematical expressions of the fundamental laws of electromagnetism. These laws are

I) *Ampère's circuital law:* The integral of the magnetic field vector over the path formed by a closed circuit is equal to the electric current that flows through the circuit. The positive directions of both the magnetic-field vector and the electric current are right-handedly related.

II) *Faraday's circuital law:* The integral of the electric field vector along a closed circuit is equal to the magnetic current that flows through the circuit. The positive directions of both the electric field vector and the magnetic current are left-handedly related. Ampère's and Faraday's laws are expressed by, respectively,

$$j = \frac{\partial D}{\partial t} = \mathrm{curl}\, H \tag{2.1}$$

and

$$\frac{\partial B}{\partial t} = -\mathrm{curl}\, E , \tag{2.2}$$

where E is the electric vector, D is the electric displacement, H is the magnetic vector, B is the magnetic induction, and j is the electric current.

From (2.1) and (2.2) it follows that

$$\frac{\partial}{\partial t}(\mathrm{div}\, D) = 0 \tag{2.3}$$

$$\frac{\partial}{\partial t}(\mathrm{div}\, B) = 0 . \tag{2.4}$$

Because isolated positive and negative magnetic poles cannot be found in nature,

$$\mathrm{div}\, B = 0 . \tag{2.5}$$

From (2.3) we obtain that the div. of the electric displacement D is independent of time; it depends only on the position of the point in question. For the case of uncharged bodies we have

$$\text{div } D = 0. \tag{2.6}$$

Equations (2.1) and (2.2) inter-relate the four quantities E, D, H, and B. For the determination of these quantities, two more relations are needed. These are provided by the behavior of the various substances under the influence of the field and are known as material equations. We will consider two forms of material equations, when the material in question is either isotropic or linearly anisotropic.

2.3 Propagation of Electromagnetic Waves in Isotropic Media

The material equations in an isotropic medium, relating the quantities D and E, and B and H, take the simple forms

$$D = KE \tag{2.7}$$

$$B = \mu H, \tag{2.8}$$

where K and μ are constants that depend on the material of the medium; they are called *dielectric coefficient* and *magnetic permeability*, respectively.

Introducing material equations (2.7) and (2.8) into Maxwell's equations (2.1) and (2.2) we obtain

$$K \frac{\partial E}{\partial t} = \text{curl } H \tag{2.9}$$

$$\mu \frac{\partial H}{\partial t} = -\text{curl } E. \tag{2.10}$$

By combining (2.9) and (2.10) and taking into account (2.6), we can conclude that the electric vector E must satisfy the differential equation

$$\frac{1}{c^2} \frac{\partial^2 E}{\partial t^2} = \nabla^2 E, \tag{2.11}$$

where

$$\frac{1}{c^2} = K\mu.$$

Equation (2.11) is the well-known wave equation and c is the velocity of the wave propagation. Experimentally, $c = 3 \times 10^{10}\,\mathrm{cm\,s^{-1}}$; this is the velocity of propagation of light or any other electromagnetic disturbance in vacuum.

If we consider the usual case of plane waves and restrict ourselves to the common case of harmonic waves, we must seek solutions of (2.11) in the form

$$E = E' e^{i\varphi} \tag{2.12}$$

with

$$\varphi = \frac{\omega}{c}(r \cdot s - ct),$$

where $r(x,y,z)$ is the position vector, $s(s_x, s_y, s_z)$ is the unit vector along the direction of light propagation, ω is the circular frequency, c the wave velocity, and t represents time.

By taking into account (2.6), we conclude that

$$E' \cdot s = 0 \tag{2.13}$$

which proves that: *The electric vector or the electric displacement is at right angles to the direction of the wave propagation and lies in the wave-front plane.*

From (2.10) and (2.12), we conclude that

$$\mu H = \frac{1}{c} e^{i\varphi} s \times E', \tag{2.14}$$

where $\times$ indicates the vector product of the corresponding vectors.

From (2.14), we obtain

$$H \cdot s = \frac{1}{\mu c} e^{i\varphi}(s \times E') \cdot s = 0, \tag{2.15}$$

that is: *The magnetic vector or the magnetic induction is at right angles to the directions of the wave propagation and lies in the plane of the wave front.*

From (2.14), we also conclude that

$$H \cdot E' = \frac{1}{\mu c} e^{i\varphi}(s \times E') \cdot E' = 0, \tag{2.16}$$

which shows that: *The electric and magnetic vectors are mutually perpendicular.*

Equations (2.13), (2.15), and (2.16) indicate that when an electromagnetic wave is propagated through an isotropic medium, the electric and magnetic field vectors are mutually perpendicular and each of them is perpendicular to the direction of the wave propagation. Both the electric and magnetic fields are propagated with the same velocity, which coincides with the velocity of propagation of the electromagnetic waves.

We can, therefore, conclude that an electromagnetic wave, propagating in an isotropic medium, may be represented either by the electric or the magnetic field vector, which are coincident with the electric displacement or the magnetic induction vectors, respectively. Which of these four vectors is chosen as the representative vector of the electromagnetic wave is a matter of definition. Subsequently, the electric displacement vector will be used as the representative vector of the electromagnetic wave.

2.4 Propagation of Electromagnetic Waves in Anisotropic Media

The material equations for the usual case of an electrically anisotropic, but magnetically isotropic, medium take the form

$$D_x = K_{xx} E_x + K_{xy} E_y + K_{xz} E_z$$
$$D_y = K_{yx} E_x + K_{yy} E_y + K_{yz} E_z \tag{2.17}$$
$$D_z = K_{zx} E_x + K_{zy} E_y + K_{zz} E_z$$

$$B = \mu H.$$

By use of the well-known Einstein's summation convention, (2.17) can be written in the form

$$D_i = K_{ij} E_j \quad (i,j = x,y,z), \tag{2.18}$$

where K_{ij} is the so-called *dielectric tensor*.

By use of energy considerations the dielectric tensor K_{ij} can be proved to be symmetric, that is

$$K_{ij} = K_{ji}. \tag{2.19}$$

The tensor K_{ij} can therefore be diagonalized; when it is referred to its principal dielectric axes, the material equations (2.17) take the simple form

$$D_x = K_x E_x, \quad D_y = K_y E_y, \quad D_z = K_z E_z. \tag{2.20}$$

The constants K_x, K_y, and K_z are called the *principal dielectric coefficients*.

Relations (2.17) and (2.20) show that the directions of D and E do not generally coincide, as in the case of isotropic materials, unless E coincides with one of the principal dielectric axes.

For a plane wave of the form of (2.12) with v the wave velocity, Maxwell's equations (2.1) and (2.2) take the form

$$-vD = s \times H$$
$$\mu v H = s \times E .$$

(2.21)

These equations show that vector H, and therefore vector B also, are perpendicular to vectors E, D, and s. Thus, vectors E, D, and s are coplanar. In addition, (2.21) show that vector s is orthogonal to vector D. Thus, vectors H and D are perpendicular to the direction of propagation s, as well as mutually perpendicular. Vector E is not, in general, perpendicular to the propagation direction, as in the case of an isotropic material. A representative picture of the position of all these vectors is shown in Fig. 2.1.

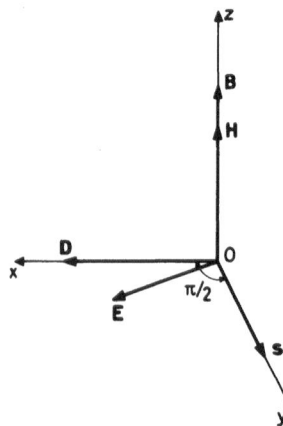

Fig. 2.1. Relative position of the wave normal s, and the characteristic electric E and D and magnetic H and B field vectors in an electrically anisotropic and magnetically isotropic medium

By eliminating vector H from (2.21), we obtain

$$D = -\frac{s \times (s \times E)}{\mu v^2} = \frac{E - s(s \cdot E)}{\mu v^2} .$$

(2.22)

By using material equations (2.20), referred to the principal dielectric axes, we obtain

$$K_k E_k = \frac{E_k - s_k(s \cdot E)}{\mu v^2} , \qquad (k = x, y, z) .$$

(2.23)

From (2.23), we have

$$E_k = \frac{s_k(s \cdot E)}{1 - \mu v^2 K_k} .$$

(2.24)

Multiplying (2.24) by s_k and adding the resulting three equations, we obtain, by dividing each member of it by $(s \cdot E)$,

$$\frac{s_x^2 v_x^2}{v_x^2 - v^2} + \frac{s_y^2 v_y^2}{v_y^2 - v^2} + \frac{s_z^2 v_z^2}{v_z^2 - v^2} = 1 , \tag{2.25}$$

in which

$$v_k = (\mu K_k)^{-1/2} . \tag{2.26}$$

Equation (2.25), when

$$s_x^2 + s_y^2 + s_z^2 = 1$$

is taken into account, gives

$$\frac{s_x^2}{v_x^2 - v^2} + \frac{s_y^2}{v_y^2 - v^2} + \frac{s_z^2}{v_z^2 - v^2} = 0 . \tag{2.27}$$

Equation (2.27) is quadratic in v^2. Therefore, two wave velocities correspond to any orientation s of the wave front. To each of these two velocities, (2.24) gives the ratios $E_x : E_y : E_z$, and then (2.20) give the corresponding ratios $D_x : D_y : D_z$. Because these ratios are real, the fields E and D are linearly polarized. We therefore obtain the basic theorem

The two monochromatic plane waves that propagate in a crystal, corresponding to a given direction of the incident wave front, are linearly polarized.

We shall show later on that the two directions of the vector D that correspond to a given direction of the wave front are mutually perpendicular.

2.4.1 The Fresnel Ellipsoid

The energy of an electric field, with vectors E and D, is proportional to the internal product

$$C = E \cdot D ,$$

which, when the material equations, expressed in the principal dielectric directions by (2.20), are taken into account, takes the form

$$C = K_x E_x^2 + K_y E_y^2 + K_z E_z^2 , \tag{2.28}$$

or

$$C = \frac{D_x^2}{K_x} + \frac{D_y^2}{K_y} + \frac{D_z^2}{K_z} , \tag{2.29}$$

where C is a constant.

When $D_x/\sqrt{C}$, $D_y/\sqrt{C}$, $D_z/\sqrt{C}$ are replaced by x, y, z, (2.29) takes the form

$$\frac{x^2}{K_x}+\frac{y^2}{K_y}+\frac{z^2}{K_z}=1\,,\tag{2.30}$$

which is the equation of an ellipsoid whose semi-axes are equal to $\sqrt{K_x}$, $\sqrt{K_y}$, and $\sqrt{K_z}$; they coincide with the principal dielectric axes. This ellipsoid will be called the *index ellipsoid* or the *Fresnel ellipsoid*.

Let us find the magnitudes and directions of the principal axes of the cross section of this ellipsoid by a diametral plane, parallel to the wave front. The equation of this plane is

$$s_x x+s_y y+s_z z=0\,.\tag{2.31}$$

The problem is reduced to finding the extreme values of the radius r, where

$$r^2=x^2+y^2+z^2\,,\tag{2.32}$$

subject to the restrictions (2.30) and (2.31).

According to Lagrange's method of undetermined multipliers, we formulate the function $f=f(x,y,z)$ as

$$f=x^2+y^2+z^2+2A\,(s_x x+s_y y+s_z z)+B\left(\frac{x^2}{K_x}+\frac{y^2}{K_y}+\frac{z^2}{K_z}-1\right),\tag{2.33}$$

where A and B are multipliers. By putting the derivatives of the function f with respect to x, y, z equal to zero, we obtain the three equations

$$x+As_x+\frac{Bx}{K_x}=0\,,\quad y+As_y+\frac{By}{K_y}=0\,,\quad z+As_z+\frac{Bz}{K_z}=0\,.\tag{2.34}$$

Multiplying (2.34) by x, y, and z, respectively, and adding, we obtain

$$r^2+B=0\,.\tag{2.35}$$

Multiplying these equations by s_x, s_y, and s_z, respectively, and adding we obtain

$$A+B\left(\frac{xs_x}{K_x}+\frac{ys_y}{K_y}+\frac{zs_z}{K_z}\right)=0\,.\tag{2.36}$$

Substituting the values of A and B from (2.35) and (2.36) in (2.34), we obtain

$$x\left(1 - \frac{r^2}{K_x}\right) + s_x r^2 \left(\frac{x s_x}{K_x} + \frac{y s_y}{K_y} + \frac{z s_z}{K_z}\right) = 0$$

$$y\left(1 - \frac{r^2}{K_y}\right) + s_y r^2 \left(\frac{x s_x}{K_x} + \frac{y s_y}{K_y} + \frac{z s_z}{K_z}\right) = 0 \qquad (2.37)$$

$$z\left(1 - \frac{r^2}{K_z}\right) + s_z r^2 \left(\frac{x s_x}{K_x} + \frac{y s_y}{K_y} + \frac{z s_z}{K_z}\right) = 0 .$$

When the values for x, x/K_x, and r^2 given by

$$x = \frac{D_x}{\sqrt{C}}, \quad \frac{x}{K_x} = \frac{E_x}{\sqrt{C}}, \quad \text{and} \quad r^2 = \frac{D^2}{C} = \frac{D^2}{E \cdot D} = \frac{1}{v^2 \mu}, \qquad (2.38)$$

where the last relation was derived from (2.22), are substituted in (2.37), the latter become

$$D_k = \frac{E_k - s_k(s \cdot E)}{\mu v^2}, \qquad (k = x, y, z). \qquad (2.39)$$

Equation (2.39) are identical with (2.23).

We therefore conclude that the two wave-propagation velocities v, corresponding to a given wave front, are inversely proportional to the lengths of the semi-axes of the elliptical cross section through the center of the ellipsoid and parallel to the wave front, and that these semi-axes coincide with the corresponding directions of the vector D. Because the semi-axes of an ellipse are mutually perpendicular, we obtain the basic theorem

The two vectors D that correspond to a given direction of propagation in a crystal are mutually perpendicular.

If we define the refractive index n along a given direction, corresponding to a wave-propagation velocity v by the relation

$$n = \frac{c}{v}, \qquad (2.40)$$

where c is the velocity of light in vacuum, then the lengths of the semi-axes of the above defined ellipsoid are proportional to the corresponding refractive indices. For this reason, this ellipsoid is called the *refractive ellipsoid*, or the *Fresnel ellipsoid*. Therefore, we conclude that

The two refractive indices that correspond to a given propagation direction are proportional to the lengths of the semi-axes of the elliptical cross section of the ellipsoid by a plane parallel to the wave front and passing through its center.

Figure 2.2 represents the graphical determination, of the two refractive indices n_1, n_2 that correspond to a given propagation direction s of light, by use of Fresnel's ellipsoid.

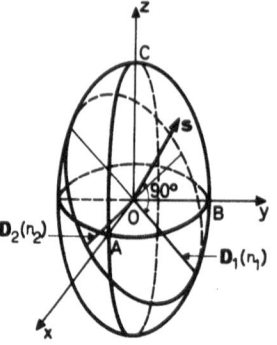

Fig. 2.2. Geometrical determination of the electric displacement vectors D_1 and D_2 and the corresponding refractive indices n_1 and n_2 for a given direction of the wave normal s by use of the Fresnel ellipsoid. The values of D_1 and D_2 or n_1 and n_2 are proportional to the lengths of the semi-axes of the ellipse formed by the intersection of the ellipsoid with a plane perpendicular to the wave normal s and through the center of the ellipsoid

There is also another simple graphical determination of the two refractive indices, based on the fact that an ellipsoid has two circular cross sections, passing through its center, whose normals are coplanar with the longest and shortest principal axes of the ellipsoid. Let the elliptical cross section of the ellipsoid, by a plane normal to a given direction, be represented by the ellipse of Fig. 2.3, whose principal axes are OA and OA'. Let OB and OB' be the intersections of this plane with the circular cross sections of the ellipsoid. Then, the lengths OB and OB' will be equal; therefore, they will also be equally inclined to the principal axes OA and OA'. If, now, we consider the intersections OX and OX' of the ellipse of this figure with the planes defined by its normal, corresponding to the direction of propagation, and the normals to the two circular cross sections of the ellipsoid, then OX and OX' must be perpendicularly oriented with respect to OB' and OB, respectively. Therefore, the directions OX and OX' are equally inclined to the two principal axes OA and OA' of the ellipse.

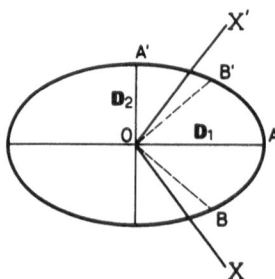

Fig. 2.3. The directions of the electric displacement vectors D_1 and D_2 corresponding to a given wave normal s are the internal and external bisectors of the angle formed by the intersections OX and OX' of the ellipse normal to the wave normal s with the planes defined by the normals to the two circular cross sections (optic axes) of the Fresnel ellipsoid and the wave normal s

Consequently, the two electric displacement directions, corresponding to a given propagation direction, can be defined as the internal or external bisectors of the angle defined by the intersection of the ellipse normal to the given direction with the planes defined by this normal and the normals to the two circular cross sections of the ellipsoid.

Besides the above-defined phase velocity v, which is along the direction s of the wave normal, we can also define the ray velocity v_r along the direction of propagation of the energy of the electromagnetic wave. This velocity v_r is normal to the vectors H and E and is related to the velocity v by

$$v = v_r \cos \alpha \, ,$$

where α is the angle subtended between the wave normal s and the velocity v_r. As the phase velocity v along the direction s and the corresponding two values of the electric displacement vector D form an orthogonal triplet of vectors, in the same manner the ray velocity v_r and the corresponding two values of the electric vector E form an orthogonal triplet of vectors. It can also be proved that the two values of the ray velocities, corresponding to a given direction of energy propagation, are proportional to the lengths of the semi-axes of the elliptical cross section of the ray ellipsoid with a plane perpendicular to the energy-propagation direction. This ellipsoid has principal axes inversely proportional to the square roots of the principal dielectric coefficients.

2.4.2 Uniaxial and Biaxial Crystals

Transparent crystals of all crystallographic systems fall into three categories according to the shapes of their Fresnel ellipsoids. If all three axes of the ellipsoid are unequal, the corresponding crystal is called *biaxial*; if two axes are equal, it is called *uniaxial*; and if all axes are equal (sphere), it is called *isotropic*. This naming of the crystal arises from the fact that an ellipsoid that has three unequal axes has two circular cross sections, whereas an ellipsoid that has two equal axes has only one circular cross section, and a sphere has all its cross sections circular. The normal to such a circular cross section of the ellipsoid is called the *optic axis* of the crystal, and has the property that, for a wave that travels along the optic axis, there is only one velocity of propagation. Thus, a biaxial crystal has two optic axes, a uniaxial crystal has one such optic axis, and in an isotropic crystal the direction of the optic axis is indeterminate.

Besides the Fresnel ellipsoid, another useful way of visualizing the variation of the phase or the ray velocity in a crystal with the direction of the wave normal or of energy propagation is provided by the so-called normal surface or ray surface, respectively. The normal surface is formed, if, along each wave normal s, two lengths proportional to the two corresponding phase velocities are taken and their end points are connected. Similarly, the ray surface is formed by taking, along each direction of energy propagation, two lengths proportional to the corresponding ray velocities.

For the most general case of a biaxial crystal the equation of the normal surface is given by (2.27). This equation can be written in the form

$$s_x^2(v_y^2 - v^2)(v_z^2 - v^2) + s_y^2(v_z^2 - v^2)(v_x^2 - v^2) + s_z^2(v_x^2 - v^2)(v_y^2 - v^2) = 0 \, . \qquad (2.41)$$

In order to get an idea of the shape of this surface, let us consider its intersections with the coordinate planes $x=0$, $y=0$, and $z=0$ of the reference system, taken along the principal dielectric axes. The intersection of the normal surface by the plane $x=0$, obtained from (2.41) by putting $s_x=0$, takes the form

$$(v_x^2 - v^2)[(s_y^2 v_z^2 + s_z^2 v_y^2) - v^2] = 0 . \tag{2.42}$$

Equation (2.42) gives

$$\begin{aligned} v^2 &= v_x^2 \\ v^2 &= s_y^2 v_z^2 + s_z^2 v_y^2 , \end{aligned} \tag{2.43}$$

and by putting $vs_y = y$, $vs_z = z$, we get

$$y^2 + z^2 = v_x^2 \quad (y^2 + z^2)^2 = v_z^2 y^2 + v_y^2 z^2 . \tag{2.44}$$

That is, the intersection of the normal surface with the plane $x=0$ is a circle and an oval. Similarly, for reasons of symmetry, the intersection with each of the other planes $y=0$ and $z=0$ will be composed of a circle and an oval. The only difference between the intersections with the three coordinate planes will be in the relative positions of the circle and the oval. According to whether

$$K_x < K_y < Z_z$$

or

$$v_x > v_y > v_z ,$$

the oval will be inside the circle in the $y-z$ plane, or outside the circle in the $x-y$ plane; it will intersect the circle in four points in the $x-z$ plane. Fig. 2.4 represents the intersection patterns of the normal surface by the three coordinate planes. A perspective view of the normal surface is shown in Fig. 2.5.

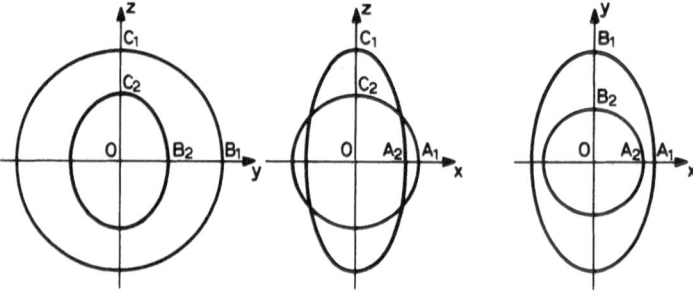

Fig. 2.4. Intersections of the normal surface by the coordinate planes of a biaxial crystal

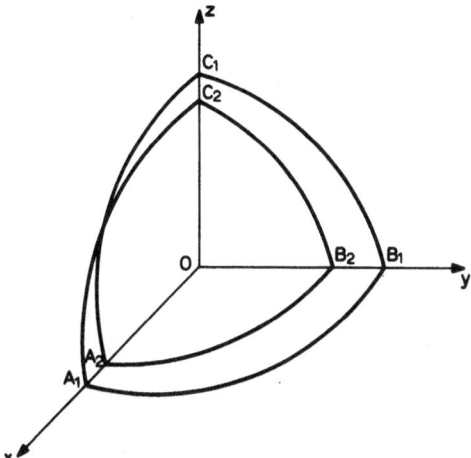

Fig. 2.5. Perspective view of the normal surface of a biaxial crystal

In a similar manner, it can be proved that the ray surface has a shape analogous to the normal surface, but with the above-considered ovals replaced by ellipses, arranged relative to the corresponding circles similarly as in the case of the normal surface.

References

2.1 M. Born, E. Wolf: *Principles of Optics*, 5th ed. (Pergamon Press, London 1975) pp. 1–18, 665–690
2.2 B. Rossi: *Optics*, 3rd ed. (Addison-Wesley, Reading, Mass. 1965)
2.3 G. N. Ramachandran, S. Ramaseshan: "Crystal Optics", in *Crystal Optics, Diffraction*, ed. by S. Flügge, Encyclopedia of Physics, Vol. 25/1 (Springer, Berlin, Göttingen, Heidelberg 1961) pp. 1–217

3. Description of Polarized Light

3.1 Introduction

According to the principles of both the electromagnetic and the wave theory of light, a polarized light beam can be represented by the light vector, which, in the electromagnetic theory, is defined as the electric- or the magnetic-field vector, whereas in the wave theory it is the vector that specifies at each instant the position of the ether particle, moving on the plane perpendicular to the direction of propagation. This simple concept of the light vector has been successfully applied in many problems of polarization optics dealing with the passage of polarized light through a number of birefringent or optically rotating media. The state of polarization of the light beam is completely determined if, at each position along the path of propagation, the light vector is known.

However, when a large number of optical elements is encountered along the light path, the resulting calculations become long, cumbersome, tedious, and error-prone. Also, the formulas obtained are complicated and it is always with great difficulty that we get the required information. More adequate methods for treating optical phenomena involving propagation of light through a series of optical elements have been invented, which are based on new descriptions of polarized light. These methods are based on the concepts of the Poincaré sphere, the *j*-circle method, and the Mueller and Jones vectors.

The sphere conceived by *Poincaré* [3.1] is a geometrical representation of every elliptical polarization form by a point on a sphere. This is a mapping procedure, by which, to each polarization state corresponds a point on the sphere and conversely, to each point on the sphere corresponds a definite polarization form. Thus, having in mind the sphere, we have a unified picture of all elliptical polarization forms and can visualize how one form can be transformed to another, by making a suitable manipulation on the sphere. Thus, we can find pictorially the kind of the optical device needed for transforming each polarization state to another; conversely, we can solve the problem of finding the emerging form from a device that has the corresponding input form. A large number of problems of polarization optics can be formulated in this way by use of the Poincaré sphere.

It is worthwhile to mention here that, instead of working with the Poincaré sphere, which is a three-dimensional device, *Kuske* [3.2] has developed the idea of *Menges* [3.3] of projecting points on the sphere onto its equatorial plane and to work with the projection vector of each point on the sphere.

This vector characterizes a particular polarization form and is called the *j vector*; the corresponding representation of each polarization form by a point inside a circle is called the *j-circle method*. This method will not be considered separately here, because any geometrical transformation in this method can be obtained from the corresponding transformation on the Poincaré sphere by taking the corresponding projections on the equatorial plane of the sphere. However, due to its importance, this method will be developed with applications to photoelasticity in Chapter 13 of this book.

Furthermore, *Stokes* [3.4] has proved that any form of partially polarized light can be characterized by four parameters. These parameters constitute the elements of a four-element vector called the Stokes vector. Stokes parameters have dimensions of intensity and can be operationally defined as the emerging light intensity from four suitable polarizing filters. Although these parameters can be defined in the more general case of partially polarized light, they will be used throughout this book only for describing completely polarized light. We shall define Stokes vector elements in terms of the electromagnetic theory of light.

The vector introduced by *Jones* [3.5] is a two-element complex vector that describes the polarization form of a light beam. This vector is restricted to dealing only with forms of polarization; it cannot be applied to unpolarized beams.

All of these three methods of describing forms of polarization are characterized by particular elegance and provide powerful means when dealing with problems of polarization optics. A corresponding calculus is available for each such method to facilitate solution of problems in which series of optical elements are encountered. These calculi will be developed in the next chapter.

In the present chapter, the conventional vectorial method, as well as the modern methods of the Poincaré sphere and the Stokes and Jones vectors, for the characterization of states of polarization will be described in detail. The particular features of each of these methods will be discussed. Furthermore, a comparison and the interrelations between these methods will be given. The use of these methods to predict the emerging polarization form from an optical device, which facilitates solution of problems dealing with a series, or "train" of optical devices, will be considered in the next chapter.

3.2 Ordinary and Polarized Light

According to the electromagnetic theory of light, *ordinary or unpolarized* light is light in which the end point of the electric (and also of the magnetic) vector moves in space irregularly and does not show any preferred directional properties. Within the concepts of the *light vector*, unpolarized light is defined as the light in which the light vector moves irregularly and shows no directional or rotational preference.

When some order is introduced into the irregular motion of the light vector, the light is called *polarized*. The end point of the light vector in polarization moves along well-defined simple curves in a definite direction. Polarized light is the simplest form of light. In terms of polarized light, unpolarized light can be defined as a mixture of all kinds of polarized light in an arbitrary and unsystematic manner.

Strictly speaking, perfectly polarized light does not exist in nature. No optical device used to transform unpolarized to polarized light is perfect; in all cases, some amount of ordinary, unpolarized, light is also present. The mixture of polarized and unpolarized light is called *partially polarized light*.

In the case of polarized light, the motion of the ether particle is not arbitrary, but well defined. This kind of light will be assumed throughout this book. In the investigation of the various phenomena, we shall consider the ideal case of perfectly polarized light, neglecting any kind of depolarization. This can be done without any significant error because, in all of the phenomena treated in this book, the effect of the presence of unpolarized light is only to weaken the pattern produced by the polarized light. Thus, the existence of an unpolarized component influences only the contrast of the obtained optical patterns.

3.3 Forms of Polarized Light

The various forms of polarized light are defined by the particular type of curve along which the end of the light vector moves. The simplest forms of curves are the straight line, the circle, and the ellipse, thus defining the linear, the circular, and the elliptical forms of polarization.

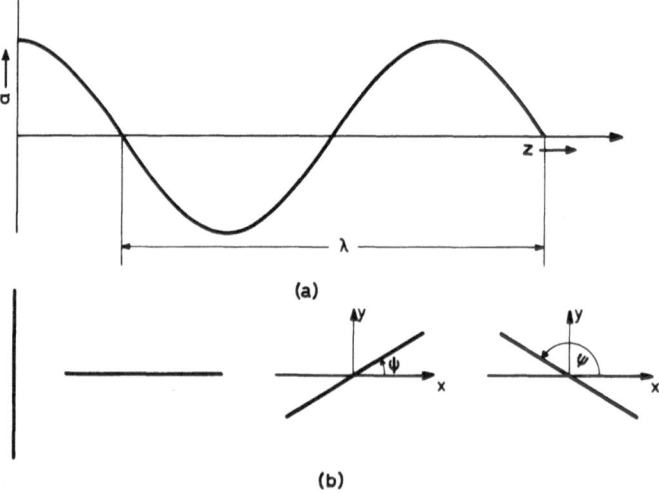

Fig. 3.1a and b. Representation of linearly polarized light by (a) a sinusoidal curve that shows the variation of the magnitude of the light vector along the direction of propagation at a single instant $t = t_0$ and by (b) the double amplitude of the light vector for different orientations on the wave front

A polarized light is said to be *linear* when the end point of the light vector moves along a straight line. In this case, the light vector does not change direction with time, but only magnitude. The largest deviation of the ether particle defines the amplitude of the linearly polarized light. Such a light can be represented by a sinusoidal curve that shows the amplitude of the wave train at a single instant, as it would be seen by an o'bserver looking along the direction normal to the light propagation (Fig. 3.1a), and by a line of length equal to twice the amplitude of the light vector, as it is seen by an observer looking along the direction of propagation, toward the light source (Fig. 3.1b). Usually, we omit the sinusoidal curve of the plane wave and represent linearly polarized light by a straight line, corresponding to a time-exposure photograph taken by a camera aimed at the light source. This part of the straight line defines the amplitude and the orientation of the linearly polarized light. Furthermore, the amplitude of the wave is taken equal to unity, so that the linear type of polarized light includes a one-parameter family, each form of linear polarization being defined by the inclination of the light vector to the horizontal.

In the *circularly polarized* light, the end point of the light vector moves on the circumference of a circle. Thus, the magnitude of the light vector remains constant, while its inclination varies continuously between 0 and 2π. The form of the light wave can be represented at a single instant in time by a cylindrical helix (Fig. 3.2a), whereas, its transverse-sectional pattern, as it is seen by an observer looking along the line of propagation of the wave, toward the light source, is a circle. By defining the amplitude of the wave as being equal to unity, two different types of circular polarization can be distinguished: the *right-circular polarization* and the *left-circular polarization*. The right-circular polarization is defined by the clockwise sence of motion of the light vector when the transverse-sectional pattern of light is seen by an observer looking along the propagation direction toward the light source, whereas the counterclockwise motion corresponds to left-circular polarization (Fig. 3.2b).

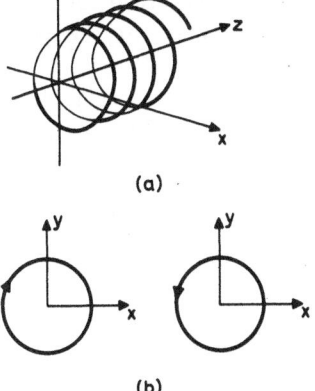

(a)

(b)

Fig. 3.2a and b. Representation of circularly polarized light by (a) a circular helix that shows the curve formed by the tips of the light vectors along the direction of propagation at a single instant and by (b) a circle that corresponds to the positions of the tips of the light vectors on the wave front at all instants. Two forms, right and left circularly polarized light, are possible

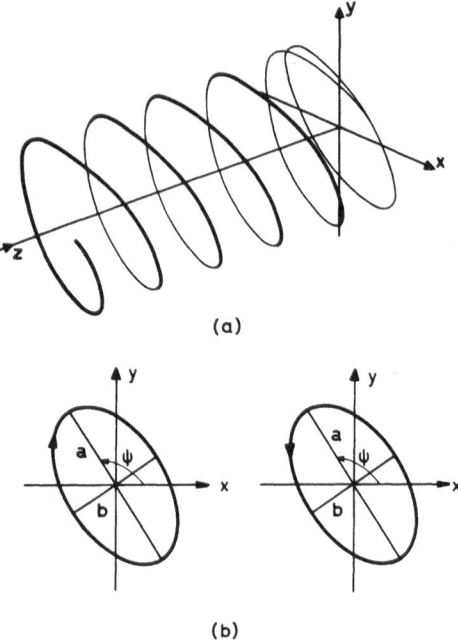

(a)

(b)

Fig. 3.3a and b. Representation of elliptically polarized light by (a) an elliptical helix that shows the curve formed by the tips of the light vectors along the direction of propagation at a single instant and by (b) an ellipse that corresponds to the positions of the tips of the light vectors on the wave front at all instants. If we are not interested in the intensity of the light, three parameters, namely the azimuth ψ, the ellipticity ω [$\tan \omega = (b/a)$], and the handedness (right or left) are needed for description of elliptically polarized light

Elliptical polarization is the most general form of polarized light used in practical applications. In this type of polarized light, the ether particle moves on an ellipse, so that the magnitude as well as the angle of inclination of the light vector vary continuously. The instantaneous view of the light wave can be represented by an elliptical helix (Fig. 3.3a); its transverse-sectional pattern is an ellipse (Fig. 3.3b). Three different parameters must be used to define completely the elliptical polarization, that is, the ratio (b/a) of the semi-axes of the ellipse, called the *ellipticity*; the inclination ψ of the major semi-axis of the ellipse with respect to the x axis, called the *azimuth*; and the direction of motion of the end point of the light vector, the so-called "handedness", defined as for circular polarization. Right- and left-elliptical forms of polarization (Fig. 3.3b) are distinguished. Ellipticity (b/a) may vary between zero and infinity; azimuth varies between 0 and 180 degrees.

Linear and circular forms of polarization can be defined as special cases of the elliptical polarization, when ellipticity (b/a) takes the values 0 and 1, respectively.

3.4 Description of Polarized Light

As has already been mentioned, light, according to the electromagnetic theory of light, consists of self-preserving electric and magnetic fields, which are produced by oscillating motions of the electric charges of the light source, and transmitted in space in the form of electric and magnetic waves. Usually, the electric-field vector is used to define the light vector. This vector, in polarized light, is well defined according to the particular type of polarized light. Thus, polarized light can be described by using vector-calculus concepts.

Although by this vectorial description of polarized light all problems involved at the passage of a light beam through a train of optical elements can be solved, the calculations are frequently cumbersome and make the solution of the problem difficult. This was the reason why other, more convenient, descriptions of polarized light have been introduced. By these modern methods the calculations involved with the passage of light through optical elements are greatly simplified. Of these new methods, we shall describe the Poincaré sphere, the j-circle method, and the Stokes and Jones vectors.

3.4.1 Vectorial Representation

Vector calculus can be used to describe the light vector, which represents the vector of the electric field, according to electromagnetic theory of light or the vector that defines the position of the ether particle at each instant, according to the wave theory. We shall now describe vectorially the linear, circular, and elliptical polarization forms.

Linear Polarization

A monochromatic linearly polarized plane wave along the x axis, generated by a disturbance, which varies sinusoidally with time t and position z along the z axis, can be represented by the vector α, defined by

$$\alpha = A \cos 2\pi \left(\frac{t}{T} + \frac{z}{\lambda} \right) i = A \cos \left(\omega t + \frac{2\pi z}{\lambda} \right) i, \tag{3.1}$$

where A is the amplitude, T the period, λ the wavelength of the monochromatic light, $\omega = (2\pi/T)$ is the angular frequency, and i is the unit vector along the x axis.

Equation (3.1), at a given instant $t = t_0$ of time, defines the wave that is propagated along the z axis (Fig. 3.1a), whereas for a given position $z = z_0$ along the z axis it defines the various positions of the light vector along the x axis (Fig. 3.1b). A more general form of (3.1), including the phase δ_x of the above-defined linearly polarized light along the x axis, can be expressed in the form

$$\alpha = A \cos \left(\omega t + \frac{2\pi z}{\lambda} + \delta_x \right) i. \tag{3.2}$$

Relation (3.2), by use of the complex variable notation, can be written as

$$\alpha = A \, \mathrm{Re} \left[e^{(\omega t + 2\pi z/\lambda + \delta_x)i} \right] i \, , \tag{3.3}$$

where Re represents the real part of the corresponding function and i is the square root of (-1).

Circular Polarization

A circularly polarized light beam can be represented by the light vector α given by

$$\alpha = \alpha_x i + \alpha_y j \, , \tag{3.4}$$

with

$$\begin{aligned} \alpha_x &= A \, \mathrm{Re} \left[e^{(\omega t + 2\pi z/\lambda + \delta_x)i} \right] \\ \alpha_y &= A \, \mathrm{Re} \left[e^{(\omega t + 2\pi z/\lambda + \delta_y)i} \right] \end{aligned} \tag{3.5}$$

and

$$\delta = \delta_y - \delta_x = \pm 90 \, \mathrm{deg.}$$

Equation (3.4) and (3.5) represent the parametric equations of a circle.

When $\delta = 90$ deg., the light is right-circularly polarized. When $\delta = -90$ deg. the light is left-circularly polarized.

Elliptical Polarization

In the more-general case of an elliptical polarization, the light vector α is given by

$$\alpha = \alpha_x i + \alpha_y j \, , \tag{3.6}$$

with

$$\begin{aligned} \alpha_x &= A_x \, \mathrm{Re} \left[e^{(\omega t + 2\pi z/\lambda + \delta_x)i} \right] \\ \alpha_y &= A_y \, \mathrm{Re} \left[e^{(\omega t + 2\pi z/\lambda + \delta_y)i} \right] . \end{aligned} \tag{3.7}$$

We shall show that (3.6) and (3.7) represent the parametric equations of an ellipse. By substitution of $v = (\omega t + 2\pi z/\lambda)$, (3.7) can be written as

$$\begin{aligned} \alpha_x &= A_x \cos (v + \delta_x) \\ \alpha_y &= A_y \cos (v + \delta_y), \end{aligned} \tag{3.8}$$

which can be further transformed to

$$\frac{\alpha_x}{A_x} = \cos v \cos \delta_x - \sin v \sin \delta_x$$

$$\frac{\alpha_y}{A_y} = \cos v \cos \delta_y - \sin v \sin \delta_y .$$

(3.9)

In order to eliminate the time factor v in (3.9), we multiply the first one by $\sin \delta_y$ and subtract from it the second one multiplied by $\sin \delta_x$. We also multiply the first one by $\cos \delta_y$ and subtract from it the second one multiplied by $\cos \delta_x$. Thus we come up with the relations:

$$\frac{\alpha_x}{A_x} \sin \delta_y - \frac{\alpha_y}{A_y} \sin \delta_x = \cos v \sin (\delta_y - \delta_x)$$

$$\frac{\alpha_x}{A_x} \cos \delta_y - \frac{\alpha_y}{A_y} \cos \delta_x = \sin v \sin (\delta_y - \delta_x) .$$

(3.10)

By squaring and adding (3.10), we obtain

$$\left(\frac{\alpha_x}{A_x}\right)^2 + \left(\frac{\alpha_y}{A_y}\right)^2 - \frac{2\alpha_x \alpha_y}{A_x A_y} \cos \delta = \sin^2 \delta$$

(3.11)

with $\delta = (\delta_y - \delta_x) .$

(3.12)

Equation (3.11) takes the form

$$\alpha_{11} x^2 + \alpha_{22} y^2 + 2\alpha_{12} xy + 2\alpha_{13} x + 2\alpha_{23} y + \alpha_{33} = 0 ,$$

which represents the more-general form of a second-degree curve. This curve represents an ellipse if

$$\alpha_{11} \alpha_{22} - \alpha_{12}^2 > 0 .$$

For (3.11) it can be deduced that this quantity is equal to

$$\left(\frac{\sin \delta}{A_x A_y}\right)^2 > 0 .$$

That is, (3.11), whose parametric form is given by (3.7), represents an ellipse.

In order to obtain the maximum or minimum of the coordinate α_x of this ellipse, we differentiate (3.11) with respect to α_y and equate this derivative to zero. Thus, we obtain

$$\frac{\alpha_x}{A_x} = \frac{\alpha_y}{A_y} (\cos \delta)^{-1} .$$

Combining this relation with (3.11), we obtain for the coordinates of the point A that represent a maximum or minimum amplitude α_x that

$$\alpha_x^A = \pm A_x$$
$$\alpha_y^A = \pm A_y \cos\delta. \tag{3.13}$$

Similarly, the coordinates of the point B with the coordinate α_y representing a maximum or a minimum, are given by

$$\alpha_x^B = \pm A_x \cos\delta$$
$$\alpha_y^B = \pm A_y. \tag{3.14}$$

From (3.13) and (3.14), we conclude that the ellipse described by (3.11) is included into a rectangle with sides parallel to the coordinate axes. This rectangle has lengths $2A_x$ and $2A_y$ and is tangent to the ellipse at the points $A\,(\pm A_x,\ \pm A_y\cos\delta)$ and $B\,(\pm A_x\cos\delta,\ \pm A_y)$ (Fig. 3.4).

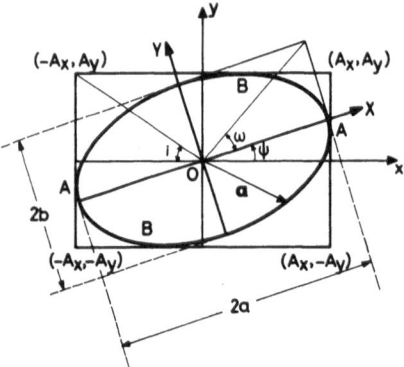

Fig. 3.4. Notation for the various quantities involved in elliptically polarized light

Let us now determine the lengths of the principal axes of the ellipse and their angles of inclination. If the principal axes of the ellipse with lengths $2a$ and $2b$ $(a \geq b)$ are along the coordinate axes OX and OY, respectively, the parametric equations of the ellipse, referred to its principal axes, are given by

$$\alpha_X = a\cos(v+\delta_0)$$
$$\alpha_Y = \pm b\sin(v+\delta_0), \tag{3.15}$$

where the plus and minus signs in α_Y are introduced to indicate the two opposite directions, either left-handed or right-handed, with which the ether particle or the corresponding light vector describes the ellipse.

If the angle between the Ox and OX-axes is denoted by ψ, $(0 \leq \psi < \pi)$ and it is called the azimuth of the light ellipse, then the coordinates α_X, α_Y and α_x, α_y are related by

$$\begin{aligned}
\alpha_X &= \alpha_x \cos \psi + \alpha_y \sin \psi \\
\alpha_Y &= -\alpha_x \sin \psi + \alpha_y \cos \psi .
\end{aligned} \tag{3.16}$$

Equations (3.16), by use of (3.9), can be written in the form

$$\begin{aligned}
\alpha_X &= A_x(\cos v \cos \delta_x - \sin v \sin \delta_x) \cos \psi + A_y(\cos v \cos \delta_y - \sin v \sin \delta_y) \sin \psi \\
\alpha_Y &= -A_x(\cos v \cos \delta_x - \sin v \sin \delta_x) \sin \psi + A_y(\cos v \cos \delta_y - \sin v \sin \delta_y) \cos \psi .
\end{aligned} \tag{3.17}$$

Equation (3.15) can also be written as

$$\begin{aligned}
\alpha_X &= a(\cos v \cos \delta_0 - \sin v \sin \delta_0) \\
\alpha_Y &= \pm b(\sin v \cos \delta_0 + \cos v \sin \delta_0) .
\end{aligned} \tag{3.18}$$

By equating the coefficients of $\cos v$ and $\sin v$ in (3.17) and (3.18), we obtain

$$\begin{aligned}
a \cos \delta_0 &= A_x \cos \delta_x \cos \psi + A_y \cos \delta_y \sin \psi \\
a \sin \delta_0 &= A_x \sin \delta_x \cos \psi + A_y \sin \delta_y \sin \psi
\end{aligned} \tag{3.19}$$

and

$$\begin{aligned}
\pm b \cos \delta_0 &= A_x \sin \delta_x \sin \psi - A_y \sin \delta_y \cos \psi \\
\pm b \sin \delta_0 &= -A_x \cos \delta_x \sin \psi + A_y \cos \delta_y \cos \psi .
\end{aligned} \tag{3.20}$$

By squaring and adding the pairs of (3.19) and (3.20), we obtain

$$\begin{aligned}
a^2 &= A_x^2 \cos^2 \psi + A_y^2 \sin^2 \psi + A_x A_y \sin 2\psi \cos \delta \\
b^2 &= A_x^2 \sin^2 \psi + A_y^2 \cos^2 \psi - A_x A_y \sin 2\psi \cos \delta .
\end{aligned} \tag{3.21}$$

By adding (3.21), we deduce that

$$a^2 + b^2 = A_x^2 + A_y^2 . \tag{3.22}$$

By multiplying the first and second equations of (3.19) and (3.20), respectively, and adding, we deduce that

$$\pm ab = A_x A_y \sin \delta , \tag{3.23}$$

whereas, by dividing these equations we obtain

$$\pm \frac{b}{a} = \frac{A_x \sin\delta_x \sin\psi - A_y \sin\delta_y \cos\psi}{A_x \cos\delta_x \cos\psi + A_y \cos\delta_y \sin\psi} = \frac{-A_x \cos\delta_x \sin\psi + A_y \cos\delta_y \cos\psi}{A_x \sin\delta_x \cos\psi + A_y \sin\delta_y \sin\psi}.$$

(3.24)

Equation (3.24) yields

$$\tan 2\psi = \frac{2A_x A_y}{A_x^2 - A_y^2} \cos\delta.$$

(3.25)

By introducing the angle i, such that

$$\tan i = \frac{A_y}{A_x}$$

(3.26)

(3.25) takes the form

$$\tan 2\psi = \tan 2i \cos\delta.$$

(3.27)

Relation (3.27) defines the angle ψ of inclination of the major axis of the light ellipse. By inserting this value of ψ into (3.21), we obtain for the lengths a and b of the semi-axes of the light ellipse that

$$a, b = \left\{ \frac{(A_x^2 + A_y^2)[1 \pm (\cos^2 2i + \sin^2 2i \cos^2\delta)^{1/2}]}{2} \right\}^{1/2}.$$

(3.28)

From (3.22) and (3.23) we deduce

$$\pm \frac{2ab}{a^2 + b^2} = \frac{2A_x A_y}{A_x^2 + A_y^2} \sin\delta$$

(3.29)

and, by introducing an angle ω such that

$$\pm \frac{b}{a} = \tan\omega \quad (-45° \leqq \omega \leqq 45°),$$

(3.30)

(3.29) takes the form

$$\sin 2\omega = \sin 2i \sin\delta.$$

(3.31)

Relations (3.22, 27, 28, 31) relate the semi-axes a and b of the ellipse and the angle of inclination ψ with the amplitudes A_x and A_y and the phase difference $\delta = (\delta_y - \delta_x)$ of the original vibrations given by (3.8).

According to the previous definition, a circular or elliptical polarization is called right-handed when an observer looking towards the light source sees the light vector rotating in the clockwise sense. It can be easily shown from (3.8), by considering two instants separated, for example, by a quarter of a period, for a right-handed elliptical polarization, that $\sin \delta > 0$, that is $0° < \delta < 180°$, whereas elliptical polarization is left-handed when $180° < \delta < 360°$.

For the case when δ takes the values $\delta = 0°$, $180°$, and $360°$, the ellipse degenerates to a straight line and we have the case of linear polarization, whereas when $A_x = A_y$ and $\delta = \pm 90$ deg., we have circularly polarized light.

Various forms of polarization that correspond to various values of the phase angle δ are shown in Fig. 3.5.

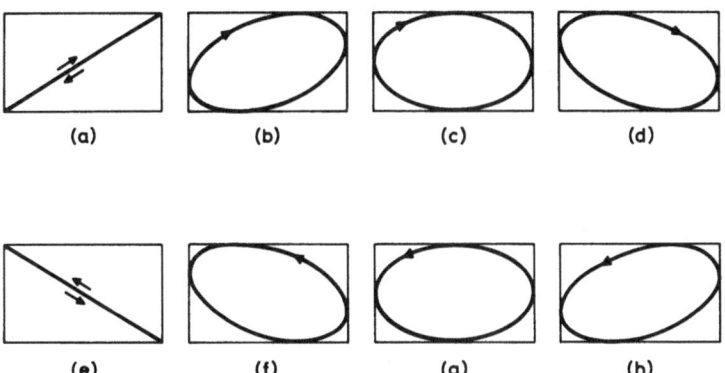

Fig. 3.5a–h. Elliptical-polarization forms that correspond to the indicated values of the phase difference δ between two mutually perpendicular components of the light vector (a) $\delta = 0$; (b) $0 < \delta < 90°$; (c) $\delta = 90°$; (d) $90° < \delta < 180°$; (e) $\delta = 180°$; (f) $180° < \delta < 270°$; (g) $\delta = 270°$; (h) $270° < \delta < 360°$

3.4.2 The Poincaré Sphere

As has already been shown, the most general form of polarization is elliptical polarization, which is represented by the light ellipse. This ellipse is characterized by three independent quantities, either the amplitudes A_x and A_y of the vibrations and their phase difference δ, or by the lengths of the major and minor semi-axes a and b and the angle ψ of inclination of the major axis of the ellipse with respect to the Ox axis. Complete characterization of the light ellipse requires also knowledge of the direction of rotation of the light vector when it describes the ellipse. As was previously shown, right-handed elliptical polarization corresponds to a phase difference δ in the interval $0 < \delta < 180°$, whereas for left-handed elliptical polarization $180° < \delta < 360°$. From (3.31) we obtain that for $0 < \delta < 180°$, $0 < \omega \leq 45°$, whereas for $180° < \delta < 360°$, $-45° \leq \omega < 0$.

In many problems that involve the passage of a light ray through a train of optical elements we are not interested in the absolute value of the light intensity. which is expressed by the sum of the squares of the amplitudes of the light ellipse $(a^2 + b^2 = A_x^2 + A_y^2)$, but on its relative value. In this way the transformation of the shape, and not the size of the light ellipse by each optical element is of interest. For such problems, two quantities can be used to describe the light ellipse: first the angle ω $(-45° \leq \omega \leq 45°)$ whose $\tan\omega$ is the ratio (b/a) of the semi-axes of the ellipse and whose sign specifies the sense of rotation of the light vector that forms the ellipse, and secondly, the angle ψ $(0 \leq \psi < \pi)$ of inclination of the major axis of the ellipse with respect to the x axis. In such cases, the light intensity may be assumed equal to unity $(a^2 + b^2 = A_x^2 + A_y^2 = 1)$, that is, the light vector is normalized to unit intensity.

The Poincaré sphere is a sphere of unit radius, each point P of which is defined by its two spherical angular coordinates (the longitude and the latitude) and represents a different polarization form. The longitude and latitude of point P are 2ψ and 2ω, respectively (Fig. 3.6).

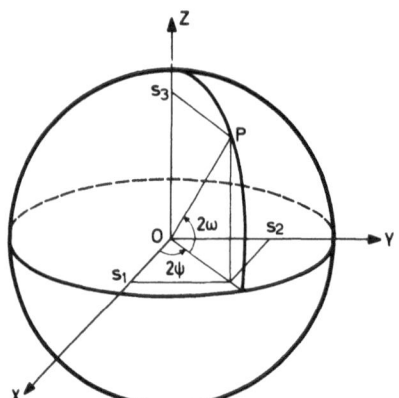

Fig. 3.6. Representation of elliptically polarized light by a point P on the Poincaré sphere. (The longitude 2ψ and latitude 2ω of the point P are equal to the double values of the azimuth ψ of the major axis and the ellipticity ω of the corresponding light ellipse.) The cartesian coordinates s_1, s_2, s_3 of the point P are the last three Stokes parameters of the polarized light

The Poincaré sphere is a convenient way of mapping, such that, to each possible polarization form corresponds a point on the sphere with spherical angular coordinates 2ψ and 2ω. Conversely, to each point of the sphere with longitude 2ψ and latitude 2ω corresponds an elliptically polarized light, whose cross-sectional pattern has azimuth ψ and ellipticity ω.

The Poincaré sphere is referred to a right-handed cartesian system of coordinates (Fig. 3.6). The angle 2ψ is measured from the $0x$ axis in the counter-clockwise sense $(0 \leq 2\psi < 360°)$. Positive values of angle ω, which represent right-elliptical polarization forms, are represented on the upper hemisphere $(0 < 2\omega \leq 90°)$. Negative values of ω, which represent left-elliptical polarization forms, are represented on the lower (southern) hemisphere $(-90° \leq 2\omega < 0)$. Therefore,

I) Each point on the equator of the sphere, defined by $0 \leq 2\psi < 360°$ and $\omega = 0$, represents a different form of linear polarization. Moreover, the point on the positive x axis represents horizontal polarization ($\psi = 0$), whereas the point on the negative x axis represents vertical polarization ($\psi = 90°$).

II) The north pole of the Poincaré sphere represents right-circular polarization ($\omega = 45°$, since $a = b$), whereas the south pole represents left-circular polarization.

III) Each point on the northern hemisphere represents a right-elliptical polarization form, whereas each point on the southern hemisphere represents left-elliptically polarized light. Furthermore, the half values of the longitude 2ψ and the latitude 2ω of each point of the sphere define the azimuth ψ and the ellipticity ω of the cross-sectional pattern of the corresponding elliptical polarization.

IV) Points on the same meridian ($\psi = $ constant) represent all forms of elliptically polarized light that have light ellipses of the same orientation, whereas points on the same parallel ($\omega = $ constant) represent all forms of elliptical polarization with the same ellipticity.

If we are interested in the absolute value of the light intensity of an elliptically polarized light, the radius of the Poincaré sphere is not taken equal to unit y, but is equal to the light intensity ($I = a^2 + b^2 = A_x^2 + A_y^2$).

We can readily show that the cartesian coordinates X, Y, Z of a point on the Poincaré sphere are

$$X = \cos 2\omega \cos 2\psi$$

$$Y = \cos 2\omega \sin 2\psi \qquad (3.32)$$

$$Z = \sin 2\omega .$$

Because unpolarized light can be defined as a mixture of all possible polarization forms, represented by points on the Poincaré sphere, the center of the sphere may be defined to represent unpolarized light.

3.4.3 The j Circle

Whereas in the Poincaré sphere a three-dimensional model is used to describe elliptically polarized light, in the j-circle method the representation is two dimensional. Indeed, the j circle is the orthographic (parallel) projection of the Poincaré sphere on its equatorial plane.

In the j-circle method, each form of polarization is represented by a vector, called the j vector, whose length is defined by

$$|j| = \frac{a^2 - b^2}{a^2 + b^2}, \qquad (3.33)$$

where a and b are the lengths of the semi-axes of the corresponding light ellipse,

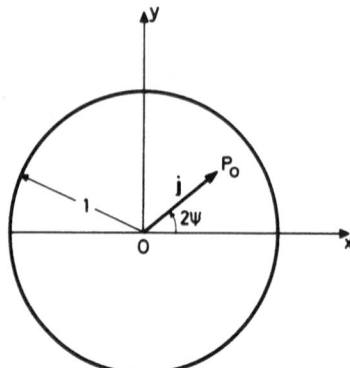

Fig. 3.7. Representation of elliptically polarized light by the j vector in the j-circle method. The magnitude of the j vector is $|j| = (a^2 - b^2)/(a^2 + b^2)$ where a and b are the semi-axes of the corresponding light ellipse, and its inclination with the Ox axis is twice the azimuth ψ of the major axis of the corresponding light ellipse

and whose direction forms an angle equal to 2ψ with the Ox axis and indicates the direction of the major axis of the light ellipse (Fig. 3.7).

We can readily show from (3.33) that the tips of all j vectors, which correspond to all forms of elliptical polarization, lie inside a circle of unit radius, called the j circle.

According to this representation of a polarization form,

I) All forms of linear polarization lie on the circumference of the j circle.

II) Circularly polarized light is represented by the origin of the coordinate axes.

III) Any form of elliptical polarization is represented by a point inside the j circle and vice-versa.

3.4.4 The Stokes Vector

The Stokes vector is a four-element vector, which completely characterizes any form of elliptical polarization. The four elements s_0, s_1, s_2, and s_3 of this vector are defined in terms of the amplitudes A_x, A_y and the phase difference δ of the α_x and α_y components of the light vector α, given by (3.8) and (3.12) in the most-general case of elliptical polarization, by

$$s_0 = A_x^2 + A_y^2$$
$$s_1 = A_x^2 - A_y^2$$
$$s_2 = 2 A_x A_y \cos \delta \qquad (3.34)$$
$$s_3 = 2 A_x A_y \sin \delta .$$

It is clear from these relations that the first element s_0 of the Stokes vector represents the intensity of the polarized light, and that all of the other three elements s_1, s_2, s_3 also have dimensions of intensity. Usually, as will be shown later in conjunction with the Mueller calculus, the four elements s_0, s_1, s_2, s_3 are arranged to form a four-element column vector

$$S = \begin{bmatrix} s_0 \\ s_1 \\ s_2 \\ s_3 \end{bmatrix}. \tag{3.35}$$

The four parameters s_0, s_1, s_2, s_3 are not independent, but satisfy the identity

$$s_0^2 = s_1^2 + s_2^2 + s_3^2. \tag{3.36}$$

By use of (3.22, 23, 26, 27, 30, 31), the Stokes-vector elements can be expressed in terms of azimuth ψ and ellipticity ω that define the inclination of the light ellipse and the ratio of its semi-axes, respectively, by

$$s_1 = s_0 \cos 2\omega \cos 2\psi$$
$$s_2 = s_0 \cos 2\omega \sin 2\psi \tag{3.37}$$
$$s_3 = s_0 \sin 2\omega.$$

By comparing (3.37) and (3.32), we can see that the elements s_1, s_2, s_3 represent the cartesian coordinates X, Y, Z of a point on the Poincaré sphere, of radius s_0.

By use of (3.34) or (3.37), the elements of the Stokes vector, corresponding to a given polarization form, can be easily determined. Thus, for linear horizontally polarized light ($A_x \neq 0$, $A_y = 0$, $\delta = 0$, or $\psi = 0$, $\omega = 0$) the four quantities s_0, s_1, s_2, s_3 are

$$s_0 = A_x^2, \quad s_1 = A_x^2, \quad s_2 = 0, \quad s_3 = 0. \tag{3.38}$$

For right-circularly polarized light ($A_x = A_y = A$, $\delta = 90°$, or $\omega = 45°$),

$$s_0 = 2A^2, \quad s_1 = 0, \quad s_2 = 0, \quad s_3 = 2A^2, \tag{3.39}$$

whereas, for left-circularly polarized light ($A_x = A_y = A$, $\delta = -90°$, or $\omega = -45°$) we have

$$s_0 = 2A^2, \quad s_1 = 0, \quad s_2 = 0, \quad s_3 = -2A^2. \tag{3.40}$$

Usually, we are interested only in the relative values of the Stokes parameters, so that we can divide all of them by the first parameter s_0. The normalized Stokes parameters, thus obtained, correspond to polarized light of unit intensity. For the previously examined three cases of polarized light, the normalized vectors are

$$\begin{bmatrix} 1 \\ 1 \\ 0 \\ 0 \end{bmatrix}, \quad \begin{bmatrix} 1 \\ 0 \\ 0 \\ 1 \end{bmatrix}, \quad \begin{bmatrix} 1 \\ 0 \\ 0 \\ -1 \end{bmatrix}. \tag{3.41}$$

Relations (3.34) or (3.37) establish a unique correspondence between a given polarization form and the Stokes parameters and vice-versa.

The Stokes vector has a fundamental property for the addition of two incoherent polarized light beams. Because all elements of the Stokes vector have dimensions of intensity, and intensity is a scalar quantity, it is easily deduced that the Stokes vector of the addition of two incoherent polarized beams of light is the sum of the corresponding Stokes vectors of the two separate beams. Thus, if two beams with Stokes vectors S' and S'', given by

$$S' = \begin{bmatrix} s'_0 \\ s'_1 \\ s'_2 \\ s'_3 \end{bmatrix}, \quad S'' = \begin{bmatrix} s''_0 \\ s''_1 \\ s''_2 \\ s''_3 \end{bmatrix}, \tag{3.42}$$

are superposed, the Stokes vector S of the combination of these two beams is

$$S = S' + S'' = \begin{bmatrix} s'_0 \\ s'_1 \\ s'_2 \\ s'_3 \end{bmatrix} + \begin{bmatrix} s''_0 \\ s''_1 \\ s''_2 \\ s''_3 \end{bmatrix} = \begin{bmatrix} s'_0 + s''_0 \\ s'_1 + s''_1 \\ s'_2 + s''_2 \\ s'_3 + s''_3 \end{bmatrix}. \tag{3.43}$$

This "additive" property of Stokes vector facilitates calculations involved in the addition of two incoherent polarized beams of light.

Table 3.1 gives the Stokes vector corresponding to some common polarization forms.

3.4.5 The Jones Vector

The Jones vector is a two-element complex column vector, whose two elements are equal to the a_x and a_y components of the light vector a. The Jones vector, in the most general case of elliptical polarization is

$$a = \begin{bmatrix} a_x \\ a_y \end{bmatrix} = \begin{bmatrix} A_x e^{i(v + \delta_x)} \\ A_y e^{i(v + \delta_y)} \end{bmatrix}, \tag{3.44}$$

3.4 Description of Polarized Light   37

Table 3.1. Stokes vectors of polarized light

	Elliptically polarized light		Circularly polarized light		Linearly polarized light		
	Light vector α $\alpha = a_x \mathbf{i} + a_y \mathbf{j}$ $a_x = A_x e^{i(v+\delta_x)}$ $a_y = A_y e^{i(v+\delta_y)}$ $\delta = \delta_y - \delta_x$	Light ellipse with azimuth ψ and ellipticity ω ($\tan\omega = b/a$) b, a semi-axes of the ellipse.	Right $A_x = A_y = A,\ \delta = (\pi/2)$ $\omega = (\pi/4)$	Left $A_x = A_y = A,\ \delta = -(\pi/2)$ $\omega = -(\pi/4)$	General at angle θ with the Ox axis $\delta = 0,\ (A_x/A) = \cos\theta,\ (A_y/A) = \sin\theta,\ \omega = 0$	Horizontal $\theta = 0$	Vertical $\theta = (\pi/2)$
Stokes vectors	$\begin{bmatrix} A_x^2 + A_y^2 \\ A_x^2 - A_y^2 \\ 2A_xA_y\cos\delta \\ 2A_xA_y\sin\delta \end{bmatrix}$	$\begin{bmatrix} s_0{}^a \\ s_0\cos 2\omega\cos 2\psi \\ s_0\cos 2\omega\sin 2\psi \\ s_0\sin 2\omega \end{bmatrix}$	$\begin{bmatrix} 2A^2 \\ 0 \\ 0 \\ 2A^2 \end{bmatrix}$	$\begin{bmatrix} 2A^2 \\ 0 \\ 0 \\ -2A^2 \end{bmatrix}$	$\begin{bmatrix} A^2 \\ A^2\cos 2\theta \\ A^2\sin 2\theta \\ 0 \end{bmatrix}$	$\begin{bmatrix} A^2 \\ A^2 \\ 0 \\ 0 \end{bmatrix}$	$\begin{bmatrix} A^2 \\ -A^2 \\ 0 \\ 0 \end{bmatrix}$
Normalized Stokes vectors	$\begin{bmatrix} 1 \\ \cos 2\omega\cos 2\psi \\ \cos 2\omega\sin 2\psi \\ \sin 2\omega \end{bmatrix}$		$\begin{bmatrix} 1 \\ 0 \\ 0 \\ 1 \end{bmatrix}$	$\begin{bmatrix} 1 \\ 0 \\ 0 \\ -1 \end{bmatrix}$	$\begin{bmatrix} 1 \\ \cos 2\theta \\ \sin 2\theta \\ 0 \end{bmatrix}$	$\begin{bmatrix} 1 \\ 1 \\ 0 \\ 0 \end{bmatrix}$	$\begin{bmatrix} 1 \\ -1 \\ 0 \\ 0 \end{bmatrix}$

[a] $s_0 = A_x^2 + A_y^2 = a^2 + b^2$

where

$$v = \omega t + \frac{2\pi z}{\lambda}.$$

This vector can be written in the form

$$e^{iv} \begin{bmatrix} A_x e^{i\delta_x} \\ A_y e^{i\delta_y} \end{bmatrix}. \tag{3.45}$$

In cases in which we are not interested in the time factor $\exp(iv)$, whose magnitude is equal to unity, the Jones vector, normalized to time factor, takes the form

$$\alpha = \begin{bmatrix} A_x e^{i\delta_x} \\ A_y e^{i\delta_y} \end{bmatrix}. \tag{3.46}$$

Usually, the Jones vector is normalized by multiplying its two elements by whatever scalar quantities are suitable to reduce the value of the corresponding light intensity to unity. Thus, the normalized form of (3.46) can be written

$$\begin{bmatrix} \cos B e^{-i(\delta/2)} \\ \sin B e^{i(\delta/2)} \end{bmatrix}, \tag{3.47}$$

where

$$B = \left| \arctan \frac{A_y}{A_x} \right|, \qquad \delta = (\delta_y - \delta_x).$$

By use of either (3.46) or (3.47), the elements of the Jones vector, corresponding to a given polarization form, can be easily determined. Thus, for linear horizontally polarized light, we have the standard and the simplified vectors

$$\begin{bmatrix} A_x e^{i\delta_x} \\ 0 \end{bmatrix}, \quad \begin{bmatrix} 1 \\ 0 \end{bmatrix}. \tag{3.48}$$

For either a right-circularly, or a left-circularly polarized light we have either of the forms

$$\begin{bmatrix} A e^{i\delta_x} \\ A e^{i(\delta_x + \pi/2)} \end{bmatrix}, \quad \begin{bmatrix} A e^{i(\delta_x + \pi/2)} \\ A e^{i\delta_x} \end{bmatrix}. \tag{3.49}$$

If we apply (3.47) we can obtain the respective normalized vectors

$$\frac{1+i}{2}\begin{bmatrix} -i \\ 1 \end{bmatrix}, \quad \frac{1-i}{2}\begin{bmatrix} i \\ 1 \end{bmatrix}.$$

Furthermore, the normalized vectors can be expressed as

$$\frac{1}{\sqrt{2}}\begin{bmatrix} -i \\ 1 \end{bmatrix} \quad \text{and} \quad \frac{1}{\sqrt{2}}\begin{bmatrix} i \\ 1 \end{bmatrix}. \tag{3.50}$$

From either (3.46) or (3.47) it can be seen that to a given polarization form there is a Jones vector corresponding to it and vice-versa.

The light intensity I of the polarized light beam, for which the Jones vector is given by (3.45), is equal to

$$I = A_x^2 + A_y^2. \tag{3.51}$$

Relation (3.51) of the light intensity can be derived as the matrix product of the matrix a and the complex conjugate $\tilde{a}$ of its transpose, called the hermitian conjugate matrix of matrix a. This matrix $\tilde{a}$ is defined by

$$\tilde{a} = [A_x e^{-i\delta_x} \quad A_y e^{-i\delta_y}]. \tag{3.52}$$

Thus, we have

$$I = \tilde{a}\, a = [A_x e^{-i\delta_x} \quad A_y e^{-i\delta_y}]\begin{bmatrix} A_x e^{i\delta_x} \\ A_y e^{i\delta_y} \end{bmatrix} = A_x^2 + A_y^2. \tag{3.53}$$

Unlike the Stokes vector, which is particularly relevant in the incoherent addition of two beams of light, this Jones vector can be used to find the result of the coherent addition of two beams of polarized light. For this purpose the Jones vectors of the two beams are added. Thus, if

$$a' = \begin{bmatrix} a'_x \\ a'_y \end{bmatrix} \quad \text{and} \quad a'' = \begin{bmatrix} a''_x \\ a''_y \end{bmatrix} \tag{3.54}$$

are the Jones vectors of the two beams of light, the Jones vector a of their

Table 3.2. Jones vectors of polarized light

Elliptically polarized light	Circularly polarized light		Linearly polarized light		
	Right	Left	General at an angle θ with the Ox axis, $\delta=0$	Horizontal	Vertical
Light vector α $\alpha = a_x i + a_y j$ $a_x = A_x e^{i\delta_x}$ $a_y = A_y e^{i\delta_y}$ $\delta = \delta_y - \delta_x$	$A_x = A_y = A$ $\delta = (\pi/2)$	$A_x = A_y = A$ $\delta = -(\pi/2)$		$\theta=0$ $A_y=0$	$\theta=(\pi/2)$ $A_x=0$
$\begin{bmatrix} A_x e^{i\delta_x} \\ A_y e^{i\delta_y} \end{bmatrix}$	$\begin{bmatrix} A e^{i\delta_x} \\ A e^{i(\delta_x+\pi/2)} \end{bmatrix}$	$\begin{bmatrix} A e^{i(\delta_x+\pi/2)} \\ A e^{i\delta_x} \end{bmatrix}$	$\begin{bmatrix} A_x e^{i\delta_x} \\ \pm A_y e^{i\delta_x} \end{bmatrix}$	$\begin{bmatrix} A_x e^{i\delta_x} \\ 0 \end{bmatrix}$	$\begin{bmatrix} 0 \\ A_y e^{i\delta_y} \end{bmatrix}$
Normalized Jones vectors					
$\begin{bmatrix} \cos B\, e^{-i(\delta/2)} \\ \sin B\, e^{i(\delta/2)} \end{bmatrix}$	$\dfrac{1+i}{2}\begin{bmatrix} -i \\ 1 \end{bmatrix},\ \dfrac{1}{\sqrt{2}}\begin{bmatrix} -i \\ 1 \end{bmatrix}$	$\dfrac{1-i}{2}\begin{bmatrix} i \\ 1 \end{bmatrix},\ \dfrac{1}{\sqrt{2}}\begin{bmatrix} i \\ 1 \end{bmatrix}$	$\begin{bmatrix} \cos\theta \\ \pm\sin\theta \end{bmatrix}$	$\begin{bmatrix} 1 \\ 0 \end{bmatrix}$	$\begin{bmatrix} 0 \\ 1 \end{bmatrix}$

^{a} $B = \left| \arctan \dfrac{A_y}{A_x} \right|$

combination will be

$$a = a' + a'' = \begin{bmatrix} a'_x \\ a'_y \end{bmatrix} + \begin{bmatrix} a''_x \\ a''_y \end{bmatrix} = \begin{bmatrix} a'_x + a''_x \\ a'_y + a''_y \end{bmatrix}. \tag{3.55}$$

Table 3.2 gives the Jones vectors that correspond to some common polarization forms.

3.5 Relationship Between the Methods of Characterization of a Polarized Light

The above-established methods of characterization of polarization form, that is, the Poincaré sphere, the j-circle method, the Stokes vector, and the Jones vector, were introduced independently in terms of the parameters necessary to define elliptical polarization, namely the azimuth, the ellipticity, and the sense of rotation of the ellipse described by the light vector. However, these methods are not independent. Each one of them may be expressed in terms of the others. We shall now establish the relations between these methods.

3.5.1 The Poincaré Sphere and the j Circle

The j vector is used to specify each polarization form by a point inside a unit circle; the magnitude of the j vector is given by (3.33), whereas its direction subtends an angle 2ψ with the Ox axis equal to twice the angle of inclination of the major axis of the light ellipse to that axis. This circle can be regarded as the orthographic projection on the equatorial plane of the point on the Poincaré sphere of unit radius, which point defines the same polarization form. The projection of this radius on the equatorial plane makes the same angle 2ψ with the Ox axis that the j vector does, and has a magnitude OP' equal to

$$OP' = \cos 2\omega, \tag{3.56}$$

which, on account of (3.31), yields

$$OP' = (\cos^2 2i + \sin^2 2i \cos^2 \delta)^{1/2}. \tag{3.57}$$

The magnitude of the j vector, given by (3.33) by use of (3.30, 31), yields the same expression (3.57). This means that the magnitude of the j vector and the projection of the corresponding unit radius of the Poincaré sphere, which represents the same polarization form on the equatorial plane, are equal.

Therefore, all $\boldsymbol{j}$ vectors that represent various polarization forms may be regarded as normal projections on the equatorial plane of the corresponding unit radius vectors of the Poincaré sphere, whose end points represent the same polarization form as the $\boldsymbol{j}$ vector.

3.5.2 The Poincaré Sphere and the Stokes Vector

By comparison of (3.32) and (3.37), the three elements of the Stokes vector s_1, s_2, s_3 can be shown to equal the cartesian coordinates of the corresponding point on the Poincaré sphere.

Relation (3.36) suggests also that when the parameters s_1, s_2, s_3 are regarded as the cartesian coordinates of a point P, then this point lies on a sphere of radius $R = s_0 = A_x^2 + A_y^2$, that is, *the geometrical representation of the Stokes vector is the Poincaré sphere.*

3.5.3 The Stokes and Jones Vectors

Because both the Stokes and the Jones vectors describe polarization forms, a relationship between these vectors must exist. In order to define this relationship, we shall introduce the hermitian conjugate matrix $\tilde{A}$ of a given matrix A by the relation

$$\tilde{A}_{ij} = \bar{A}_{ji} \tag{3.58}$$

where $\bar{A}_{ji}$ denotes the complex conjugate of A_{ji}.

Then, if $\boldsymbol{a}$ is the Jones vector of a given polarization form, defined by

$$\boldsymbol{a} = \begin{bmatrix} A_x e^{i\delta_x} \\ A_y e^{i\delta_y} \end{bmatrix}, \tag{3.59}$$

and S is the corresponding Stokes vector of the same polarization form, defined by

$$S = \begin{bmatrix} s_0 \\ s_1 \\ s_2 \\ s_3 \end{bmatrix} = \begin{bmatrix} A_x^2 + A_y^2 \\ A_x^2 - A_y^2 \\ 2A_x A_y \cos\delta \\ 2A_x A_y \sin\delta \end{bmatrix}, \quad \delta = (\delta_y - \delta_x), \tag{3.60}$$

the following relations can readily be proved:

$$\begin{aligned}
s_0 &= \tilde{a}\,Q_1\,a & s_2 &= \tilde{a}\,Q_3\,a \\
s_1 &= \tilde{a}\,Q_2\,a & s_3 &= \tilde{a}\,Q_4\,a\,, & \text{where}
\end{aligned} \tag{3.61}$$

$$Q_1 = \begin{bmatrix} 1 & 0 \\ 0 & 1 \end{bmatrix}, \quad Q_2 = \begin{bmatrix} 1 & 0 \\ 0 & -1 \end{bmatrix}, \quad Q_3 = \begin{bmatrix} 0 & 1 \\ 1 & 0 \end{bmatrix}, \quad Q_4 = \begin{bmatrix} 0 & -i \\ i & 0 \end{bmatrix}.$$

(3.62)

Matrices Q_2, Q_3, Q_4 are closely related to Pauli spin matrices [3.6].

3.6 Discussion of the Methods of Characterization of Polarized Light

The effectiveness, powerfulness, and superiority of the new methods of characterization of polarization forms over the older methods of using the light vector will be shown in the next chapter, where the optical transformations of polarized light when it passes through a train of optical elements will be discussed.

Although only completely polarized light will be considered throughout this book, the Poincaré sphere can also be used to describe partially polarized light; the Stokes vector is also convenient for partially polarized, as well as for unpolarized light. Thus, the Stokes parameters were first introduced by the following relations based on concepts of the electromagnetic theory of light

$$\begin{aligned}
s_0 &= \langle A_x^2 + A_y^2 \rangle \\
s_1 &= \langle A_x^2 - A_y^2 \rangle \\
s_2 &= \langle 2 A_x A_y \cos \delta \rangle \\
s_3 &= \langle 2 A_x A_y \sin \delta \rangle,
\end{aligned}$$

(3.63)

where the angle brackets indicate time averages of the corresponding quantities.

Thus, for the case of unpolarized light, for which there is no time-averaged preference between A_x and A_y, and in the phase difference, the four Stokes parameters take the values

$$s_0 = \langle 2 A_x^2 \rangle, \quad s_1 = s_2 = s_3 = 0.$$

Thus, we conclude that the normalized Stokes vector

$$\begin{bmatrix} 1 \\ 0 \\ 0 \\ 0 \end{bmatrix}$$

(3.64)

represents unpolarized light.

Besides this definition of Stokes parameters in terms of the electromagnetic theory of light, these parameters can also be introduced operationally. This approach considers passage of the light through a set of four different polarizing filters and measurement of the intensity of the light that passes through each separate filter. It corresponds to the historical development of the Stokes parameters. Both approaches can be proved to be equivalent. The operational introduction of Stokes parameters provides further evidence that these parameters can be used to describe any kind of light, regardless of whether it is polarized, unpolarized, partially polarized, monochromatic, or polychromatic.

Unlike the Stokes vector, the Jones vector is used to describe only polarized light. This vector, which contains only two elements, describes completely polarized light with great simplicity and elegance.

Also, unlike the Stokes vector, which is particularly relevant in the incoherent addition of two beams of light, the Jones vector is especially suitable for immediately determining the combination of two coherent polarized light beams.

The importance of Stokes and Jones vectors will be shown in the next chapter, in conjunction with Mueller and Jones calculi, respectively, in predicting the effect of an optical element on polarized light.

While Stokes and Jones vectors are used to compute the effects of optical elements on polarized light, the Poincaré sphere and *j*-circle methods are especially convenient for graphical prediction of the same effects. Application of the properties of the Poincaré sphere may yield short-cut solutions to photoelastic problems, at least qualitatively, whose solutions by conventional methods may be difficult.

References

3.1 H. Poincaré: *Théorie Mathématique de la Lumière*, Vol. 2 (Gauthiers-Villars, Paris 1892) Chap. 12
3.2 A. Kuske: Exp. Mech. **6**, 218 (1966)
3.3 H. J. Menges: Z. angew. Math. Mech. **20**, 210 (1940)
3.4 G. G. Stokes: Trans. Cambridge Phil. Soc. **9**, 399 (1852)
3.5 R. C. Jones: J. Opt. Soc. Am. **31**, 488 (1941)
3.6 C. Whitney: J. Opt. Soc. Am. **61**, 1207 (1971)

4. Passage of Polarized Light Through Optical Elements

4.1 Introduction

In the previous chapter the modern methods of describing polarization form, based on the Poincaré sphere and the Stokes and Jones vectors, were described. All of these methods are characterized by extreme elegance and compactness; they provide, in a unified manner, for all forms of polarized light. Thus, any kind of polarization, regardless of whether elliptical or linear, can be represented by a point on the Poincaré sphere, which provides a clear and well-defined picture of all particular polarization forms. Similarly, by the Stokes vector, each kind of light, regardless of whether it is totally or partially polarized, can be characterized by a four-element vector. The elements of the vector completely define the degree and the state of polarization (azimuth, ellipticity, and handedness of the light ellipse) of each particular form of light. In the same way every form of elliptically polarized light can be completely described by the two-element complex Jones vector.

However, the main advantage of these modern methods is not merely uniform representation of each separate polarization form, but that they provide suitable means for facilitating solutions of the problem of finding the polarization form of the output light from an optical device, for any form of input polarization and any type of device.

Thus, in the Poincaré-sphere method, because all forms of polarization are represented by particular points on the sphere and because each input polarization form is transformed by any interposed optical device into a definite output polarization form, the transition from the input to the output form can be defined by a transformation on the sphere. As will be shown later, this transformation, for the case of any device, regardless of whether linear or elliptical, consists of rotating the sphere about a given axis, through an appropriate angle, with the suitable sense of rotation. Similarly, the light intensity, transmitted through any elliptical polarizer, is expressed in terms of the arc that connects the relevant points of the input polarization form and the eigenvector of one of the two principal axes of the polarizer.

This simple and uniform manner of handling problems that involve passage of light through birefringent media and polarizers enables us to solve qualitatively the corresponding problem by having in mind the sphere and imagining appropriate rotations of it. Furthermore, useful theorems of polarization optics can be formulated and proved, dealing with polarizers and retarders.

The Poincaré-sphere method [4.1] was applied by many investigators to problems of polarization optics during the first decades of our century [4.2–13]. *Jerrard* [4.14] gave a brief account of the theory of the Poincaré sphere and constructed a simple model of the sphere. Suitable spheres with degree scales were also provided by *Koester* [4.15]. A special "Wulff-net" type of paper for handling problems of geometrical manipulations of the sphere was described by *Hartshorne* and *Stuart* [4.16] and used in [4.14]. *Wright* [4.17] used a stereographic projection of the Poincaré sphere and converted all transformations of the sphere into corresponding transformations on a plane. A comprehensive and thorough discussion of all problems connected with the Poincaré sphere and its various applications to polarization-optics problems, dealing with completely or partially polarized light, was given by *Ramachandran* and *Ramaseshan* [4.18]. *Pancharatnam* [4.19, 20] and *Koester* [4.15] used the Poincaré sphere in their investigations for the design of an achromatic, multilayer retarder. *Pancharatnam* [4.21, 22] used also the Poincaré sphere for the study of the interference of two coherent beams of light and the resolution of any polarization form into two prescribed polarization forms. Discussions and applications of the Poincaré sphere can also be found in [4.23–39].

Kuske, in a series of papers [4.40–43] studied some polarization-optics problems with applications to two- and three-dimensional photoelasticity by using the orthographic projection of the Poincaré sphere on its equatorial plane; he was thus able to work with a two-dimensional instead of the three-dimensional model that is required by the method of the Poincaré sphere. With each point of the Poincaré sphere *Kuske* associated its projection on the equatorial plane of the sphere; this produced a vector, called the *j* *vector*, which characterizes each particular polarization form. All geometrical transformations in the *j* circle can be derived from the corresponding transformations on the sphere; thus the *j-circle method* need not be considered separately from the Poincaré sphere method. *Kuske* has applied this useful idea of the *j*-circle method to a series of problems in two- and three-dimensional photoelasticity.

According to the method of the Stokes vector, each particular form of light, whether partially or completely polarized, is fully characterized by four real quantities, which constitute the components of the Stokes vector. The use of the Stokes vector in the solution of polarization-optics problems was introduced by *Mueller* [4.44, 45]. *Mueller* based his calculus on the fact that, if the four Stokes parameters are considered as the elements of a four-element column vector, then, because both the input and output light beams to and from an optical device are represented by such vectors, the device itself must be represented by a four-by-four real matrix. Indeed, the four-by-four matrix is the suitable operator that connects the two four-element column matrices. The complete characterization of an optical device by such a matrix, whose elements have appropriate values, according to the particular device considered, leads to an elegant and unified prediction of the polarization form of the light beam that emerges from a device, when the polarization form of the incident light is known. The only thing necessary is to multiply the column vector of the incident light by the matrix of the device; the result will be the Stokes vector of the emerging light. The multiplication process can be applied for any number of optical devices interposed in the light path. Besides the Stokes-vector concept, particular help for the formulation of the Mueller calculus was provided by the previous works of *Soleillet* and *Perrin*. Thus, *Soleillet* [4.46] proved that the Stokes vector is transformed linearly when light passes through an optical device, whereas *Perrin* [4.47] showed that this transformation can be expressed in matrix form; *Perrin* was therefore the first to conceive the idea of the Mueller calculus.

Parke [4.48–51] used the Mueller calculus to deal with problems of unpolarized light and related it to the electromagnetic theory of light. A brief description of Mueller calculus and its relation to the Poincaré sphere was published by *Walker* [4.52]. *Fano* [4.53, 54] and *McMaster* [4.55, 56] gave a quantum-mechanics representation of the calculus and applied it to the scattering of gamma radiation. The Mueller calculus was used by *Billings* [4.57–59] in the solution of polarization-optics problems. Other interesting accounts, dealing with the Mueller calculus and its various applications can be found in [4.60–62].

The Jones calculus is based on the corresponding Jones vector, which characterizes a given elliptical polarization form, in the same manner in which the Mueller calculus is based on the Stokes vector. However, the Jones calculus can be applied to only elliptically polarized light, whereas the Mueller calculus may be used for any form of light. Because any elliptically polarized light is represented by a two-element complex column vector, and the matrix operator that connects two such vectors is a two-by-two complex matrix, any optical device must be represented by such a two-by-two matrix. This matrix is called the *Jones matrix* of the device, and the corresponding calculus, which leads to short-cut solutions of polarization-optics problems, is called the *Jones calculus*. This calculus was conceived by *Jones* and was presented in a series of papers [4.63–71]. *Hsü* et al. [4.72] have used the Jones calculus for the calculation of the intensity of a light beam passing through a train of retardation plates placed between two polarizers. For the same purpose, the Jones calculus was used by *Dawson* and *Young* [4.73] and *Evans* [4.74]. Other pertinent works, using the Jones calculus for the solution of polarization or quantum-optics problems, or related to this calculus, can be found in [4.75, 77]. Also, in [4.48–51] both the Mueller and Jones calculi are discussed and interrelated.

A different approach to the Jones calculus was provided by *Aben* [4.39, 78–80], who derived the Jones matrices of birefringent and rotating media by using Ginsburg's electromagnetic equations. This approach will be described in Chapter 8.

In the following sections, all of these three methods of handling polarization-optics problems, based on the Poincaré sphere, and the Mueller and Jones calculi, will be described and discussed. Particular attention will be given to the problem of predicting the output polarization form from an optical device, when the input form is given. The relationship between the Mueller and Jones matrices will be indicated and established. At the end of the chapter some useful theorems related to the behavior of a train of optical elements to a polarized light beam will be discussed.

4.2 Definition of Optical Elements

Before proceeding to the prediction of the effect of an optical element on a given light beam, we will define the most common optical elements that will be discussed in this book. We shall deal only with the mode of operation of these elements for specified light beams; we shall not become involved with the construction of such optical elements, for which the interested reader is referred to appropriate books of optics [4.18, 37, 81]. The optical transformations that occur when a polarized light passes through an optical element will be established by use of the conventional vectorial description of polarized light, the Poincaré sphere, and the Mueller and Jones calculi.

4.2.1 Orthogonal Forms

Two forms of elliptical polarization are called orthogonal if their corresponding light ellipses have the same ellipticity, their azimuths differ by 90°, and they are described in opposite directions. From this definition it follows that two perpendicular linear polarizations are orthogonal, and right- and left-circularly polarized beams are also orthogonal pairs. Orthogonal forms of linear, circular, and elliptical polarizations are shown in Fig. 4.1.

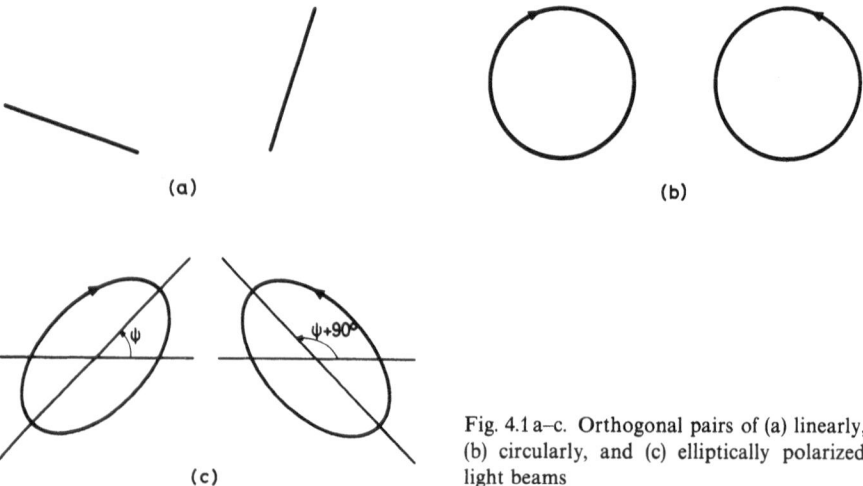

Fig. 4.1 a–c. Orthogonal pairs of (a) linearly, (b) circularly, and (c) elliptically polarized light beams

4.2.2 Polarizers

A *polarizer* is an optical element that divides the incident light beam into two orthogonal forms and transmits them with different intensities.

The action of a polarizer is described by its two eigenvectors and its two transmittance coefficients. An eigenvector is a polarization form that, when it impinges on a polarizer, emerges unchanged. An eigenvector of a polarizer can be depicted by a point on the Poincaré sphere or by a Stokes or a Jones vector. The transmittance coefficient of a polarizer along an eigenvector is defined as the ratio of the intensities of the transmitted to the incident beams, both of these two beams being described by the eigenvectors of the polarizer. The transmittance coefficients are designated k_1 and k_2; for the polarizers considered in this book, they will be assumed to be $k_1 = 1$ and $k_2 = 0$. When $k_1 = 1$ and $k_2 = 0$, a polarizer is defined as the optical element that produces a certain polarization form. If the polarization form that is produced is linear, circular, or elliptical, the polarizer is called linear, circular, or elliptical, respectively.

Thus, a linear polarizer whose eigenvector corresponds to a horizontal linear polarization transmits, without any loss of intensity, all of the horizontally polarized light and extinguishes all of the vertically polarized light.

4.2.3 Retarders

A *retarder* or a *birefringent plate* is an optical element that divides an incident monochromatic polarized light beam into two orthogonal polarization forms and produces in one of them a certain optical retardation relative to the other.

The two polarization forms that are conserved by a retarder are called the eigenvectors of the retarder, and according to whether they are linear, circular, or elliptical, the retarder is called linear, circular, or elliptical, respectively. The eigenvector of a retarder that corresponds to the smaller refractive index is called the *fast eigenvector*; the other is called the *slow eigenvector*. A retarder is characterized by its two eigenvectors and its optical retardation. A circular retarder is frequently called an *optically active device*, or a *rotator*; a linear retarder that has retardation equal to 90° or 180° is called a *quarter-* or a *half-wave plate*, respectively.

4.3 The Passage of a Polarized Beam Through a Polarizer or a Retarder

We establish now the transformation of a polarized light beam when it passes through a polarizer or a retarder. The cases of a linear polarizer and a linear retarder will be treated in detail, because these are the most common devices used in photoelasticity. However, for purposes of completeness, the influence of the more-general cases of an elliptical polarizer and an elliptical retarder on a monochromatic light beam will also be given. The derivation of the influence of a polarizer or a retarder on a given polarized light will be studied by using either the conventional vector calculus, or the new methods of the Poincaré sphere, the j-circle method, and the Mueller and Jones calculi.

4.3.1 Vector Calculus

We shall treat separately the cases of a linear polarizer and a linear retarder.

The Linear Polarizer

Let us consider the more-general case of a linear polarizer, whose eigenvector is a linearly polarized light that makes an angle θ with the x axis; that is, a linearly polarized light at an angle θ with the Ox axis passes without any change through the polarizer. Suppose also that a linearly polarized light, of amplitude A at an angle φ with the Ox axis, impinges on the polarizer. Then, the incident light can be represented by the light vector $\boldsymbol{\alpha}$ given by

$$\boldsymbol{\alpha} = a_x \boldsymbol{i} + a_y \boldsymbol{j} = A \cos\varphi\, \boldsymbol{i} + A \sin\varphi\, \boldsymbol{j}. \tag{4.1}$$

The polarizer transmits only the component of the $\boldsymbol{\alpha}$ vector along its axis at angle θ to the Ox direction. The amplitude W of this component is given by

$$W = a_x \cos\theta + a_y \sin\theta. \tag{4.2}$$

When W is analyzed along the Ox and Oy axes,

$$\begin{aligned}
a'_x &= W\cos\theta = a_x \cos^2\theta + a_y \sin\theta \cos\theta \\
a'_y &= W\sin\theta = a_x \sin\theta \cos\theta + a_y \sin^2\theta.
\end{aligned} \tag{4.3}$$

Relations (4.3) give the components along the Ox and Oy axes of the light vector $\boldsymbol{\alpha'} = a'_x \boldsymbol{i} + a'_y \boldsymbol{j}$, transmitted through the polarizer.

The Linear Retarder

Consider now the light vector $\boldsymbol{\alpha}$, given by (4.1), which impinges on a retarder whose fast axis is inclined at an angle θ with the Ox axis and whose retardation is equal to δ.

The components a_f, a_s of the light vector $\boldsymbol{\alpha}$ along the fast and slow axes of the retarder at the entrance of the light beam are given by

$$\begin{aligned}
a_f &= a_x \cos\theta + a_y \sin\theta \\
a_s &= -a_x \sin\theta + a_y \cos\theta.
\end{aligned} \tag{4.4}$$

After passing through the retarder, the a_s component is retarded by an amount equal to the retardation δ of the retarder relative to the a_f component, so that at the exit of the retarder the light components along the fast (a_f') and slow (a_s') axes of the retarder are given by

$$a_f' = a_f e^{i\delta}$$
$$a_s' = a_s . \tag{4.5}$$

The components a_x' and a_y' of the emerging light from the retarder along the Ox and Oy axes are given by

$$a_x' = a_f' \cos\theta - a_s' \sin\theta = (e^{i\delta} \cos^2\theta + \sin^2\theta) a_x + (e^{i\delta} - 1) \sin\theta\cos\theta\, a_y$$
$$a_y' = a_f' \sin\theta + a_s' \cos\theta = (e^{i\delta} - 1) \sin\theta\cos\theta\, a_x + (e^{i\delta} \sin^2\theta + \cos^2\theta) a_y . \tag{4.6}$$

Relations (4.6) give the components along the Ox and Oy axes of the light vector $\alpha' = a_x' \boldsymbol{i} + a_y' \boldsymbol{j}$ that emerges from the retarder.

4.3.2 The Jones Calculus

Relations (4.3) for the case of a polarizer and (4.6) for the case of a retarder establish a linear relationship between the components along the Ox and Oy axes of the incident and the emerging light beams. If these light beams are described by Jones vectors of the forms

$$\begin{bmatrix} a_x \\ a_y \end{bmatrix} = \begin{bmatrix} A_x e^{i\delta_x} \\ A_y e^{i\delta_y} \end{bmatrix}, \quad \begin{bmatrix} a_x' \\ a_y' \end{bmatrix} = \begin{bmatrix} A_x' e^{i\delta_x'} \\ A_y' e^{i\delta_y'} \end{bmatrix}, \tag{4.7}$$

then (4.3) and (4.6) can be written in matrix form as

$$\begin{bmatrix} A_x' e^{i\delta_x'} \\ A_y' e^{i\delta_y'} \end{bmatrix} = \begin{bmatrix} \cos^2\theta & \sin\theta\cos\theta \\ \sin\theta\cos\theta & \sin^2\theta \end{bmatrix} \begin{bmatrix} A_x e^{i\delta_x} \\ A_y e^{i\delta_y} \end{bmatrix} \tag{4.8}$$

for the linear polarizer at angle θ with the Ox axis, and as

$$\begin{bmatrix} A_x' e^{i\delta_x'} \\ A_y' e^{i\delta_y'} \end{bmatrix} = \begin{bmatrix} e^{i\delta} \cos^2\theta + \sin^2\theta & (e^{i\delta} - 1) \sin\theta\cos\theta \\ (e^{i\delta} - 1) \sin\theta\cos\theta & e^{i\delta} \sin^2\theta + \cos^2\theta \end{bmatrix} \begin{bmatrix} A_x e^{i\delta_x} \\ A_y e^{i\delta_y} \end{bmatrix} \tag{4.9}$$

for the linear retarder with retardation δ, whose fast axis subtends an angle θ with the Ox axis.

The two-by-two matrices in (4.8) and (4.9), which characterize a given optical element (polarizer or retarder), are called the Jones matrices and the corresponding calculus, the Jones calculus. By knowing the Jones matrix of an optical device, the Jones vector of the light beam leaving the device can be calculated by simple multiplication of the Jones matrix of the device by the Jones vector of the entering light beam.

Thus, if we denote by a, a', and J the Jones vectors of the entering and the emerging light beams and the Jones matrix of the optical device, respectively,

$$a' = J a .$$ (4.10)

Relation (4.10) constitutes the key formula of the Jones calculus. Thus, if we have two devices in series with Jones matrices J_1 and J_2 and the Jones vectors of the original beam of light, the beam between the devices and the beam after the second device are a_1, a_2, and a_3 respectively, then

$$a_2 = J_1 a_1 \quad \text{and} \quad a_3 = J_2 a_2$$

and, by combination,

$$a_3 = J_2 (J_1 a_1) = J_2 J_1 a_1 .$$ (4.11)

In (4.11) we have applied the associative property of matrices.

Similarly, if we consider a train of optical devices $1, 2, \ldots, n$, whose Jones matrices are $J_1, J_2, \ldots, J_n$, respectively, and that a light beam whose Jones vector is a_0 enters first the device numbered 1, and then consequtively all of the other devices until the last (n) in that order, the Jones vector a of the light beam that emerges from the series of devices $(1, 2, \ldots, n)$ will be given by

$$a = J_n \ldots J_2 J_1 a_0 .$$ (4.12)

It can easily be shown that

$$\begin{bmatrix} \cos^2 \theta & \sin \theta \cos \theta \\ \sin \theta \cos \theta & \sin^2 \theta \end{bmatrix} = \begin{bmatrix} \cos \theta & -\sin \theta \\ \sin \theta & \cos \theta \end{bmatrix} \begin{bmatrix} 1 & 0 \\ 0 & 0 \end{bmatrix} \begin{bmatrix} \cos \theta & \sin \theta \\ -\sin \theta & \cos \theta \end{bmatrix},$$ (4.13)

where the matrix $J_0 = \begin{bmatrix} 1 & 0 \\ 0 & 0 \end{bmatrix}$ represents the Jones matrix of a polarizer with a horizontal optical axis (This matrix results from the general matrix of a polarizer with its axis subtending an angle θ with the Ox axis, by putting $\theta = 0$).

Thus, (4.13) can be written in the form

$$J_\theta = R(-\theta) J_0 R(\theta),$$ (4.14)

where J_θ is the Jones matrix of the polarizer with its axis at an angle θ with the Ox axis, J_0 is the corresponding matrix with $\theta = 0$, and $R(\theta)$ is the rotation matrix given by

$$R = \begin{bmatrix} \cos\theta & \sin\theta \\ -\sin\theta & \cos\theta \end{bmatrix}. \tag{4.15}$$

Similarly, it can be shown that

$$\begin{bmatrix} e^{i\delta}\cos^2\theta + \sin^2\theta & (e^{i\delta}-1)\sin\theta\cos\theta \\ (e^{i\delta}-1)\sin\theta\cos\theta & e^{i\delta}\sin^2\theta + \cos^2\theta \end{bmatrix}$$

$$= \begin{bmatrix} \cos\theta & -\sin\theta \\ \sin\theta & \cos\theta \end{bmatrix} \begin{bmatrix} e^{i\delta} & 0 \\ 0 & 1 \end{bmatrix} \begin{bmatrix} \cos\theta & \sin\theta \\ -\sin\theta & \cos\theta \end{bmatrix}, \tag{4.16}$$

where the matrix $J_0 = \begin{bmatrix} e^{i\delta} & 0 \\ 0 & 1 \end{bmatrix}$ represents the matrix of a linear retarder whose fast axis coincides with the Ox axis. Thus, (4.16) can be put in the form of (4.14).

Table 4.1. Jones matrices for polarizers

Type of polarizer	Jones matrix
Ideal elliptical, producing elliptically polarized light with azimuth ψ and ellipticity ω such that $\tan 2\psi = \tan 2\theta \cos\delta$ $\sin 2\omega = \sin 2\theta \sin\delta$	$\begin{bmatrix} \cos^2\theta & e^{-i\delta}\sin\theta\cos\theta \\ e^{i\delta}\sin\theta\cos\theta & \cos^2\theta \end{bmatrix}$
Ideal right-circular, ($\theta = 45°$, $\delta = 90°$)	$\dfrac{1}{2}\begin{bmatrix} 1 & -i \\ i & 1 \end{bmatrix}$
Ideal left-circular, ($\theta = 45°$, $\delta = -90°$)	$\dfrac{1}{2}\begin{bmatrix} 1 & i \\ -i & 1 \end{bmatrix}$
Ideal linear, with its axis at angle θ ($\delta = 0$)	$\begin{bmatrix} \cos^2\theta & \sin\theta\cos\theta \\ \sin\theta\cos\theta & \sin^2\theta \end{bmatrix}$
Partially linear, with principal intensity transmission coefficients k_1 and k_2 along its principal directions 1 and 2.	$\begin{bmatrix} \sqrt{k_1} & 0 \\ 0 & \sqrt{k_2} \end{bmatrix}$

Remark. The Jones matrix $J(\theta + \varphi)$ of any ideal polarizer whose principal axis subtends an angle $(\theta + \varphi)$ with the Ox axis is obtained from the corresponding Jones matrix $J(\varphi)$ of the same polarizer with its principal axis at an angle φ by use of

$$J(\theta + \varphi) = R(-\theta)J(\varphi)R(\theta),$$

where $R(\theta)$ is the rotation Jones matrix

$$R(\theta) = \begin{bmatrix} \cos\theta & \sin\theta \\ -\sin\theta & \cos\theta \end{bmatrix}.$$

Table 4.2. Jones matrices for retarders

Type of retarder	Jones matrix
Elliptical, giving two orthogonal elliptical polarization forms with azimuth ψ and ellipticity ω such that $\tan 2\psi = \tan 2\theta \cos \Delta$ $\sin 2\omega = \sin 2\theta \sin \Delta$ where Δ and θ are the elements of the constituent vibrations of the ellipses (see Sec. 3.4.1) and a retardation δ between them.	$\begin{bmatrix} e^{i\delta}\cos^2\theta + \sin^2\theta & (e^{i\delta}-1)e^{-i\Delta}\sin\theta\cos\theta \\ (e^{i\delta}-1)e^{i\Delta}\sin\theta\cos\theta & e^{i\delta}\sin^2\theta + \cos^2\theta \end{bmatrix}$
Right-circular, $(\theta = 45°, \Delta = 90°)$	$\begin{bmatrix} \cos\dfrac{\delta}{2} & \sin\dfrac{\delta}{2} \\ -\sin\dfrac{\delta}{2} & \cos\dfrac{\delta}{2} \end{bmatrix}$
Left-circular, $(\theta = 45°, \Delta = -90°)$	$\begin{bmatrix} \cos\dfrac{\delta}{2} & -\sin\dfrac{\delta}{2} \\ \sin\dfrac{\delta}{2} & \cos\dfrac{\delta}{2} \end{bmatrix}$
Linear, with its fast axis at an angle θ and retardation δ $(\Delta = 0, \psi = \theta)$	$\begin{bmatrix} e^{i\delta}\cos^2\theta + \sin^2\theta & (e^{i\delta}-1)\sin\theta\cos\theta \\ (e^{i\delta}-1)\sin\theta\cos\theta & e^{i\delta}\sin^2\theta + \cos^2\theta \end{bmatrix}$
Half-wave plate, with its fast axis at an angle θ $(\Delta = 0, \psi = \theta, \delta = 180°)$	$\begin{bmatrix} -\cos 2\theta & -\sin 2\theta \\ -\sin 2\theta & \cos 2\theta \end{bmatrix}$
Quarter-wave plate, with its fast axis at an angle θ $(\Delta = 0, \psi = \theta, \delta = 90°)$	$\begin{bmatrix} i\cos^2\theta + \sin^2\theta & (i-1)\sin\theta\cos\theta \\ (i-1)\sin\theta\cos\theta & i\sin^2\theta + \cos^2\theta \end{bmatrix}$

Remark. The Jones matrix $J(\theta + \varphi)$ of any retarder whose principal axis subtends an angle $(\theta + \varphi)$ with the Ox axis is obtained from the corresponding Jones matrix $J(\varphi)$ of the same retarder with its principal axis at an angle φ by use of the relation

$$J(\theta + \varphi) = R(-\theta)J(\varphi)R(\theta),$$

where $R(\theta)$ is the rotation Jones matrix

$$R(\theta) = \begin{bmatrix} \cos\theta & \sin\theta \\ -\sin\theta & \cos\theta \end{bmatrix}.$$

Similarly, it can be shown that

$$J(\theta+\varphi)=R(-\theta)J(\varphi)R(\theta),\tag{4.17}$$

where $J(\theta+\varphi)$ represents the Jones matrix of an optical device with orientation $(\theta+\varphi)$ and $J(\varphi)$ the relevant matrix of the same device at an orientation φ.

Thus, (4.17) can be used to transform the Jones matrix for an optical device with a given orientation φ into the Jones matrix for the same device with a new orientation $(\theta+\varphi)$.

The Jones matrices for the most common types of polarizers and retarders used in this book are listed in Tables 4.1 and 4.2, respectively.

4.3.3 An Independent Introduction of Jones Matrices

The Jones matrices were previously introduced by use of vector calculus and of the basic properties of the corresponding optical device. However, these matrices can be introduced independently by employing only the linearity principle of an optical device, that is, the fact that the components of a beam leaving the device are linear combinations of the components of the beam that entered the device. This principle was experimentally established at various optical frequencies. However, nonlinear effects are introduced at electric fields of very high strength, such as result, for example, from a laser source. By use of the linearity principle and the properties of the particular optical element, the Jones vector of this element can be easily obtained.

Let us apply this procedure to the derivation of the Jones matrix of a retarder with retardation δ and fast axis subtending an angle θ with the Ox axis. The linearity between the incident and emerging light beams can be expressed in the form

$$\begin{bmatrix} A'_x e^{i\varphi'_x} \\ A'_y e^{i\varphi'_y} \end{bmatrix} = \begin{bmatrix} J_{11} & J_{12} \\ J_{21} & J_{22} \end{bmatrix} \begin{bmatrix} A_x e^{i\varphi_x} \\ A_y e^{i\varphi_y} \end{bmatrix},\tag{4.18}$$

where

$$a = \begin{bmatrix} A_x e^{i\varphi_x} \\ A_y e^{i\varphi_y} \end{bmatrix}, \qquad a' = \begin{bmatrix} A'_x e^{i\varphi'_x} \\ A'_y e^{i\varphi'_y} \end{bmatrix}$$

are the Jones vectors for the incident and emerging light beams and $J_{11}, J_{12}, J_{21}, J_{22}$ are the elements of the Jones matrix of the retarder.

Let us consider a linearly polarized light beam along the Ox axis with Jones vector

$$\begin{bmatrix} A \\ 0 \end{bmatrix}.$$

The components of this beam along the principal axes of the retarder can be found by multiplying this vector by the rotation matrix, that is

$$
\begin{bmatrix} \cos\theta & \sin\theta \\ -\sin\theta & \cos\theta \end{bmatrix} \begin{bmatrix} A \\ 0 \end{bmatrix} = \begin{bmatrix} A\cos\theta \\ -A\sin\theta \end{bmatrix}. \tag{4.19}
$$

The component along the slow axis of the retarder is retarded by δ, so that we have for the emerging light beam the vector

$$
\begin{bmatrix} A\,e^{i\delta}\cos\theta \\ -A\sin\theta \end{bmatrix}.
$$

By multiplying it by the matrix $\mathbf{R}(-\theta)$, we can find the components of this beam along the Ox and Oy axes. Thus, we obtain

$$
\begin{bmatrix} \cos\theta & -\sin\theta \\ \sin\theta & \cos\theta \end{bmatrix} \begin{bmatrix} A\,e^{i\delta}\cos\theta \\ -A\sin\theta \end{bmatrix} = \begin{bmatrix} (e^{i\delta}\cos^2\theta + \sin^2\theta)A \\ (e^{i\delta}-1)\sin\theta\cos\theta\,A \end{bmatrix}. \tag{4.20}
$$

Applying now (4.18) for

$$
\mathbf{a} = \begin{bmatrix} A \\ 0 \end{bmatrix} \quad \text{and} \quad \mathbf{a}' = \begin{bmatrix} (e^{i\delta}\cos^2\theta + \sin^2\theta)A \\ (e^{i\delta}-1)\sin\theta\cos\theta\,A \end{bmatrix}, \tag{4.21}
$$

we obtain

$$
J_{11} = e^{i\delta}\cos^2\theta + \sin^2\theta \quad \text{and} \quad J_{21} = (e^{i\delta}-1)\sin\theta\cos\theta.
$$

Similarly, by considering a linearly polarized light beam along the Oy axis, we obtain

$$
J_{12} = (e^{i\delta}-1)\sin\theta\cos\theta, \qquad J_{22} = e^{i\delta}\sin^2\theta + \cos^2\theta,
$$

that is, the matrix of a linear retarder found previously in (4.9).

By using the linearity principle and the particular properties of a given optical device, we can easily find the Jones matrices that correspond to the more-general case of an elliptical polarizer or retarder. These matrices are shown in Tables 4.1 and 4.2 for optical polarizers and retarders, respectively.

4.3.4 The Mueller Calculus

In the Mueller calculus, the Stokes vector that characterizes a polarized light beam is used. This vector, as has already been indicated, can be described by a four-element column matrix. Because the incident and emerging light beams are described by column matrices with four elements, the optical device should be described by a four-by-four matrix. This matrix, which completely characterizes an optical device, is called the *Mueller matrix*, and the calculus based on Mueller matrices and Stokes vectors is called the *Mueller calculus*.

The Mueller matrix of a certain optical element can be readily found by using the linearity principle and the characteristic properties of the optical element considered. This principle can be expressed in the form

$$
\begin{bmatrix} s'_0 \\ s'_1 \\ s'_2 \\ s'_3 \end{bmatrix} = \begin{bmatrix} m_{11} & m_{12} & m_{13} & m_{14} \\ m_{21} & m_{22} & m_{23} & m_{24} \\ m_{31} & m_{32} & m_{33} & m_{34} \\ m_{41} & m_{42} & m_{43} & m_{44} \end{bmatrix} \begin{bmatrix} s_0 \\ s_1 \\ s_2 \\ s_3 \end{bmatrix},
\tag{4.22}
$$

where the column matrix with primes on the left-hand side of this relation represents the Stokes vector of the emerging light beam, that on the right-hand side represents the corresponding Stokes vector of the incident beam, and the four-by-four matrix is the Mueller matrix of the optical device. Relation (4.22) can be written in the form

$$
S' = MS,
\tag{4.23}
$$

where:

$$
S' = \begin{bmatrix} s'_0 \\ s'_1 \\ s'_2 \\ s'_3 \end{bmatrix} \quad S = \begin{bmatrix} s_0 \\ s_1 \\ s_2 \\ s_3 \end{bmatrix} \quad M = \begin{bmatrix} m_{11} & m_{12} & m_{13} & m_{14} \\ m_{21} & m_{22} & m_{23} & m_{24} \\ m_{31} & m_{32} & m_{33} & m_{34} \\ m_{41} & m_{42} & m_{43} & m_{44} \end{bmatrix}.
\tag{4.24}
$$

We will now derive the Mueller matrices for a linear polarizer and a linear retarder.

The Linear Polarizer

Let us consider a linear polarizer whose axis subtends an angle θ with the Ox axis. We will calculate the coefficients m_{ij} $(i,j = 1,2,3,4)$ of the matrix M, by using the basic properties of the linear polarizer:

I) An unpolarized light that passes through a linear polarizer becomes linearly polarized at an angle θ with the Ox axis. Thus if

$$
S = \begin{bmatrix} 2 \\ 0 \\ 0 \\ 0 \end{bmatrix}, \quad \text{then} \quad S' = \begin{bmatrix} 1 \\ \cos 2\theta \\ \sin 2\theta \\ 0 \end{bmatrix}.
\tag{4.25}
$$

II) A linearly polarized light at an angle θ with the Ox axis remains unaffected; that is, if

$$S = \begin{bmatrix} 1 \\ \cos 2\theta \\ \sin 2\theta \\ 0 \end{bmatrix}, \quad \text{then} \quad S' = \begin{bmatrix} 1 \\ \cos 2\theta \\ \sin 2\theta \\ 0 \end{bmatrix}. \tag{4.26}$$

III) A linearly polarized light along the Ox axis becomes linearly polarized at an angle θ with the Ox axis. If the intensity of the incident light is equal to unity, then the Stokes parameters of the emerging light will be $s_0 = \cos^2 \theta$, $s_1 = \cos^2 \theta \cos 2\theta$, $s_2 = \cos^2 \theta \sin 2\theta$, $s_3 = 0$; that is, if

$$S = \begin{bmatrix} 1 \\ 1 \\ 0 \\ 0 \end{bmatrix}, \quad \text{then} \quad S' = \begin{bmatrix} \cos^2 \theta \\ \cos^2 \theta \cos 2\theta \\ \cos^2 \theta \sin 2\theta \\ 0 \end{bmatrix}. \tag{4.27}$$

IV) A right-circular light of unit intensity, passing through the linear polarizer, becomes linearly polarized at an angle θ with the Ox axis with an intensity equal to half the intensity of the incident light; that is, if

$$S = \begin{bmatrix} 1 \\ 0 \\ 0 \\ 1 \end{bmatrix}, \quad \text{then} \quad S' = \frac{1}{2} \begin{bmatrix} 1 \\ \cos 2\theta \\ \sin 2\theta \\ 0 \end{bmatrix}. \tag{4.28}$$

By applying (4.22), with Stokes vectors S and S' given by (4.25) it follows that

$$m_{11} = \frac{1}{2}, \quad m_{21} = \frac{\cos 2\theta}{2}, \quad m_{31} = \frac{\sin 2\theta}{2}, \quad m_{41} = 0. \tag{4.29}$$

Similarly, from (4.22) and (4.26) it follows that

$$\frac{1}{2} + m_{12} \cos 2\theta + m_{13} \sin 2\theta = 1$$

$$\frac{\cos 2\theta}{2} + m_{22} \cos 2\theta + m_{23} \sin 2\theta = \cos 2\theta$$

$$\frac{\sin 2\theta}{2} + m_{32} \cos 2\theta + m_{33} \sin 2\theta = \sin 2\theta$$

$$m_{42} \cos 2\theta + m_{43} \sin 2\theta = 0.$$

$$\tag{4.30}$$

From (4.22) and (4.27) it follows that

$$\frac{1}{2}+m_{12}=\cos^2\theta \qquad\qquad m_{12}=\frac{1}{2}\cos 2\theta$$

$$\frac{\cos 2\theta}{2}+m_{22}=\cos^2\theta\cos 2\theta \qquad\qquad m_{22}=\frac{1}{2}\cos^2 2\theta$$

or   (4.31)

$$\frac{\sin 2\theta}{2}+m_{32}=\cos^2\theta\sin 2\theta \qquad\qquad m_{32}=\frac{1}{2}\sin 2\theta\cos 2\theta$$

$$m_{42}=0 \qquad\qquad m_{42}=0.$$

With these values of m_{12}, m_{22}, m_{32} and m_{42}, (4.30) give

$$m_{13}=\frac{1}{2}\sin 2\theta,\quad m_{23}=\frac{1}{2}\sin 2\theta\cos 2\theta,\quad m_{33}=\frac{1}{2}\sin^2 2\theta,\quad m_{43}=0.$$
(4.32)

Finally, from (4.22) and (4.28), it follows that

$$\frac{1}{2}+m_{14}=\frac{1}{2} \qquad\qquad m_{14}=0$$

$$\frac{\cos 2\theta}{2}+m_{24}=\frac{\cos 2\theta}{2} \qquad\qquad m_{24}=0$$

or   (4.33)

$$\frac{\sin 2\theta}{2}+m_{34}=\frac{\sin 2\theta}{2} \qquad\qquad m_{34}=0$$

$$m_{44}=0.$$

Relations (4.29, 31–33) give the Mueller matrix P_θ of the linear polarizer whose principal axis subtends an angle θ with the Ox axis:

$$P_\theta=\frac{1}{2}\begin{bmatrix} 1 & \cos 2\theta & \sin 2\theta & 0 \\ \cos 2\theta & \cos^2 2\theta & \sin 2\theta\cos 2\theta & 0 \\ \sin 2\theta & \sin 2\theta\cos 2\theta & \sin^2 2\theta & 0 \\ 0 & 0 & 0 & 0 \end{bmatrix}.$$
(4.34)

The Mueller matrices for the most-common polarizers used in polarization optics are listed in Table 4.3.

Table 4.3. Mueller matrices

Type of polarizer	Mueller matrix
Ideal elliptical, producing elliptically polarized light with azimuth ψ and ellipticity ω such that $\tan 2\psi = \tan 2\theta \cos \delta$ $\sin 2\omega = \sin 2\theta \sin \delta$	$\dfrac{1}{2}\begin{bmatrix} 1 & \cos 2\theta & \sin 2\theta \cos \delta & \sin 2\theta \sin \delta \\ \cos 2\theta & \cos^2 2\theta & \sin 2\theta \cos 2\theta \cos \delta & \sin 2\theta \cos 2\theta \sin \delta \\ \sin 2\theta \cos \delta & \sin 2\theta \cos 2\theta \cos \delta & \sin^2 2\theta \cos^2 \delta & \sin^2 2\theta \sin \delta \cos \delta \\ \sin 2\theta \sin \delta & \sin 2\theta \cos 2\theta \sin \delta & \sin^2 2\theta \sin \delta \cos \delta & \sin^2 2\theta \sin^2 \delta \end{bmatrix}$
Ideal right-circular, $(\theta = 45°, \delta = 90°)$	$\dfrac{1}{2}\begin{bmatrix} 1 & 0 & 0 & 1 \\ 0 & 0 & 0 & 0 \\ 0 & 0 & 0 & 0 \\ 1 & 0 & 0 & 1 \end{bmatrix}$
Ideal left-circular, $(\theta = 45°, \delta = -90°)$	$\dfrac{1}{2}\begin{bmatrix} 1 & 0 & 0 & -1 \\ 0 & 0 & 0 & 0 \\ 0 & 0 & 0 & 0 \\ -1 & 0 & 0 & 1 \end{bmatrix}$
Ideal linear, with its axis at angle θ $(\delta = 0)$	$\dfrac{1}{2}\begin{bmatrix} 1 & \cos 2\theta & \sin 2\theta & 0 \\ \cos 2\theta & \cos^2 2\theta & \sin 2\theta \cos 2\theta & 0 \\ \sin 2\theta & \sin 2\theta \cos 2\theta & \sin^2 2\theta & 0 \\ 0 & 0 & 0 & 0 \end{bmatrix}$
Partial linear, with principal intensity transmission coefficients k_1 and k_2 along its principal axes 1 and 2	$\dfrac{1}{2}\begin{bmatrix} k_1 + k_2 & k_1 - k_2 & 0 & 0 \\ k_1 - k_2 & k_1 + k_2 & 0 & 0 \\ 0 & 0 & 2\sqrt{k_1 k_2} & 0 \\ 0 & 0 & 0 & 2\sqrt{k_1 k_2} \end{bmatrix}$

Remark. The Mueller matrix $P(\theta + \varphi)$ of an ideal polarizer whose principal axis subtends an angle $(\theta + \varphi)$ with the Ox axis is obtained from the corresponding Mueller matrix $P(\varphi)$ of the same polarizer with its principal axis at an angle φ by use of the relation

$$P(\theta + \varphi) = T(-2\theta) P(\varphi) T(2\theta),$$

where $T(2\theta)$ is the rotation Mueller matrix

$$T(2\theta) = \begin{bmatrix} 1 & 0 & 0 & 0 \\ 0 & \cos 2\theta & \sin 2\theta & 0 \\ 0 & -\sin 2\theta & \cos 2\theta & 0 \\ 0 & 0 & 0 & 1 \end{bmatrix}.$$

The Linear Retarder

We shall now find the Mueller matrix of a linear retarder with a retardation δ, whose fast axis makes an angle θ with the Ox axis, by using the linearity principle expressed by (4.22) and the basic properties of the retarder.

I) An unpolarized light beam, passing through a linear retarder, remains unaffected; that is, if

$$S = \begin{bmatrix} 1 \\ 0 \\ 0 \\ 0 \end{bmatrix}, \quad \text{then} \quad S' = \begin{bmatrix} 1 \\ 0 \\ 0 \\ 0 \end{bmatrix}. \tag{4.35}$$

II) A linearly polarized light beam passing through the fast axis of the linear retarder remains unaffected; that is, if

$$S = \begin{bmatrix} 1 \\ \cos 2\theta \\ \sin 2\theta \\ 0 \end{bmatrix}, \quad \text{then} \quad S' = \begin{bmatrix} 1 \\ \cos 2\theta \\ \sin 2\theta \\ 0 \end{bmatrix}. \tag{4.36}$$

III) The components a'_x, a'_y of a beam of unit amplitude linearly polarized along the Ox axis with $a_x = e^{i\delta_x}, a_y = 0$, after passing through the linear retarder, take the following form, according to (4.6),

$$a'_x = (e^{i\delta} \cos^2 \theta + \sin^2 \theta) a_x, \quad a'_y = (e^{i\delta} - 1) \sin \theta \cos \theta \, a_x.$$

These give the Stokes parameters of the emergent light beam:

$$s_0 = 1 \quad s_1 = \cos^2 2\theta + \sin^2 2\theta \cos \delta \quad s_2 = (1 - \cos \delta) \sin 2\theta \cos 2\theta$$
$$s_3 = \sin 2\theta \sin \delta.$$

That is, if

$$S = \begin{bmatrix} 1 \\ 1 \\ 0 \\ 0 \end{bmatrix}, \quad \text{then} \quad S' = \begin{bmatrix} 1 \\ \cos^2 2\theta + \sin^2 2\theta \cos \delta \\ (1 - \cos \delta) \sin 2\theta \cos 2\theta \\ \sin 2\theta \sin \delta \end{bmatrix}. \tag{4.37}$$

IV) Similarly, for a right-circularly polarized light beam S, passing through a linear retarder,

$$S = \begin{bmatrix} 1 \\ 0 \\ 0 \\ 1 \end{bmatrix}, \quad \text{and} \quad S' = \begin{bmatrix} 1 \\ -\sin 2\theta \sin \delta \\ \cos 2\theta \sin \delta \\ \cos \delta \end{bmatrix}. \tag{4.38}$$

From (4.22) and (4.35) it follows that

$$m_{11} = 1, \quad m_{21} = 0, \quad m_{31} = 0, \quad m_{41} = 0. \tag{4.39}$$

Similarly, from (4.22) and (4.36) it follows that

$$1 + m_{12} \cos 2\theta + m_{13} \sin 2\theta = 1$$
$$m_{22} \cos 2\theta + m_{23} \sin 2\theta = \cos 2\theta$$
$$m_{32} \cos 2\theta + m_{33} \sin 2\theta = \sin 2\theta \tag{4.40}$$
$$m_{42} \cos 2\theta + m_{43} \sin 2\theta = 0.$$

From (4.22) and (4.37) it follows that

$$m_{12} = 0$$
$$m_{22} = \cos^2 2\theta + \sin^2 2\theta \cos \delta$$
$$m_{32} = (1 - \cos \delta) \sin 2\theta \cos 2\theta \tag{4.41}$$
$$m_{42} = \sin 2\theta \sin \delta.$$

Then, (4.40) give

$$m_{13} = 0$$
$$m_{23} = (1 - \cos \delta) \sin 2\theta \cos 2\theta$$
$$m_{33} = \sin^2 2\theta + \cos^2 2\theta \cos \delta \tag{4.42}$$
$$m_{43} = -\cos 2\theta \sin \delta.$$

Finally, (4.22) and (4.38) yield

$$m_{14} = 0$$
$$m_{24} = -\sin 2\theta \cos \delta$$
$$m_{34} = \cos 2\theta \sin \delta \tag{4.43}$$
$$m_{44} = \cos \delta.$$

Relations (4.39, 41–43) give the Mueller matrix of a linear retarder with retardation δ, whose fast axis subtends an angle θ with the Ox axis,

$$
R_\theta = \begin{bmatrix}
1 & 0 & 0 & 0 \\
0 & \cos^2 2\theta + \sin^2 2\theta \cos\delta & (1-\cos\delta)\sin 2\theta \cos 2\theta & -\sin 2\theta \sin\delta \\
0 & (1-\cos\delta)\sin 2\theta \cos 2\theta & \sin^2 2\theta + \cos^2 2\theta \cos\delta & \cos 2\theta \sin\delta \\
0 & \sin 2\theta \sin\delta & -\cos 2\theta \sin\delta & \cos\delta
\end{bmatrix}.
$$

$$(4.44)$$

The Mueller matrices for the more-general cases of an elliptical polarizer or an elliptical retarder can be found in an analogous manner. The Mueller matrices for the more-common retarders used in polarization optics are listed in Table 4.4.

As in the case of the Jones calculus, it can be proved that if we have a train of optical devices $1, 2, ..., n$, with Mueller matrices $M_1, M_2, ..., M_n$, respectively, and consider that a light beam with Stokes vector S_0 enters first the device numbered 1, then that numbered 2 and so on, then the Stokes vector S of the light beam that emerges from the last device n, will be given by

$$S = M_n ... M_2 M_1 S_0. \tag{4.45}$$

It can also be proved, as in the case of the Jones calculus, that the Mueller matrix of an optical device, whose principal axis is at an angle $(\theta + \varphi)$ with the Ox axis, is related to the Mueller matrix of the same device when its principal axis is at an angle φ with the Ox axis, through the relation

$$M(\theta + \varphi) = T(-2\theta) M(\varphi) T(2\theta), \tag{4.46}$$

where $T(2\theta)$ is the well-known rotator matrix,

$$
T(2\theta) = \begin{bmatrix}
1 & 0 & 0 & 0 \\
0 & \cos 2\theta & \sin 2\theta & 0 \\
0 & -\sin 2\theta & \cos 2\theta & 0 \\
0 & 0 & 0 & 1
\end{bmatrix}. \tag{4.47}
$$

4.3.5 The Poincaré Sphere

As has already been indicated in Section 3.4.2, the Poincaré sphere can be considered to be the geometrical representation of the Stokes vector, whose components s_1, s_2, s_3 express the cartesian coordinates of a point on the Poincaré sphere. Thus, the appropriate transformations on the Poincaré sphere, for finding the influence of a polarizer or a retarder on a beam of

Table 4.4. Mueller matrices

Type of retarder	Mueller matrix
Elliptical, giving two orthogonal elliptical polarization forms with azimuth ψ and ellipticity ω and retardation δ between them	$$\begin{bmatrix} 1 & 0 & 0 & 0 \\ 0 & A_1^2 - A_2^2 - A_3^2 + A_4^2 & 2(A_1 A_2 + A_3 A_4) & -2(A_1 A_3 + A_2 A_4) \\ 0 & 2(A_1 A_2 - A_3 A_4) & -A_1^2 + A_2^2 - A_3^2 + A_4^2 & 2(A_1 A_4 - A_2 A_3) \\ 0 & -2(A_1 A_3 - A_2 A_4) & -2(A_1 A_4 + A_2 A_3) & -A_1^2 - A_2^2 + A_3^2 + A_4^2 \end{bmatrix}$$ with: $$A_1 = \cos 2\omega \cos 2\psi \sin \frac{\delta}{2} \qquad A_2 = \cos 2\omega \sin 2\psi \sin \frac{\delta}{2}$$ $$A_3 = \sin 2\omega \sin \frac{\delta}{2} \qquad A_4 = \cos \frac{\delta}{2}$$
Right-circular, $(\omega = 45°)$	$$\begin{bmatrix} 1 & 0 & 0 & 0 \\ 0 & \cos\delta & \sin\delta & 0 \\ 0 & -\sin\delta & \cos\delta & 0 \\ 0 & 0 & 0 & 1 \end{bmatrix}$$
Left-circular, $(\omega = -45°)$	$$\begin{bmatrix} 1 & 0 & 0 & 0 \\ 0 & \cos\delta & -\sin\delta & 0 \\ 0 & \sin\delta & \cos\delta & 0 \\ 0 & 0 & 0 & 1 \end{bmatrix}$$
Linear, with its fast axis at an angle θ and retardation δ $(\omega = 0, \psi = \theta)$	$$\begin{bmatrix} 1 & 0 & 0 & 0 \\ 0 & \cos^2 2\theta + \sin^2 2\theta \cos\delta & (1-\cos\delta)\sin 2\theta \cos 2\theta & -\sin 2\theta \sin\delta \\ 0 & (1-\cos\delta)\sin 2\theta \cos 2\theta & \sin^2 2\theta + \cos^2 2\theta \cos\delta & \cos 2\theta \sin\delta \\ 0 & \sin 2\theta \sin\delta & -\cos 2\theta \sin\delta & \cos\delta \end{bmatrix}$$
Half-wave plate, with its fast axis at an angle θ $(\omega = 0, \psi = \theta, \delta = 180°)$	$$\begin{bmatrix} 1 & 0 & 0 & 0 \\ 0 & \cos 4\theta & \sin 4\theta & 0 \\ 0 & \sin 4\theta & -\cos 4\theta & 0 \\ 0 & 0 & 0 & -1 \end{bmatrix}$$
Quarter-wave plate, with its fast axis at an angle θ $(\omega = 0, \psi = \theta, \delta = 90°)$	$$\begin{bmatrix} 1 & 0 & 0 & 0 \\ 0 & \cos^2 2\theta & \sin 2\theta \cos 2\theta & -\sin 2\theta \\ 0 & \sin 2\theta \cos 2\theta & \sin^2 2\theta & \cos 2\theta \\ 0 & \sin 2\theta & -\cos 2\theta & 0 \end{bmatrix}$$

elliptically polarized light, can be readily established from the corresponding transformations of the Stokes vector, by using the Mueller calculus. Indeed, in this calculus the matrices that represent the various optical elements, found previously, can be used for the calculation of the Stokes vectors of the emerging light beams, from the corresponding vectors of the incident beams. By transforming the s_1, s_2, s_3 components of Stokes vectors for the incident and the emerging light beam into the cartesian coordinates of points on the Poincaré sphere, we obtain, on the sphere, two points that represent the incident and the emerging polarization forms. Thus, the appropriate manipulations on the Poincaré sphere, for transferring from the point that represents the incident polarization to that which represents the emerging polarization on the sphere, can be readily established.

However, the transformations on the Poincaré sphere, required for finding the result of an optical element on a polarized light beam, will be established independently. We will consider separately the cases of a linear polarizer and a linear retarder.

The Linear Polarizer

Consider a linear polarizer with, as eigenvector, a linearly polarized light beam at an angle θ with the Ox axis and an elliptically polarized light beam of unit intensity, incident on the linear polarizer. If we consider that the incident light is specified by the light ellipse with orientation ψ and semi-axes a and b, then the intensity I of the emerging light beam will be given by

$$I = a^2 \cos^2(\theta - \psi) + b^2 \sin^2(\theta - \psi). \tag{4.48}$$

When (3.30) is taken into account, and when the intensity of the incident light is unity, that is $a = \cos\omega$ and $b = \sin\omega$, (4.48) takes the form

$$I = \frac{1}{2} + \frac{1}{2}\cos 2\omega \cos 2(\psi - \theta). \tag{4.49}$$

Remark. The Mueller matrix $M(\theta + \varphi)$ of any retarder whose principal axis subtends an angle $(\theta + \varphi)$ with the Ox axis is obtained from the corresponding Mueller matrix $M(\varphi)$ of the same retarder with its principal axis at an angle φ by use of the relation

$$M(\theta + \varphi) = T(-2\theta)M(\varphi)T(2\theta),$$

where $T(2\theta)$ is the rotation Mueller matrix

$$T(2\theta) = \begin{bmatrix} 1 & 0 & 0 & 0 \\ 0 & \cos 2\theta & \sin 2\theta & 0 \\ 0 & -\sin 2\theta & \cos 2\theta & 0 \\ 0 & 0 & 0 & 1 \end{bmatrix}.$$

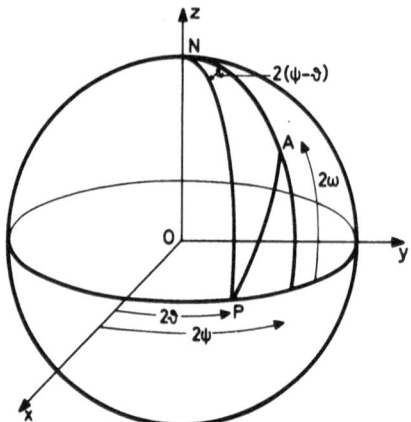

Fig. 4.2. Representation on the Poincaré sphere of the action of a linear polarizer P on an elliptically polarized light beam A. The intensity of the light transmitted through the polarizer is $I = \cos^2(\widehat{AP}/2)$

Let us now represent the eigenvector of the linear polarizer and the polarization form of the incident light beam on the Poincaré sphere (Fig. 4.2). Because the polarizer is linear with its axis at an angle θ with the Ox axis, its eigenvector will be represented by a point P on the equatorial plane of the Poincaré sphere, defined by the longitude 2θ. The incident elliptically polarized light is represented by a point A with longitude 2ψ and latitude 2ω. From the spherical triangle NPA of Fig. 4.2 it follows that

$$\cos \widehat{AP} = \cos 2\omega \cos 2(\psi - \theta),$$

so (4.49) takes the form

$$I = \cos^2 \frac{\widehat{AP}}{2}. \tag{4.50}$$

Relation (4.50) shows that the intensity of the light with polarization state A, which is transmitted by a linear polarizer whose eigenvector is defined by the polarization state P, is equal to $\cos^2(\widehat{AP}/2)$, where $\widehat{AP}$ is the length of the arc joining points A and P on the Poincaré sphere. This result is valid for the most general case, when the polarizer is elliptical and is represented on the Poincaré sphere by a point $P(2\theta, 2\varphi)$ not lying on the equatorial plane (Fig. 4.3).

Another interesting result, obtained from (4.50), is that the locus on the Poincaré sphere of the polarization states for which the emerging intensity from a given polarizer is constant is a small circle on the sphere, the pole of which is at the point that represents the eigenvector of the polarizer (Fig. 4.4).

Relation (4.50) has a number of important applications and can be used to prove many laws involving polarizers. Thus, the intensity transmitted by a pair of "crossed" polarizers, that is polarizers whose eigenvectors represent orthogonal polarization forms, is zero, because orthogonal forms are represented on the Poincaré sphere by two opposite points P and A, for which

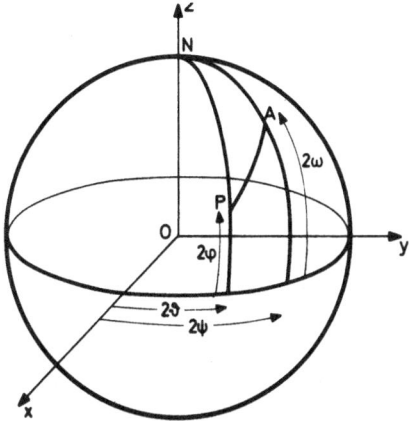

Fig. 4.3. Representation on the Poincaré sphere of the action of an elliptical polarizer P on an elliptically polarized light beam A. The intensity of the light transmitted through the polarizer is $I = \cos^2(\widehat{AP}/2)$

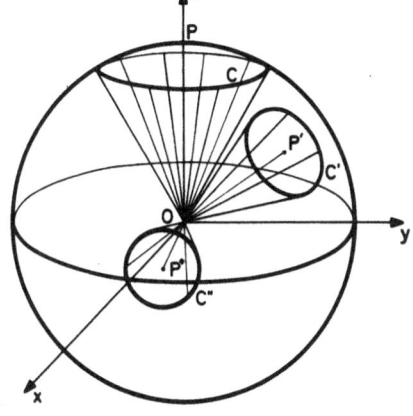

Fig. 4.4. Circles C, C', and C'' that represent on the Poincaré sphere the loci of polarization states for which the emerging light intensity from the polarizers with eigenvectors P, P', and P'' is constant. These circles, C, C', and C'' are normal to the directions OP, OP', and OP''

$\widehat{PA} = \pi$. We can also see that a polarizer of state A completely transmits a light of the same state A, because for $\widehat{AA} = 0$, $I = 1$. Also, a linear polarizer P of any azimuth transmits half the intensity of a circularly polarized beam A (N or S), because for every point P on the equatorial plane $\widehat{PA} = \pi/2$, for which (4.50) gives $I = 1/2$.

The Linear Retarder

We shall now establish the geometrical operations on the Poincaré sphere for finding the effect of a linear retarder of retardation δ, whose fast axis makes an angle θ with the Ox axis on an elliptically polarized light beam.

Let us first consider the simple case in which the incident light is linearly polarized at an angle λ with the Ox axis and the fast and slow axes of the retardation plate are along the Ox and Oy axes, that is $\theta = 0$. If the intensity of the incident light is equal to unity, then the semi-axes of the light ellipse of the emerging light beam would be $\cos \omega$, $\sin \omega$. Resolving the vibration represented by the light ellipse along the Ox and Oy axes, we can find the two amplitudes of the emerging light beam to be

$$
\begin{aligned}
A_1 &= \cos \omega \cos \psi + i \sin \omega \sin \psi \\
A_2 &= \cos \omega \sin \psi - i \sin \omega \cos \psi .
\end{aligned}
\tag{4.51}
$$

Because

$$A_1 = \cos \lambda, \qquad A_2 = \sin \lambda, \tag{4.52}$$

it follows from (4.51) and (4.52) that

$$\cos^2 \lambda = \cos^2 \omega \cos^2 \psi + \sin^2 \omega \sin^2 \psi$$
$$\sin^2 \lambda = \cos^2 \omega \sin^2 \psi + \sin^2 \omega \cos^2 \psi,$$

from which it follows that

$$\cos 2\lambda = \cos 2\omega \cos 2\psi. \tag{4.53}$$

From (3.27) and (3.31), it follows that the phase difference δ is connected with ω and ψ through the relation

$$\tan \delta = \frac{\tan 2\omega}{\sin 2\psi}. \tag{4.54}$$

Let us now represent by the point P_0 on the Poincaré sphere (Fig. 4.5) the linearly polarized light that is incident on the linear retarder, such that $\widehat{FP_0} = 2\lambda$, where F is a horizontally linear polarized light, and represent the emerging light beam by the point P, such that $\widehat{FA} = 2\psi$ and $\widehat{AP} = 2\omega$. From the right-angled spherical triangles FAP and FP_0P, it follows from (4.53) and (4.54) that

$$\widehat{FP} = \widehat{FP_0} \quad \text{and} \quad \widehat{PFP_0} = \delta.$$

That is, the point P of the emerging light is obtained from point P_0 of the incident light by rotating the Poincaré sphere about the axis FS, which con-

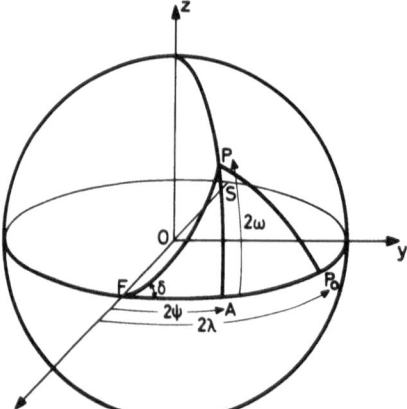

Fig. 4.5. Representation on the Poincaré sphere of the effect of a linear retarder with retardation δ, whose eigenvectors OF, OS are along the positive and negative Ox axis, on an initially linearly polarized light beam P_0. Point P of the emerging polarization form is obtained from point P_0 of the incident form by rotating the sphere about the axis FS, by an angle $\widehat{P_0P} = \delta$

nects the points that represent the polarization forms of the fast and slow eigenvectors of the linear retarder, by an angle δ that is equal to the retardation of the retarder. The rotation of the sphere should be in the counterclockwise sense for an observer looking from F to S.

Because the definition on the Poincaré sphere of the horizontally linear polarized light beam, based on the position of the Ox axis, is arbitrary, the above result can easily be generalized for the case in which the principal axes of the retarder are not along the Ox axis. Similarly, this result is still valid for the case in which the incident light beam is not linearly polarized, when we have the most-general case of an elliptical retarder. Thus, for the prediction of the effect of an elliptical retarder on an elliptically polarized light beam, the following operations on the Poincaré sphere should be performed (Fig. 4.6):

I) We mark on the sphere the point P_0 that represents the incident light beam, and the points F and S that represent the eigenvectors of the fast and slow axes of the retarder. These points are diametrically opposite, because they represent a pair of orthogonal polarization forms.

II) We rotate the sphere about the axis FS, through an angle equal to the retardance δ. The sense of rotation is counterclockwise for an observer looking from F to S.

III) We find the new position of point P, which represents the light beam emerging from the retarder.

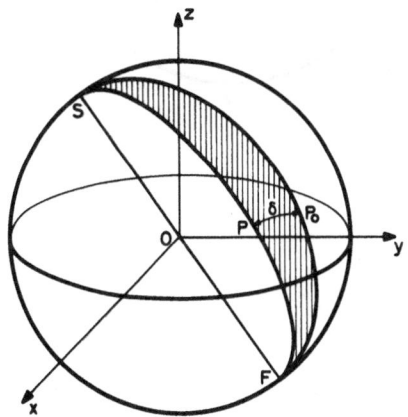

Fig. 4.6. The effect of an elliptical retarder with retardation δ on elliptically polarized light. The fast and slow eigenvectors of the retarder are represented by the points F and S and the elliptically polarized light by point P_0 on the Poincaré sphere. The output polarization form is obtained by rotating the sphere about the axis FS, through an angle δ, in a counterclockwise direction, for an observer looking from F to S

The geometrical representation of the effect of a linear retarder on a polarization state P_0 is shown in Fig. 4.7. The polarization state P of the emerging light is found by rotating the sphere about the axis FS, where F and S are the fast and slow eigenvectors of the retarder, through an angle δ.

This very important result has a large number of useful applications and makes the Poincaré sphere a unique and elegant means for predicting the effect of a retarder on a polarized light beam. Such predictions can be made

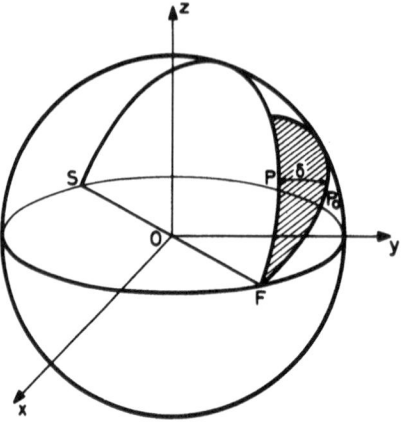

Fig. 4.7. As in Fig. 3.6 for a linear retarder

Fig. 4.8. Representation on the Poincaré sphere of the effect of a circular retarder. The output polarization forms P or P' obtained from the input forms P_0 or P'_0 by rotating the sphere around the axis NS have the same ellipticity and handedness as the input forms

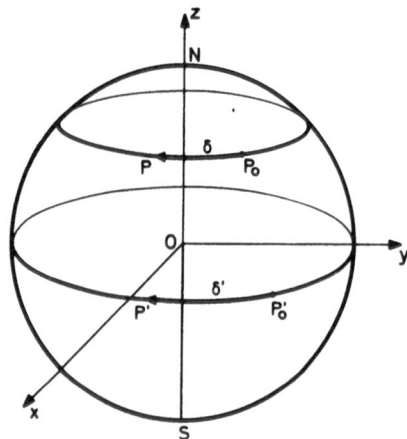

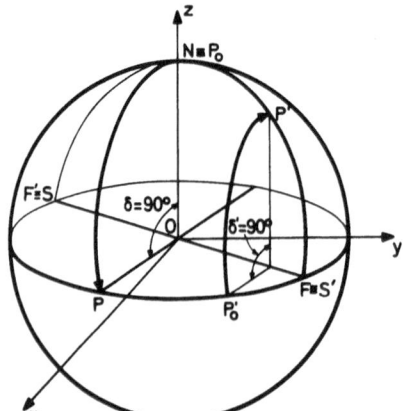

Fig. 4.9. Conversion of circularly polarized light $N \equiv P_0$ into linearly polarized light P by passage through a quarter-wave plate with eigenvectors F and S on the Poincaré sphere. The inverse relation is not generally true (P'_0, P')

by having in mind only the properties of the Poincaré sphere and performing the suitable transformations on it.

Many useful problems that involve retarders can be immediately solved by use of the Poincaré sphere. For example, a circular retarder, whose eigenvectors are represented by the poles of the Poincaré sphere, converts any elliptical polarization form to another that has the same handedness and the same ellipticity (Fig. 4.8). Thus, every linear polarization is transformed into another linear polarization by a circular retarder. Circularly polarized light, represented by a pole of the sphere, is converted by means of a quarter-wave plate ($\delta = 90°$) into linearly polarized light, because each pole, rotated about a horizontal axis through 90°, gives a point on the equator of the sphere (Fig. 4.9). The inverse is not always true; that is, linearly polarized light is not always converted into circularly polarized light by a quarter-wave plate (Fig. 4.9). To accomplish this, the points that represent the axes of the plate and the polarization form of the incident light on the Poincaré sphere must differ by 90°, that is, the direction of polarization must make an angle 45° with the axes of the quarter-wave plate.

An important and useful result may be directly derived from the Poincaré sphere, that any polarization form can be converted into any other by inserting into the first form a 180° retarder (generally elliptical) (Fig. 4.10). Indeed, by tracing the great circle that is defined by the entrance and exit points and bisecting the angle between those points, the two eigenvectors of the retarder can be derived. Then, the 180° rotation about the axis defined by the eigenvector of the required 180° retarder brings the first point to the other. If both states are linearly polarized, a 180° linear retarder is required (Fig. 4.11).

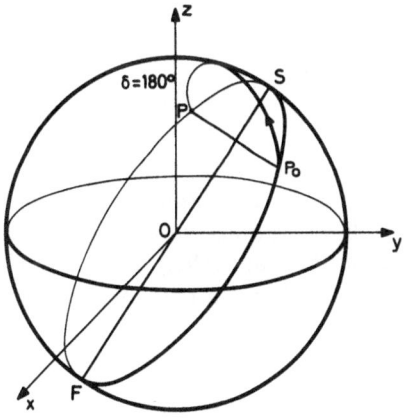

Fig. 4.10. Conversion of a given polarization form P_0 into another form P by use of a 180° retarder with eigenvectors F and S on the Poincaré sphere

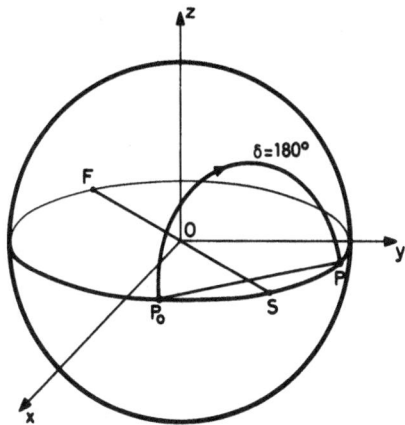

Fig. 4.11. Conversion of a given linear polarization form P_0 into another linear form P by use of a 180° linear retarder with eigenvectors F and S on the Poincaré sphere

When we have a train of retarders interposed in a polarized light beam, the emerging light from the last retarder can be found from the incident light beam by defining the eigenvectors of the retarders on the Poincaré sphere and performing successive rotations of the sphere about axes that are defined by the particular eigenvectors of each retarder. Each required angle of rotation is equal to the retardance of the corresponding retarder (Fig. 4.12). The whole series of retarders is equivalent to a single retarder, because a single arc can always be found on the surface of the sphere such as to bring the initial point on the sphere that represents the incident light beam into the final point on the sphere that represents the emerging light beam.

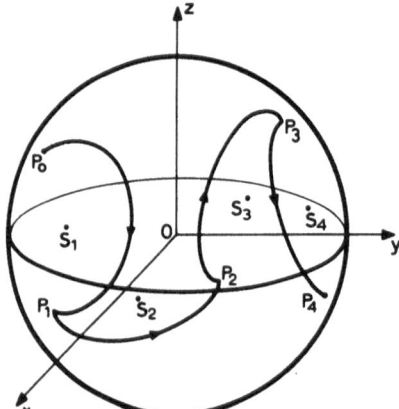

Fig. 4.12. Representation on the Poincaré sphere of the effect of a pile of retarders with slow eigenvectors S_1, S_2, S_3, S_4 on an incident polarization form P_0

4.3.6 The j-Circle Method

Because the tip of the j vector, which represents a given polarization form, can be regarded as the orthographic projection of the corresponding point of the Poincaré sphere on its equatorial plane, all operations with the j-circle method can easily be deduced from those on the Poincaré sphere. The method of this j circle with a number of applications to photoelasticity will be developed in Chap. 13 of this book.

4.4 Interrelation Between the Mueller and Jones Matrices

Because the Mueller and Jones matrices represent optical devices and the corresponding Stokes and Jones vectors of the incident and the emerging light beams are closely related (see Sec. 3.5.3), it follows that a relationship must exist between the Mueller and Jones matrices.

Let S and S' be the incident and the emerging Stokes vectors, a and a' the corresponding Jones vectors, and M and J the Mueller and Jones matrices of the optical device between S and S', or between a and a', respectively.

Then,

$$S' = \begin{bmatrix} s'_0 \\ s'_1 \\ s'_2 \\ s'_3 \end{bmatrix} = \begin{bmatrix} m_{11} & m_{12} & m_{13} & m_{14} \\ m_{21} & m_{22} & m_{23} & m_{24} \\ m_{31} & m_{32} & m_{33} & m_{34} \\ m_{41} & m_{42} & m_{43} & m_{44} \end{bmatrix} \begin{bmatrix} s_0 \\ s_1 \\ s_2 \\ s_3 \end{bmatrix} = M \cdot S \tag{4.55}$$

$$a' = \begin{bmatrix} A'_x e^{i\delta'_x} \\ A'_y e^{i\delta'_y} \end{bmatrix} = \begin{bmatrix} j_{11} & j_{12} \\ j_{21} & j_{22} \end{bmatrix} \begin{bmatrix} A_x e^{i\delta_x} \\ A_y e^{i\delta_y} \end{bmatrix} = J a . \tag{4.56}$$

From (4.55) it follows that

$$s'_0 = m_{11} s_0 + m_{12} s_1 + m_{13} s_2 + m_{14} s_3 , \tag{4.57}$$

which, when (3.61) are taken into account yields

$$\tilde{a}' Q_1 a' = m_{11}(\tilde{a} Q_1 a) + m_{12}(\tilde{a} Q_2 a) + m_{13}(\tilde{a} Q_3 a) + m_{14}(\tilde{a} Q_4 a)$$

or

$$\tilde{a}' Q_1 a' = \tilde{a}(m_{11} Q_1 + m_{12} Q_2 + m_{13} Q_3 + m_{14} Q_{14}) a . \tag{4.58}$$

When the values of Q_i $(i=1,2,3,4)$ from (3.62) are substituted, (4.58) takes the form

$$\tilde{a}' Q_1 a' = \tilde{a} \begin{bmatrix} m_{11} + m_{12} & m_{13} - i m_{14} \\ m_{13} + i m_{14} & m_{11} - m_{12} \end{bmatrix} a . \tag{4.59}$$

Similarly,

$$\tilde{a}' Q_2 a' = \tilde{a} \begin{bmatrix} m_{21} + m_{22} & m_{23} - i m_{24} \\ m_{23} + i m_{24} & m_{21} - m_{22} \end{bmatrix} a , \tag{4.60}$$

$$\tilde{a}' Q_3 a' = \tilde{a} \begin{bmatrix} m_{31} + m_{32} & m_{33} - i m_{34} \\ m_{33} + i m_{34} & m_{31} - m_{32} \end{bmatrix} a , \tag{4.61}$$

$$\tilde{a}' Q_4 a' = \tilde{a} \begin{bmatrix} m_{41} + m_{42} & m_{43} - i m_{44} \\ m_{43} + i m_{44} & m_{41} - m_{42} \end{bmatrix} a . \tag{4.62}$$

From (4.56), by taking the hermitians on both sides of this relation and using the well-known property of matrices,

$$(\widetilde{AB}) = \tilde{B} \cdot \tilde{A},$$

where $\tilde{A}, \tilde{B}$ are the hermitian conjugate matrices of the matrices A and B, respectively, we can deduce that

$$\tilde{a}' = \tilde{a}\, \tilde{J}. \tag{4.63}$$

Thus, we obtain

$$\tilde{a}' Q_1 a' = \tilde{a}(\tilde{J} Q_1 J) a,$$
$$\tilde{a}' Q_2 a' = \tilde{a}(\tilde{J} Q_2 J) a,$$
$$\tilde{a}' Q_3 a' = \tilde{a}(\tilde{J} Q_3 J) a, \tag{4.64}$$
$$\tilde{a}' Q_4 a' = \tilde{a}(\tilde{J} Q_4 J) a.$$

By equating the right-hand members of (4.59) and the first of (4.64), we obtain

$$m_{11} + m_{12} = \bar{j}_{11} j_{11} + \bar{j}_{21} j_{21},$$
$$m_{13} - i m_{14} = \bar{j}_{11} j_{12} + \bar{j}_{21} j_{22},$$
$$m_{13} + i m_{14} = \bar{j}_{12} j_{11} + \bar{j}_{22} j_{21}, \tag{4.65}$$
$$m_{11} - m_{12} = \bar{j}_{12} j_{12} + \bar{j}_{22} j_{22}.$$

Similarly, from (4.60) and the second of (4.64) we obtain

$$m_{21} + m_{22} = \bar{j}_{11} j_{11} - \bar{j}_{21} j_{21},$$
$$m_{23} - i m_{24} = \bar{j}_{11} j_{12} - \bar{j}_{21} j_{22},$$
$$m_{23} + i m_{24} = \bar{j}_{12} j_{11} - \bar{j}_{22} j_{21}, \tag{4.66}$$
$$m_{21} - m_{22} = \bar{j}_{12} j_{12} - \bar{j}_{22} j_{22}.$$

From (4.61) and the third of (4.62) it follows that

$$m_{31} + m_{32} = \bar{j}_{11} j_{21} + \bar{j}_{21} j_{11},$$
$$m_{33} - i m_{34} = \bar{j}_{11} j_{22} + \bar{j}_{21} j_{12},$$
$$m_{33} + i m_{34} = \bar{j}_{12} j_{21} + \bar{j}_{22} j_{11}, \tag{4.67}$$
$$m_{31} - m_{32} = \bar{j}_{12} j_{22} + \bar{j}_{22} j_{12}.$$

Finally, from (4.62) and the last of (4.64) it follows that

$$m_{41} + m_{42} = i(\bar{j}_{21}j_{11} - \bar{j}_{11}j_{21}),$$
$$m_{43} - im_{44} = i(\bar{j}_{21}j_{12} - \bar{j}_{11}j_{22}),$$
$$m_{43} + im_{44} = i(\bar{j}_{22}j_{11} - \bar{j}_{12}j_{21}),$$
$$m_{41} - m_{42} = i(\bar{j}_{22}j_{12} - \bar{j}_{12}j_{22}).$$

(4.68)

In (4.65–68), bars indicate complex conjugates of the corresponding quantities.

From (4.65–68), we obtain the following equations, which express the elements m_{ij} $(i,j = 1,2,3,4)$ of the Mueller matrix in terms of the elements $j_{\kappa l}$ $(\kappa, l = 1,2)$ of the corresponding Jones matrix:

$$2m_{11} = \bar{j}_{11}j_{11} + \bar{j}_{21}j_{21} + \bar{j}_{12}j_{12} + \bar{j}_{22}j_{22},$$
$$2m_{12} = \bar{j}_{11}j_{11} + \bar{j}_{21}j_{21} - \bar{j}_{12}j_{12} - \bar{j}_{22}j_{22},$$
$$2m_{13} = \bar{j}_{11}j_{12} + \bar{j}_{21}j_{22} + \bar{j}_{12}j_{11} + \bar{j}_{22}j_{21},$$
$$2m_{14} = i(\bar{j}_{11}j_{12} + \bar{j}_{21}j_{22} - \bar{j}_{12}j_{11} - \bar{j}_{22}j_{21}),$$
$$2m_{21} = \bar{j}_{11}j_{11} + \bar{j}_{12}j_{12} - \bar{j}_{21}j_{21} - \bar{j}_{22}j_{22},$$
$$2m_{22} = \bar{j}_{11}j_{11} + \bar{j}_{22}j_{22} - \bar{j}_{21}j_{21} - \bar{j}_{12}j_{12},$$
$$2m_{23} = \bar{j}_{12}j_{11} + \bar{j}_{11}j_{12} - \bar{j}_{22}j_{21} - \bar{j}_{21}j_{22},$$
$$2m_{24} = i(\bar{j}_{11}j_{12} + \bar{j}_{22}j_{21} - \bar{j}_{21}j_{22} - \bar{j}_{12}j_{11}),$$
$$2m_{31} = \bar{j}_{11}j_{21} + \bar{j}_{21}j_{11} + \bar{j}_{12}j_{22} + \bar{j}_{22}j_{12},$$
$$2m_{32} = \bar{j}_{11}j_{21} + \bar{j}_{21}j_{11} - \bar{j}_{12}j_{22} - \bar{j}_{22}j_{12},$$
$$2m_{33} = \bar{j}_{11}j_{22} + \bar{j}_{21}j_{12} + \bar{j}_{12}j_{21} + \bar{j}_{22}j_{11},$$
$$2m_{34} = i(\bar{j}_{11}j_{22} + \bar{j}_{21}j_{12} - \bar{j}_{12}j_{21} - \bar{j}_{22}j_{11}),$$
$$2m_{41} = i(\bar{j}_{21}j_{11} + \bar{j}_{22}j_{12} - \bar{j}_{11}j_{21} - \bar{j}_{12}j_{22}),$$
$$2m_{42} = i(\bar{j}_{21}j_{11} + \bar{j}_{12}j_{22} - \bar{j}_{11}j_{21} - \bar{j}_{22}j_{12}),$$
$$2m_{43} = i(\bar{j}_{21}j_{12} + \bar{j}_{22}j_{11} - \bar{j}_{11}j_{22} - \bar{j}_{12}j_{21}),$$
$$2m_{44} = \bar{j}_{22}j_{11} + \bar{j}_{11}j_{22} - \bar{j}_{12}j_{21} - \bar{j}_{21}j_{12}.$$

(4.69)

Combining (4.65–68), we can express the elements $j_{\kappa l}$ $(\kappa, l = 1,2)$ of the Jones matrix of a given optical device in terms of the elements m_{ij} $(i,j = 1,2,3,4)$ of the Mueller matrix of the same optical device. Elements $j_{\kappa l}$ are complex and can be expressed in their polar form,

$$j_{\kappa l} = |j_{\kappa l}| e^{i\theta_{\kappa l}}.$$

By adding the first equations in (4.65) and (4.66) we obtain

$$2|j_{11}|^2 = m_{11} + m_{12} + m_{21} + m_{22}.$$

(4.70)

By subtracting the same equations, we obtain

$$2|j_{21}|^2 = m_{11} + m_{12} - m_{21} - m_{22}. \tag{4.71}$$

By adding and subtracting the last equations in (4.65) and (4.66), we obtain

$$2|j_{12}|^2 = m_{11} - m_{12} + m_{21} - m_{22}, \tag{4.72}$$

$$2|j_{22}|^2 = m_{11} - m_{12} - m_{21} + m_{22}. \tag{4.73}$$

Relations (4.70–73) give the absolute values $|j_{\kappa l}|$ of the complex numbers $j_{\kappa l}$ of the Jones matrix in terms of the elements m_{ij} of the Mueller matrix.

In order to find the polar angles $\theta_{\kappa l}$ of the complex numbers $j_{\kappa l}$, we use (4.69). Thus, we obtain

$$m_{13} + m_{23} = \bar{j}_{11} j_{12} + \bar{j}_{12} j_{11} = 2|j_{11}||j_{12}| \cos(\theta_{11} - \theta_{12}),$$

or

$$\cos(\theta_{11} - \theta_{12}) = \frac{m_{13} + m_{23}}{\sqrt{(m_{11} + m_{21})^2 - (m_{12} + m_{22})^2}}. \tag{4.74}$$

From the same equations, we obtain

$$\sin(\theta_{11} - \theta_{12}) = \frac{m_{14} + m_{24}}{\sqrt{(m_{11} + m_{21})^2 - (m_{12} + m_{22})^2}}. \tag{4.75}$$

Similarly, we obtain

$$\cos(\theta_{21} - \theta_{11}) = \frac{m_{31} + m_{32}}{\sqrt{(m_{11} + m_{12})^2 - (m_{21} + m_{22})^2}},$$

$$\sin(\theta_{21} - \theta_{11}) = \frac{m_{41} + m_{42}}{\sqrt{(m_{11} + m_{12})^2 - (m_{21} + m_{22})^2}},$$

$$\cos(\theta_{22} - \theta_{11}) = \frac{m_{33} + m_{44}}{\sqrt{(m_{11} + m_{22})^2 - (m_{21} + m_{12})^2}},$$

$$\tag{4.76}$$

$$\sin(\theta_{22} - \theta_{11}) = \frac{m_{43} - m_{34}}{\sqrt{(m_{11} + m_{22})^2 - (m_{21} + m_{12})^2}}.$$

Relations (4.74–76) completely determine the angles $(\theta_{11} - \theta_{12})$, $(\theta_{21} - \theta_{11})$, $(\theta_{22} - \theta_{11})$ in the interval $0 < (\theta_{ij} - \theta_{\kappa l}) < 2\pi$.

Thus, from (4.70–76) the elements $j_{\kappa l}$ of the Jones matrices are determined in terms of the elements $m_{\kappa l}$ of the corresponding Mueller matrices.

4.5 Some Useful Theorems

We shall now prove some theorems related to a train of retarders, which will be used frequently in subsequent chapters. The theorems will be proved by use of the Jones calculus.

Let us consider a polarized light beam that has initially a Jones vector a_0, which passes through a train of optical retarders $1, 2, \ldots, (n-1), n$ whose, Jones matrices are $J_1, J_2, \ldots, J_{n-1}, J_n$, respectively. The Jones vector a of the light beam that emerges from the train of devices is given by (4.12),

$$a = J_n J_{n-1} \ldots J_2 J_1 a_0 . \tag{4.77}$$

As was shown in Section 4.3.2, the Jones matrix $J(\theta)$ of a retarder can be written in the form

$$J(\theta) = R(-\theta) J(0) R(\theta) , \tag{4.78}$$

where θ indicates the orientation of the fast axis of the retarder with respect to the Ox axis, and $R(\theta)$ is the rotation matrix.

If the orientations of the fast axes of retarders $1, 2, \ldots, (n-1), n$, whose Jones matrices are $J_1, J_2, \ldots, J_{n-1}, J_n$ with respect to the Ox axis, are $\theta_1, \theta_2, \ldots, \theta_{n-1}, \theta_n$, respectively, then we have for the Jones vector

$$a = [R(-\theta_n) J_n(0) R(\theta_n)] [R(-\theta_{n-1}) J_{n-1}(0) R(\theta_{n-1})]$$
$$\ldots [R(-\theta_2) J_2(0) R(\theta_2)] [R(-\theta_1) J_1(0) R(\theta_1)] a_0 , \tag{4.79}$$

and, by using the relation

$$R(\theta_1 + \theta_2) = R(\theta_1) R(\theta_2) ,$$

we obtain

$$a = R(-\theta_n) [J_n(0) R(\theta_n - \theta_{n-1}) J_{n-1}(0) R(\theta_{n-1} - \theta_{n-2}) \ldots J_2(0) R(\theta_2 - \theta_1)$$
$$\cdot J_1(0) R(\theta_1 - \theta_n)] R(\theta_n) a_0 = R(-\theta_1) [R(\theta_1 - \theta_n) J_n(0) R(\theta_n - \theta_{n-1}) \tag{4.80}$$
$$\cdot J_{n-1}(0) R(\theta_{n-1} - \theta_{n-2}) \ldots J_2(0) R(\theta_2 - \theta_1) J_1(0)] R(\theta_1) a_0 .$$

Equation (4.80) can be written in the form

$$a = R(-\theta_n) J R(\theta_n) a_0 = R(-\theta_1) J' R(\theta_1) a_0 , \tag{4.81}$$

where matrices J and J' depend only on the relative orientations of the elements of the optical system and are independent of the orientation of the optical system as a whole. Relation (4.81) indicates the important theorem:

The matrix that represents a train of retarders is expressed in terms of two matrices, the first of which depends only on the nature of the retarders and on their relative orientation, whereas the second matrix depends only on the orientation of the optical system as a whole with respect to Ox axis.

This theorem is the generalization, for a train of retarders, of the result for a single retarder, expressed by (4.78).

Before proceeding to other relations, we shall define a unitary matrix A. A unitary matrix is a matrix whose product with its hermitian conjugate satisfies the relation

$$\tilde{A}A = A\tilde{A} = I,\tag{4.82}$$

where I denotes the unit matrix, and $\tilde{A}$ denotes the hermitian conjugate of A.

It can easily be shown from (4.16) that the matrix of a retarder is unitary, because it is the product of unitary matrices. Indeed, matrices $J(0)$ and $R(\theta)$ are unitary matrices.

Because the product of unitary matrices is also a unitary matrix, from (4.80) the theorem can be derived:

The Jones matrix that represents a train of retarders is a unitary matrix.

Using this theorem and the well-known theorem of matrix algebra, that any unitary matrix can be considered to be the product of a unitary matrix and a rotation matrix, we obtain the useful theorem:

An optical system that contains any number of retarders is optically equivalent to a system that contains one retarder and one rotator.

This theorem is referred to as the "*equivalence theorem*" in polarization optics.

From the nature of an optical rotator, such as a crystal or a solution that exhibits optical activity, it follows that the rotation of the plane of polarization of a light beam, introduced by the rotator, changes sign when light passes through the rotator in the reverse direction. Thus, if

$$R(\theta) = \begin{bmatrix} \cos\theta & \sin\theta \\ -\sin\theta & \cos\theta \end{bmatrix}\tag{4.83}$$

is the rotation matrix for light that passes through the rotator in one direction, then the matrix for light that passes through the rotator in the reverse direction is

$$R(-\theta) = \begin{bmatrix} \cos\theta & -\sin\theta \\ \sin\theta & \cos\theta \end{bmatrix}. \tag{4.84}$$

The transpose of this matrix is equal to $R(-\theta)$,

$$\tilde{R}(\theta) = R(-\theta). \tag{4.85}$$

Unlike the optical rotator, the Jones matrix of either a retarder or a polarizer does not change, if the light passes through the retarder or the polarizer in the reverse direction. From (4.13) and (4.16) it follows that the transpose matrices of the matrices that represent either a polarizer or a retarder are equal to the original matrices.

According to the equivalence theorem, the Jones matrix $J(\theta)$ of any system that includes optical retarders can be expressed by

$$J(\theta) = J_i(\theta) R_i(\theta), \tag{4.86}$$

where $J_i(\theta)$ and $R_i(\theta)$ are the Jones matrices of the equivalent retarder and rotator, respectively.

If the same optical system is traversed by a light beam in the opposite direction, then its Jones matrix $J^*(\theta)$, according to the previously established fundamental properties of the Jones matrices for polarizers or retarders on the one hand and a rotator on the other hand, will be given by

$$J^*(\theta) = R_i(-\theta) J_i(\theta). \tag{4.87}$$

When (4.85) is taken into account and the fact that the transpose matrix $\tilde{J}_i(\theta)$ of the matrix $J_i(\theta)$ is equal to $J_i(\theta)$, (4.87) can be written in the form

$$\tilde{J}^*(\theta) = \tilde{R}_i(\theta) \tilde{J}_i(\theta). \tag{4.88}$$

By taking the transpose matrices of both sides of (4.86) we deduce that

$$\tilde{J}(\theta) = \tilde{R}_i(\theta) \tilde{J}_i(\theta). \tag{4.89}$$

From (4.88) and (4.89) it follows that

$$J^*(\theta) = \tilde{J}(\theta). \tag{4.90}$$

Relation (4.90) expresses the theorem:

The Jones matrix of an optical system that is traversed by a light beam in a certain direction must be transposed in order to obtain the Jones matrix of the same optical system when it is traversed by the light beam in the reverse direction.

This theorem is known as the reversibility theorem of polarization optics. The reversibility theorem is valid only in systems that contain rotators due to the ordinary optical activity and not those due to the Faraday effect, for which the direction of rotation of polarized light depends on the direction of the magnetic field and not on the direction of the light that passes through the rotator. For the Faraday effect, matrix $J^*(\theta)$ is not expressed by (4.90).

References

4.1   H. Poincaré: *Théorie Mathématique de la Lumière*, Vol. 2 (Gauthiers-Villars, Paris 1892) Chap. 12
4.2   J. Walker: *Analytical Theory of Light* (Cambridge, at the University Press 1904)
4.3   J. Becquerel: Communs. Phys. Lab. Univ. Leiden, No. 191 C, 19 (1928)
4.4   J. Becquerel: Communs. Phys. Lab. Univ. Leiden, No. 211 A 1 (1930)
4.5   C. A. Skinner: J. Opt. Soc. Am. **10**, 491 (1925)
4.6   M. L. Chaumont: C.R. Acad. Sci. **150**, 1604 (1913)
4.7   M. L. Chaumont: Ann. Phys. **4**, 101 (1915)
4.8   G. Bruhat, P. Grivet: J. de Physique **6**, 12 (1935)
4.9   Y. Björnstahl: Z. Phys. **40**, 437 (1939)
4.10 Y. Björnstahl: Z. Instrumk. **59**, 425 (1939)
4.11 O. Snellman, Y. Björnstahl: Kolloid Beih. **52**, 403 (1941)
4.12 M. F. Bokstein: J. Techn. Phys. (USSR) **18**, 673 (1948)
4.13 F. Pockels: *Lehrbuch der Kristalloptik*, Vol. 11–13 (Teubner, Leipzig 1906) pp. 267–283
4.14 H. G. Jerrard: J. Opt. Soc. Am. **44**, 634 (1954)
4.15 C. J. Koester: J. Opt. Soc. Am. **49**, 405 (1959)
4.16 N. H. Hartshorne, A. Stuart: *Crystals and the Polarizing Microscope*, 2nd ed. (Arnold Press, London 1950)
4.17 F. E. Wright: J. Opt. Soc. Am. **20**, 529 (1930)
4.18 G. N. Ramachandran, S. Ramaseshan: "Crystal Optics", in *Crystal Optics, Diffraction*, ed. by S. Flügge, Encyclopedia of Physics, Vol. 25/1 (Springer, Berlin, Göttingen, Heidelberg 1961) pp. 1–217
4.19 S. Pancharatnam: Proc. Indian Acad. Sci. A**41**, 130 (1955)
4.20 S. Pancharatnam: Proc. Indian Acad. Sci. A**41**, 137 (1955)
4.21 S. Pancharatnam: Proc. Indian Acad. Sci. A**44**, 247 (1956)
4.22 S. Pancharatnam: Proc. Indian Acad. Sci. A**44**, 398 (1956)
4.23 G. N. Ramachandran, S. Ramaseshan: J. Opt. Soc. Am. **42**, 49 (1952)
4.24 G. N. Ramachandran, V. Chandrasekharan: Proc. Indian Acad. Sci. A**33**, 199 (1951)
4.25 S. Ramaseshan, V. Chandrasekharan: Current Sci. (India) **20**, 150 (1951)
4.26 S. Ramaseshan: Proc. Indian Acad. Sci. A**34**, 32 (1951)
4.27 A. Robert: Int. J. Solids Struct. **6**, 423 (1970)
4.28 A. J. Robert: Exp. Mech. **7**, 224 (1967)
4.29 R. Mark: AIAA J. **2**, 150 (1964)
4.30 J. D. Riera, R. Mark: Exp. Mech. **9**, 9 (1969)
4.31 C. Whitney: J. Opt. Soc. Am. **61**, 1207 (1971)
4.32 R. Plechata: Acta Techn. **2**, 230 (1957)
4.33 H. J. Woods: J. Text. Inst. **55**, 243 (1964)
4.34 H. Schwieger: Exp. Mech. **9**, 67 (1969)
4.35 A. Robert: *Polarimétrie et Photoélasticimétrie* (Serv. Techn. Const. Armes Navales, Paris 1972)
4.36 W. A. Shurcliff, S. S. Ballard: *Polarized Light* (Van Nostrand, New Jersey 1964)
4.37 W. A. Shurcliff: *Polarized Light. Production and Use* (Harvard University Press, Cambridge, Mass. 1962)

4.38 J. W. Simmons, M. J. Gutmann: *States, Waves and Photons: A Modern Introduction to Light* (Addison-Wesley, Reading, Mass. 1970)

4.39 H. K. Aben: *Integrated Photoelasticity* (VALGUS, Tallin, USSR 1975)

4.40 A. Kuske: Exp. Mech. **6**, 218 (1966)

4.41 A. Kuske: Int. Spannungsopt. Symp. Berlin, 11.–15.4.1961 (Akademie Verlag, Berlin 1962)

4.42 A. Kuske: Optik **19**, 261 (1962)

4.43 A. Kuske: Rev. Frans. de Méc. **9**, 49 (1964)

4.44 H. Mueller: Rep. No 2 of the OSRD Project OEMsr-576, Nov. 15 (1943)

4.45 H. Mueller: J. Opt. Soc. Am. **38**, 661 (1948)

4.46 P. Soleillet: Ann. Phys. **12**, 23 (1929)

4.47 F. Perrin: J. Chem. Phys. **10**, 415 (1942)

4.48 N. G. Parke: Ph. D. Thesis, Dept. of Physics, M.I.T., May 1 (1948)

4.49 N. G. Parke: Tech. Rep. No 70, Research Lab. of Electr., M.I.T., June 30 (1948)

4.50 N. G. Parke: Tech. Rep. No 95, Research Lab. of Electr., M.I.T., January 31 (1949)

4.51 N. G. Parke: Tech. Rep. No 119, Research Lab. of Electr., M.I.T., June 15 (1949)

4.52 M. J. Walker: Am. J. Phys. **22**, 170 (1954)

4.53 U. Fano: J. Opt. Soc. Am. **39**, 859 (1949)

4.54 U. Fano: Rev. Mod. Phys. **29**, 74 (1957)

4.55 W. H. McMaster: Am. J. Phys. **22**, 351 (1954)

4.56 W. H. McMaster: Rev. Mod. Phys. **33**, 8 (1961)

4.57 B. H. Billings, E. H. Land: J. Opt. Soc. Am. **38**, 819 (1948)

4.58 B. H. Billings: J. Opt. Soc. Am. **41**, 966 (1951)

4.59 B. H. Billings: J. Opt. Soc. Am. **42**, 12 (1952)

4.60 P. Roman: Nuovo Cimento **13**, 974 (1959)

4.61 R. W. Schmieder: J. Opt. Soc. Am. **59**, 297 (1969)

4.62 D. W. Weeks: J. Math. Phys. **13**, 380 (1957)

4.63 R. C. Jones: J. Opt. Soc. Am. **31**, 488 (1941)

4.64 H. Hurwitz, R. C. Jones: J. Opt. Soc. Am. **31**, 493 (1941)

4.65 R. C. Jones: J. Opt. Soc. Am. **31**, 500 (1941)

4.66 R. C. Jones: J. Opt. Soc. Am. **32**, 486 (1942)

4.67 R. C. Jones: J. Opt. Soc. Am. **37**, 107 (1947)

4.68 R. C. Jones: J. Opt. Soc. Am. **37**, 110 (1947)

4.69 R. C. Jones: J. Opt. Soc. Am. **38**, 671 (1948)

4.70 R. C. Jones: J. Opt. Soc. Am. **46**, 126 (1956)

4.71 R. C. Jones: J. Opt. Soc. Am. **46**, 528 (1956)

4.72 H. Y. Hsü, M. Richartz, Y. K. Liang: J. Opt. Soc. Am. **37**, 99 (1947)

4.73 E. F. Dawson, N. O. Young: J. Opt. Soc. Am. **50**, 170 (1960)

4.74 J. W. Evans: J. Opt. Soc. Am. **48**, 142 (1958)

4.75 M. Richartz, H. Y. Hsü: J. Opt. Soc. Am. **39**, 136 (1949)

4.76 J. Cernosek: J. Opt. Soc. Am. **61**, 324 (1971)

4.77 G. B. Parrent, P. Roman: Nuovo Cimento **15**, 370 (1960)

4.78 H. K. Aben: Proc. Conf. on Exp. Methods of Investigating Stress and Strain in Structures, (Prague 1965), pp. 33–42

4.79 H. K. Aben: Exp. Mech. **10**, 97 (1970)

4.80 H. K. Aben: Proc. 4th Int. Conf. Stress Anal. (Cambridge 1970), pp. 175–182

4.81 D. Clarke, J. F. Grainger: *Polarized Light and Optical Measurement* (Pergamon Press, Oxford 1971)

5. Measurement of Elliptically Polarized Light

5.1 Introduction

The measurement of the state of polarization of a light beam is of particular importance for the problem of the photoelastic determination of the stresses induced in a two- or three-dimensional body. In these problems, the state of stress of the body in question modifies the polarization form of the incident light beam, so that the emerging light contains enough information for determination of the stresses of the body. The complete characterization of the state of polarization of the output light constitutes the main part of the problem.

Generally speaking the complete characterization of the state of polarization of a beam of light necessitates the determination of four independent quantities. The four Stokes parameters may be considered as such quantities, since, for an arbitrary light beam, they are not restricted to satisfy any relation. However, in the case of a perfectly polarized light beam, three independent quantities characterize completely the state of its polarization. Thus, when polarized light is defined in terms of the Stokes parameters, only three of them are independent, because the four parameters satisfy an initial condition. These three independent Stokes parameters can be used to specify the polarization form of the light beam. In terms of the Jones vector these three quantities may represent the two amplitudes and the phase difference of the components of the light vector along two arbitrary orthogonal directions. When the light ellipse is used to define the polarization state and we are not interested in the intensity of the light beam, the inclination of the major axis of the ellipse (azimuth), the ratio of its axes (ellipticity), and the sense of rotation of the vector whose end point describes the ellipse (handedness) are convenient parameters with which we specify the state of polarization. In the graphical method of the Poincaré-sphere representation the two spherical coordinates of the point on the sphere and the position of the point on the north or the south hemisphere constitute the three relevant parameters that determine the state of polarization.

Because the most relevant quantity, which can be easily determined experimentally, is the intensity of the light, all methods for measurement of the state of polarization are based on optical transformations of the light in question, by insertion of optical devices into the light beam and measuring the intensity of the emerging light. As such optical devices, retardation plates of various retardances, and linear polarizers, are usually used. These optical elements can be combined in various manners, so that many particular methods can be developed.

In the following, the basic principles of the main methods of measuring the state of polarization of a perfectly polarized light beam will be developed. At the end of this chapter, we shall also develop methods for the determination of the characteristic quantities of an optical element, defined either by the four-by-four Mueller matrix, or by the corresponding two-by-two Jones matrix.

5.2 General Considerations

As just mentioned, all convenient methods for measurement of the state of polarization of a light beam are based on the same principle, that is, insertion of retardation plates and polarizers into the light beam and measurement of the intensity of the emerging light. Because the light that emerges from a retardation plate is generally elliptically polarized and its light vector varies continuously with time, the final optical element of any measuring method must always be a linear polarizer, so that the light vector of the outcoming beam has a definite direction. Thus, a sinusoidal time variation of light intensity will emerge, which can be easily detected by any intensity measuring device. However, although the use of a linear polarizer (called the *analyzer*, due to its particular function) is always necessary, it does not suffice for the measurement of any elliptical polarization form. Only linearly polarized light beams can be detected by an analyzer alone. In all other cases, a retardation plate must also be interposed in the light path.

In order to get a better understanding of the idea of measurement of an elliptical polarization form, by insertion of a retardation plate and an analyzer, let us consider the problem quantitatively, by use of the Mueller calculus.

Let the Stokes vector S of the light beam to be measured be

$$S = \begin{bmatrix} s_0 \\ s_1 \\ s_2 \\ s_3 \end{bmatrix}. \tag{5.1}$$

Let us insert into this beam a retardation plate with retardance δ, with its fast axis at an angle β from the Ox axis of a reference frame $Oxyz$, and a linear analyzer, whose pass axis subtends an angle γ with the Ox axis (Fig. 5.1). Then, the Mueller matrices $R_\beta(\delta)$ and P_γ of the retarder and the polarizer, respectively, will be (see Tables 4.3 and 4.4)

$$R_\beta(\delta) = \begin{bmatrix} 1 & 0 & 0 & 0 \\ 0 & \cos^2 2\beta + \sin^2 2\beta \cos\delta & (1-\cos\delta)\sin 2\beta \cos 2\beta & -\sin 2\beta \sin\delta \\ 0 & (1-\cos\delta)\sin 2\beta \cos 2\beta & \sin^2 2\beta + \cos^2 2\beta \cos\delta & \cos 2\beta \sin\delta \\ 0 & \sin 2\beta \sin\delta & -\cos 2\beta \sin\delta & \cos\delta \end{bmatrix},$$

(5.2)

$$P_\gamma = \frac{1}{2} \begin{bmatrix} 1 & \cos 2\gamma & \sin 2\gamma & 0 \\ \cos 2\gamma & \cos^2 2\gamma & \sin 2\gamma \cos 2\gamma & 0 \\ \sin 2\gamma & \sin 2\gamma \cos 2\gamma & \sin^2 2\gamma & 0 \\ 0 & 0 & 0 & 0 \end{bmatrix}.$$

(5.3)

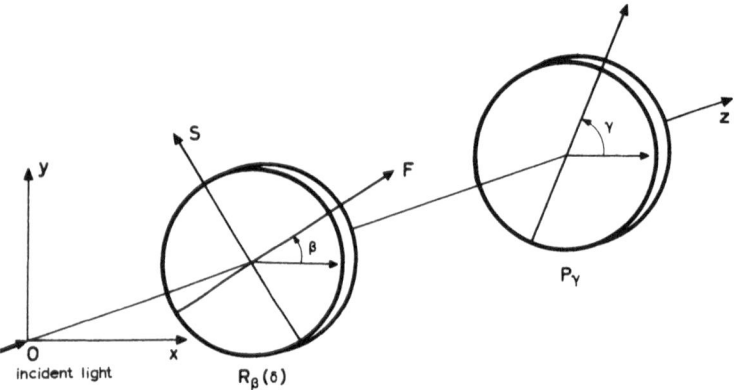

Fig. 5.1. Generalized polarimetric arrangement consisting of a retarder $R_\beta(\delta)$ and a polarizer P_γ for measurement of the state of polarization of elliptically polarized light

The Stokes vector S' of the light that emerges from the analyzer will be

$$S' = P_\gamma R_\beta(\delta) S.$$

(5.4)

By performing the matrix multiplications in (5.4), we obtain for the first element of the Stokes vector s_0' of the emerging light beam, which also represents the outcoming intensity $I(\beta,\gamma,\delta)$,

$$2I(\beta,\gamma,\delta) = 2s_0' = s_0 + [(s_1 \cos 2\beta + s_2 \sin 2\beta)\cos 2(\beta-\gamma)]$$
$$+ [(s_1 \sin 2\beta - s_2 \cos 2\beta)\cos\delta - s_3 \sin\delta]\sin 2(\beta-\gamma).$$

(5.5)

Equation (5.5) indicates that the intensity of the emerging light beam from the analyzer depends on the Stokes parameters $s_0, s_1, s_2,$ and s_3 of the incident light beam, as well as on the orientations β and γ of the retardation plate and the analyzer and also on the retardance δ of the retardation plate. Equation (5.5) therefore indicates that the values of the Stokes parameters s_0, s_1, s_2, and s_3 of the light in question, which for the case of perfect polarization satisfy

$$s_0^2 = s_1^2 + s_2^2 + s_3^2 , \tag{5.6}$$

can be determined by assigning three different sets of values to the parameters β, γ, and δ and measuring the corresponding values of the output light intensity.

However, great simplification in the determination of the Stokes parameters s_0, s_1, s_2, and s_3 may be obtained if only one of the three parameters β, γ, and δ is varied, while the other two are kept constant. Thus, all methods for the determination of the state of polarization of a light beam may be classified into the following three categories:

I) Those that use a rotating analyzer (γ variable, β and δ constant)

II) Those that use a rotating retardation plate (β variable, γ and δ constant)

III) Those that use a retardation plate of varying retardance or that use a series of retardation plates (δ variable, β and γ constant)

5.3 Determination of the Elements of the Stokes and Jones Vectors

As an application of the principles outlined, let us consider the determination of the elements of the Stokes and Jones vectors of a light beam. For the case of the Stokes vector, (5.5) will be used for the determination of the intensity of the emerging light. Let us interpose in the light path the indicated optical elements and compute the corresponding light intensities I of the beam.

I) A single analyzer, with its pass plane parallel to the Ox axis. Then, we get from (5.5)

$$I_1 = I(0,0,0) = \frac{s_0 + s_1}{2} . \tag{5.7}$$

II) A single analyzer with its pass plane parallel to the Oy axis. Then, we have

$$I_2 = I(0,90°,0) = \frac{s_0 - s_1}{2} . \tag{5.8}$$

From (5.7) and (5.8), we obtain for the Stokes parameters s_0 and s_1

$$s_0 = I_1 + I_2, \quad s_1 = I_1 - I_2 . \tag{5.9}$$

III) A single analyzer with its pass plane at $45°$ to the Ox axis. Then, we have

$$I_3 = I(0,45°,0) = \frac{s_0 + s_2}{2},$$
(5.10)

and

$$s_2 = 2I_3 - (I_1 + I_2).$$
(5.11)

IV) A quarter-wave plate ($\delta = 90°$) with its fast axis along the Ox axis ($\beta = 0$), followed by an analyzer whose pass plane is at $45°$ to the Ox axis ($\gamma = 45°$). Then, we have

$$I_4 = I(0,45°,90°) = \frac{s_0 + s_3}{2},$$
(5.12)

and

$$s_3 = 2I_4 - (I_1 + I_2).$$
(5.13)

The four Stokes parameters $s_0, s_1, s_2,$ and s_3 are determined from (5.9), (5.11), and (5.13), by use of the measured values of the intensities I_1, I_2, I_3, and I_4 of the light that emerges from the specified optical devices.

Usually, in measuring devices, only relative values of the light intensities $I_1, I_2, I_3,$ and I_4 are measured, that is, the values of (I_2/I_1), (I_3/I_1), and (I_4/I_1). This makes easier and more accurate the experimental determination of the Stokes parameters of the light in question and does not necessitate any calibration of the measuring device.

The same procedure may be followed when the elements of the Jones vector of the polarized light beam have to be determined. Thus, by use of the same optical elements, arranged similarly, we obtain for the Jones vector

I) For the light transmitted through an analyzer whose pass plane is parallel to the Ox axis (see Tables 4.1 and 4.2)

$$a' = J a = \begin{bmatrix} 1 & 0 \\ 0 & 0 \end{bmatrix} \begin{bmatrix} a_x \\ a_y \end{bmatrix} = \begin{bmatrix} a_x \\ 0 \end{bmatrix},$$
(5.14)

where

$$a = \begin{bmatrix} a_x \\ a_y \end{bmatrix} = \begin{bmatrix} A_x e^{i\delta_x} \\ A_y e^{i\delta_y} \end{bmatrix} = e^{i\delta_x} \begin{bmatrix} A_x \\ A_y e^{i\delta} \end{bmatrix}, \qquad \delta = (\delta_y - \delta_x)$$

is the Jones vector of the light in question.

The intensity I_1 of the transmitted light is

$$I_1 = [A_x \quad 0] \begin{bmatrix} A_x \\ 0 \end{bmatrix} = A_x^2.$$ (5.15)

II) For the light transmitted through an analyzer whose pass plane is vertical

$$\mathbf{a}' = \mathbf{J}\mathbf{a} = \begin{bmatrix} 0 & 0 \\ 0 & 1 \end{bmatrix} \begin{bmatrix} a_x \\ a_y \end{bmatrix} = \begin{bmatrix} 0 \\ a_y \end{bmatrix}.$$ (5.16)

The intensity I_2 of this beam is

$$I_2 = [0 \quad A_y e^{-i\delta}] \begin{bmatrix} 0 \\ A_y e^{i\delta} \end{bmatrix} = A_y^2.$$ (5.17)

III) For the light transmitted through an analyzer whose pass plane is at $45°$ to the Ox axis

$$\mathbf{a}' = \mathbf{J}\mathbf{a} = \frac{1}{2} \begin{bmatrix} 1 & 1 \\ 1 & 1 \end{bmatrix} \begin{bmatrix} a_x \\ a_y \end{bmatrix} = \frac{1}{2} \begin{bmatrix} a_x + a_y \\ a_x + a_y \end{bmatrix}.$$ (5.18)

The intensity I_3 of this light is

$$I_3 = \frac{1}{4} [A_x + A_y e^{-i\delta} \quad A_x + A_y e^{-i\delta}] \begin{bmatrix} A_x + A_y e^{i\delta} \\ A_x + A_y e^{i\delta} \end{bmatrix}$$

$$= \frac{1}{2} (A_x^2 + A_y^2 + 2 A_x A_y \cos \delta).$$ (5.19)

IV) For the light transmitted through an analyzer whose pass plane is at $45°$ to the Ox axis, preceded by a quarter-wave plate whose fast axis is along the Ox axis

$$\mathbf{a}' = \mathbf{J}_2 \mathbf{J}_1 \mathbf{a} = \frac{1}{2} \begin{bmatrix} 1 & 1 \\ 1 & 1 \end{bmatrix} \begin{bmatrix} i & 0 \\ 0 & 1 \end{bmatrix} \begin{bmatrix} a_x \\ a_y \end{bmatrix} = \frac{1}{2} \begin{bmatrix} i a_x + a_y \\ i a_x + a_y \end{bmatrix}.$$ (5.20)

The intensity I_4 of this light is

$$I_4 = \frac{1}{4} [-i A_x + A_y e^{-i\delta} \quad -i A_x + i A_y e^{-i\delta}] \begin{bmatrix} i A_x + A_y e^{i\delta} \\ i A_x + A_y e^{i\delta} \end{bmatrix}$$

$$= \frac{1}{2} (A_x^2 + A_y^2 + 2 A_x A_y \sin \delta).$$ (5.21)

By measuring the light intensities I_1 and I_2, we obtain the amplitudes A_x and A_y, whereas from the intensities I_3 and I_4 we determine the phase angle δ.

5.4 Photoelectric Methods

5.4.1 General Considerations

As mentioned in Section 5.2, determination of the state of polarization of a light beam by interposing in the light path a retardation plate, followed by an analyzer, is greatly simplified if one of the characteristic parameters of the system is varied, while the other two are kept constant. The light-intensity formula (5.5) for the first and the third methods of Section 5.2, where either a rotating analyzer, or a retardation plate with variable retardance is used, can be put in the general form

$$I(\varphi) = A_1 + A_2 \cos(\varphi + A_3),\tag{5.22}$$

whereas for the second method, in which a rotating retardation plate is used, the light-intensity formula (5.5) can be put in the form

$$I(\varphi) = A_1 + A_2 \cos(\varphi + A_3) + A_4 \cos 2(\varphi + A_5).\tag{5.23}$$

In (5.22) and (5.23), φ is the variable parameter of the system, that is, φ is related to each quantity γ, β, and δ for the first, second, or third method, respectively, whereas A_1, A_2, A_3, A_4, and A_5 are constants that depend on the Stokes parameters of the light beam in question, as well as on the values of the parameters β, γ, and δ that are kept constant for each method. Thus, measurements of the constants A_1, A_2, A_3, A_4, and A_5 in (5.22) and (5.23) permit determination of the Stokes parameters s_0, s_1, s_2, and s_3 of a light beam.

In order to get a better understanding of the meaning of constants A_1, A_2, A_3, A_4, and A_5 in (5.23), let us consider the particular case in which a quarter-wave plate is used ($\delta = 90°$) and the pass axis of the analyzer is along the Ox axis ($\gamma = 0$). Then, we obtain from (5.5)

$$2I(\beta) = \left(s_0 + \frac{s_1}{2}\right) - s_3 \sin 2\beta + \frac{(s_1^2 + s_2^2)^{1/2}}{2} \cos\left(4\beta - \tan^{-1}\frac{s_2}{s_1}\right).\tag{5.24}$$

If it is known that the light beam, whose state of polarization is to be measured, is perfectly polarized and we make absolute measurements of the intensity of the outcoming light, then either (5.22) or (5.23) suffices for the determination of the three unknown quantities of the light beam from the measured values of the constants A_1, A_2, A_3 or A_1, A_2, A_3, A_4, and A_5.

However, when it is not known if the light is perfectly polarized, or when relative values of the output intensities are measured, the first and the third of the methods leading to (5.22), do not suffice alone to determine the state of polarization. In such cases, the second method, which uses a rotating retardation plate and employs (5.23), must be used.

When an instrument for recording light intensity is used, the emergent signal $S(\varphi)$ is of the form

$$S(\varphi) = g\, I(\varphi),$$ (5.25)

where g characterizes the response of the instrument to unit intensity. For the usual case, in which only relative values of the output intensities are recorded, there is no need to evaluate g.

The signal of the detector can be processed to give the constants A_1, A_2, A_3, A_4, and A_5 in (5.22) or (5.23). All of the techniques used for the determination of these constants can be classified into three categories:

I) In the first category belong all techniques in which two intensity measurements are performed and the value of the phase $(\varphi + A_3)$ that corresponds to some particular feature of the emerging signal is determined. Thus, for example, we determine the value of the angle φ for which $S(\varphi)$ becomes maximum or minimum. For such positions $(\varphi + A_3) = n\pi$, thus leading to the value for A_3, which is equal to $A_3 = (n\pi - \varphi)$.

II) The second category includes all methods in which the light intensities at three relevant values of the angle φ are measured.

III) Finally, the third category comprises methods in which the parameter of the system that varies linearly with time, as well as the constant term A_1, and the amplitudes A_2, A_4 and the corresponding phases A_3 and A_5 of the oscillatory components of the emerging signal are separately determined. This method, when applied to the general case of elliptically polarized light, is self-sufficient and gives the values of all of the required parameters. Thus, as can be observed from (5.24), the emerging signal consists of a constant term and two oscillatory components with frequencies 2 and 4. By applying a Fourier-series expansion to the outcoming signal, the amplitudes and the phases of the oscillatory components may be determined. This leads to the determination of the values of the parameters A_1, A_2, A_3, A_4, and A_5.

The mode for detecting the emerging signal, the inherent errors, and the accuracy of determination of the characteristics of the state of polarization by each technique, belonging to one of the three general categories, are dependent on the particular method and therefore require lengthy investigations. Description of the particular characteristics of the detecting process is outside the scope of this book, which is concerned with only the general principles for measurement of states of polarization. Therefore, such details will not be given here. We shall, however, present the main characteristics of some techniques that are widely used for the photoelectric determination of the state of polarization of a light beam.

5.4.2 Special Techniques for Determining the State of Polarization

Kent and *Lawson* [5.1] have suggested a photoelectric technique that is based on the fact that, when circularly polarized light is passed through a rotating analyzer, the emerging signal does not have any oscillatory component. Otherwise, an oscillatory component would exist, with a frequency equal to twice the frequency of rotation of the analyzer. Thus, the method consists of converting any given elliptical polarization into circular polarization, by use of a suitably selected and oriented retardation plate, and afterwards passing the outcoming light through a rotating analyzer.

In order to find the proper values of retardance and the orientation of the retardation plate, so that the input elliptically polarized light is converted into circularly polarized light, let us use the Poincaré sphere (Fig. 5.2). The incident polarized light is represented on the sphere by point A that has longitude 2ψ and latitude 2ω. In order to bring point A to point N, representing right-circularly polarized light, the Poincaré sphere must be rotated about an axis BC that lies in the equatorial plane with points B and C at longitudes $2\psi'$ equal to $(270° + 2\psi)$ and $(90° + 2\psi)$, respectively, through an angle φ such that

$$|2\omega| + \varphi = 90° .$$

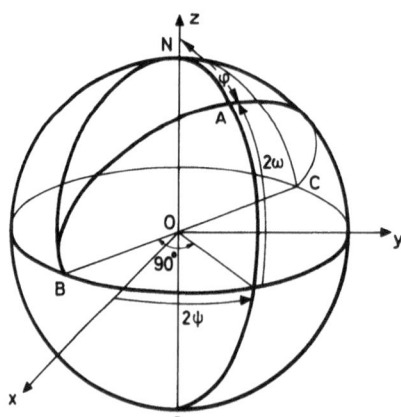

Fig. 5.2. Determination of the orientation of the principal axes and the retardance of a retardation plate that converts a given elliptical polarization into circular polarization, by the Poincaré-sphere method. Point A represents the incident polarization form, points B and C the eigenvectors, and φ the retardance of the retardation plate

Thus, we obtain for the longitude 2ψ and latitude 2ω of the original beam that

$$\psi = \psi' - 45°$$

$$|\omega| = 45° - \frac{\varphi}{2} ,$$

(5.26)

where $2\psi'$ is the longitude of point C.

However, the handedness of the light ellipse cannot be determined, because there is no way to distinguish whether the induced circularly polarized light is right or left (either point N or S in Fig. 5.2).

A double-beam technique was developed by *Archard* et al. [5.2]. In this method, a double-image polarizer was used to separate the two mutually perpendicular linear vibrations into two 90° out-of-phase components. When the two fields thus obtained are of equal intensity, the axes of the rotating polarizer are at 45° to the principal axes of the light ellipse and hence the azimuth may be determined. Then, the polarizer is rotated through 45°, so that its axes now lie along the major and minor axes of the ellipse. The ratio of the intensities of the two signals obtained gives the ratio of the principal axes of the ellipse, that is, its ellipticity. The handedness of the ellipse can be determined by inserting a retarder and measuring the intensity of the output light.

Robert [5.3–10] has used a rotating analyzer, placed in the light path, and detected the characteristics of the outcoming signal. In this case, (5.5) takes the form

$$I(0,\gamma,0) = \frac{1}{2}\left[s_0 + (s_1^2 + s_2^2)^{1/2} \cos\left(2\gamma - \tan^{-1}\frac{s_2}{s_1} \right) \right] \tag{5.27}$$

and consists of a constant term and an oscillatory one, with frequency equal to twice the frequency of the rotating analyzer. By measuring the relative values of the amplitudes of the constant and the oscillatory terms, as well as the phase of the oscillatory term, two out of the three Stokes parameters can be measured. From these two parameters, the azimuth and the ellipticity of the light ellipse can be immediately determined. For determination of the handedness of the ellipse, a quarter-wave plate, with its fast axis parallel to the major axis of the light ellipse, was placed before the analyzer and the signal thus obtained was detected. The phase of its oscillatory component gives the handedness. Based on this method of measuring the state of polarization of a beam of light, *Robert* has constructed an automatic photoelasticimeter for the photoelastic determination of two- and three-dimensional stress states.

Sekera [5.11] has used a polarimeter that detects the three amplitudes and the two phase differences of (5.24) for the complete characterization of a state of polarization.

The idea of using a retardation plate of variable birefringence (method III of Sec. 5.2) has also been used for measuring the state of polarization of a beam of light. In this case, the retardation can be varied sinusoidally with time, by applying a sinusoidally varying voltage to the plate. The parameters of the elliptically polarized light can be determined by measuring the constant and the varying terms of the output intensity [5.12,13]. The same principle was also used in the polarimetric method proposed by *Takasaki* et al. [5.14,15].

5.5 Visual Methods for Measuring the State of Polarization

Besides photoelectric methods for measuring the state of polarization of a light beam, which are based on the insertion of optical elements in the light path and measurement of the intensity of the output beam, there exist also visual methods, which do not necessitate any intensity detection. These methods use, as do photoelectric methods, optical elements inserted in the light path. They are based on a suitable arrangement of these elements such that the intensity of the field of view becomes either maximum or minimum, or the intensities of two fields of view are made equal. Such "end-point" decisions can be made only by visual means, and do not necessitate any detecting device. Thus, the corresponding methods are called "visual methods". However, when great accuracy is required, photoelectric recording devices can also be used in conjunction with visual methods.

Photoelectric methods present the advantage over visual methods that they can also be used in the nonvisible region of the spectrum, where visual methods cannot. However, photoelectric methods need more elaborate instrumentation than do visual methods. Visual methods are based on some ingenious principles, which will be described. We shall study separately some techniques for determining the azimuth, the ellipticity, and the handedness of the light ellipse. For a more thorough study of these methods, the reader is referred to a paper by *Richartz* and *Hsü* [5.16].

Besides the classification of methods for measuring the state of polarization as photoelectric and visual, they can also be distinguished according to the method of measuring the state of polarization into *direct* and *indirect methods*. In direct methods, the optical elements are arranged to measure some characteristic features of the light ellipse, whereas in indirect methods, the light ellipse is transformed, by adjusting the optical elements, to measure some known polarization forms.

5.5.1 Measurement of Azimuth

The azimuth of the light ellipse, representing an elliptically polarized light, can be measured by use of a rotating analyzer. When the pass axis of the analyzer is parallel to the major or the minor axes of the ellipse, the intensity of the light transmitted becomes maximum or minimum, respectively. This result can be immediately deduced from the Poincaré sphere. If the state of polarization is represented by the point P on the sphere (Fig. 5.3) and the linear analyzer is represented by a point A on the equator of the sphere, then the light intensity I, transmitted by the analyzer, is given by (Sec. 4.3.5)

$$I = \cos^2 \frac{\widehat{PA}}{2}. \tag{5.28}$$

The intensity I passes through a maximum or a minimum, when point A lies in the same meridian as point P, specifically at positions $A_{\max}$ or $A_{\min}$, respectively (Fig. 5.3). Because the position for the minimum value of the light intensity I is more-easily detected than the position for the maximum, we obtain for the azimuth ψ_P of the light ellipse

$$\psi_P = \psi_{A_{min}} - \frac{\pi}{2}, \tag{5.29}$$

where $\psi_{A_{min}}$ is the azimuth of the analyzer for minimum value of the transmitted intensity.

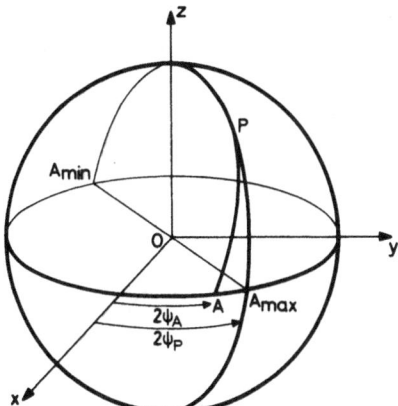

Fig. 5.3. Determination of the positions A_{max} and A_{min} of an analyzer that transmits the maximum and minimum light intensity of an incident polarization form that is represented by the point P on the Poincaré sphere

However, in practice, the exact position of the analyzer for minimum transmitted intensity cannot be determined accurately. Furthermore, the accuracy is strongly reduced when the two semi-axes of the ellipse are nearly equal. For these reasons, all visual methods for determining the azimuth are based on separation of the field of view into two parts and comparison of the intensities of these two parts. These intensities depend on the orientation of some optical device, whose correct position is that for which the two fields of view appear equally bright. The larger the difference between the intensities of the two fields of view for a slight deviation of the analyzer from its correct position, the greater the accuracy in determining the azimuth.

Many special techniques, using various optical devices, have been conceived for applying the principle of dividing the field of view into two fields. In these techniques, the most commonly used devices are

I) *The double-field analyzer* [5.17–20], which operates like a group of two analyzers, whose respective pass axes are inclined at an angle 2η to each other. Each analyzer corresponds to one of the fields of view. The light intensities of the two fields of view will be equal when either the major or the minor axis of the light ellipse is parallel to the internal bisector of the pass planes of the two analyzers. The situation in which the minor axis of the ellipse is parallel to this bisector is more sensitive than when the bisector is parallel to the major axis, because it corresponds to the more easily detected minimum of the light intensity. For such a situation, the azimuth ψ of the light ellipse is given by (5.29), where $\psi_{A_{min}}$ represents the azimuth of the internal bisector of the pass axes of the two analyzers. The procedure of applying the double-field analyzer is shown schematically in Fig. 5.4.

Many optical devices have been conceived, which apply the principle of the double-field analyzer. These are described in detail in [5.17–20].

II) *The rotating biplate*, which consists of two pure rotators, each of which corresponds to one of the fields of view. The light ellipse of the beam incident on the two rotators, undergoes rotations of the same amount, but of opposite senses. Thus, if η and $-\eta$ are the powers of the two rotators, the azimuths of the output-light ellipses will form an angle 2η. If this assembly is backed by an analyzer, then the intensities of the two halves of the field of view will be equal and will take their minimum values when the axis of the analyzer is parallel to the minor axis of the light ellipse. The manner of action of the rotating biplate is shown in Fig. 5.5.

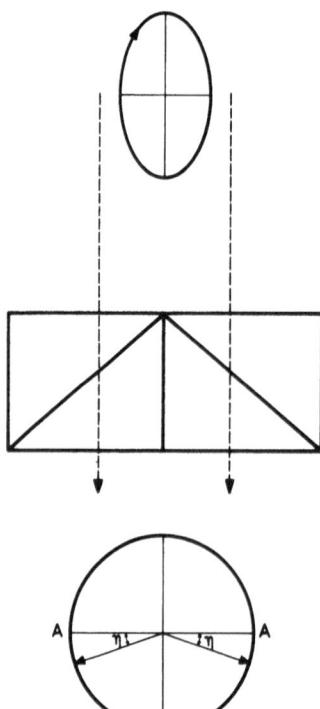

Fig. 5.4. Action of a double-field analyzer on incident elliptically polarized light

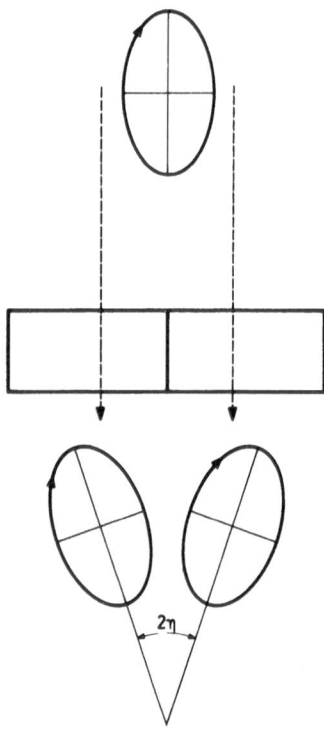

Fig. 5.5. Action of a rotating biplate on incident elliptically polarized light

For details of arrangements of rotating biplates the reader is referred to [5.21–23].

III) *The half-shadow plate.* The half-shadow method is based on the fact that when an elliptically polarized light is incident on a half-wave plate whose principal axes are parallel to the principal axes of its light ellipse, then the light ellipse of the outcoming light has the same azimuth and ellipticity, but opposite handedness with respect to the incident-light ellipse. This is shown in Fig. 5.6. The incident light is represented by the point P on the Poincaré sphere; the fast and slow axes of the half-wave plate are represented by points M and M' on the equator of the sphere, with the point M lying on the same meridian as point P. Then, the output light is obtained by rotating the sphere about the axis MM', by an angle 180°. With this rotation, point P comes to point P', having the same longitude and latitude as point P, but lying in the hemisphere opposite to point P. Thus, the output light ellipse has the same azimuth and ellipticity, but opposite handedness with respect to the input-light ellipse.

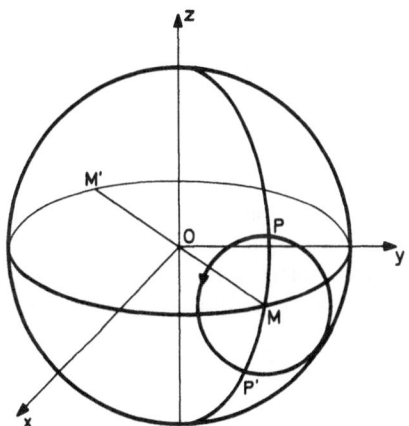

Fig. 5.6. Geometrical representation on the Poincaré sphere of the action of a half-wave plate on elliptically polarized light, whose light ellipse has principal axes that are parallel to the principal axes of the plate. The output polarization form P' that results from rotating the sphere about the axis MM' through angle 180° has the same azimuth and ellipticity but handedness opposite to that of the incident form P

According to this principle, in one-half of the field of view a half-wave plate is placed, followed by an analyzer that covers both fields of view. The half-wave plate and the analyzer are rotated together. Then, when the principal axes of the half-wave plate are parallel to the principal axes of the light ellipse, the intensities of both fields of view are equal. Thus, the azimuth of the light ellipse of the incident light is determined. This method was first conceived by *Chauvin* [5.24] and improved by *Chaumont* [5.25].

5.5.2 Measurement of Ellipticity

All methods for measuring the ellipticity of polarized light are based on the use of a retardation plate (either half- or quarter-wave, or any retardance plate) interposed in the light beam, which is to be measured. Two different approaches are used, concerning the retardation plate. In the first approach, the retardance of the retardation plate is constant and the ellipticity is measured by varying the orientation of the plate, whereas in the second approach the retardation plate has a fixed orientation, but a variable retardance. According to the definition given at the end of Section 5.5, the first arrangement is a direct method of measuring an elliptical vibration; the second is an indirect method. The methods of the first category are based on the Senarmont principle; those of the second category use various types of compensators.

Generally speaking, the methods for measuring ellipticity presume knowledge of the azimuth of the light ellipse, although some methods determine both the azimuth and ellipticity simultaneously. The measurement of ellipticity is of particular importance for the determination of the retardation of a birefringent medium. In photoelastic phenomena, optically isotropic media become birefringent, when they are subjected to stress.

The measurement of birefringence in photoelastic analysis yields the state of stress in the specimens under study, which study is the purpose of this book.

In the following, we shall develop in detail the direct methods, based on the Senarmont principle; afterwards, we shall describe the various types of compensators used in photoelastic analysis.

Direct Methods

The direct methods are based on the Senarmont principle, according to which, when a quarter-wave plate is placed with its principal axes parallel to the principal axes of the light ellipse of an elliptically polarized light, the outcoming light will be linearly polarized at an azimuth ω with respect to the slow axis of the incident light, where ω is the ellipticity of the incident light. This result is shown in Fig. 5.7. The incident light is represented by the point P on the

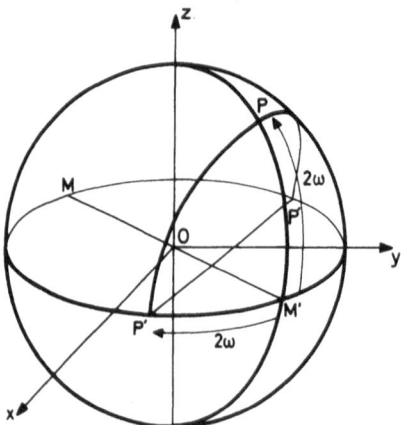

Fig. 5.7. Geometrical representation of the Senarmont principle on the Poincaré sphere. Elliptically polarized light P, whose light ellipse has axes parallel to the axes M, M' of a quarter-wave plate, is converted by the plate into linearly polarized light P' at an azimuth ω to the slow axis of the plate, equal to the ellipticity of the elliptically polarized light

Poincaré sphere; the fast and slow axes of the quarter-wave plate are represented by the points M and M' on the equator of the sphere, where point M' lies on the same meridian as point P. Then, the emerging light is obtained by rotating the sphere about the axis MM' through $90°$. This rotation brings point P to point P' on the equator of the sphere, with $\widehat{M'P} = \widehat{M'P'}$, that is, the azimuth of the output plane-polarized light with respect to the slow axis of the quarter-wave plate is equal to the ellipticity of the incident light.

For the determination of the ellipticity of a polarized light according to the Senarmont principle, the azimuth of the incident-light ellipse must be determined and the principal axes of the quarter-wave plate must be arranged parallel to the principal axes of the incident-light ellipse. Then the azimuth of the output plane-polarized light, giving the ellipticity of the incident light, is determined by a rotating analyzer.

The direct application of the Senarmont principle in measuring the ellipticity is strongly dependent on the correct orientation of the quarter-wave plate, which must be parallel to the principal axes of the incident-light ellipse. If the azimuth of the incident light is not accurately determined, then the emerging light from the quarter-wave plate will not be linearly polarized and complete extinction will not be achieved by the analyzer.

For overcoming this disadvantage of the Senarmont method *Stokes* [5.26] and *MacCullagh* [5.27] have developed a method of successive approximations, based on the Senarmont principle. They have determined by a trial and error technique the suitable setting of the quarter-wave plate for which the output light is completely extinguished at some position of an elliptical analyzer.

However, neither the direct application of the Senarmont principle nor the Stokes-MacCullagh method of successive approximations uses the half-shade principle. To incorporate the half-shade principle into the Senarmont method and thus increase the accuracy of the determination of the ellipticity, various special methods have been developed [5.28–31]. From these methods, we shall describe the method developed by *Richartz* [5.32].

Richartz's method is a combination of the rotating-biplate method, for determining the azimuth and the Senarmont method in conjunction with the half-shade principle for determination of the ellipticity of an elliptically polarized light. The principle of the method is shown in Fig. 5.8. One half of the field of view is covered by a quarter-wave plate Q and a rotating biplate B; the other half is covered by only a rotating biplate. The whole arrangement is followed by an analyzer A. First, the azimuth of the light ellipse is determined by the system of the rotating biplate and the analyzer. This is done as in the rotating-biplate method, by rotating the biplate and analyzer to produce equal brightness in the two quarters of the whole field of view. Then, the quarter-wave plate is oriented with its axes parallel to the determined axes of the light ellipse and again the analyzer is rotated until the other two quarters of the field of view are equally bright. Thus, the ellipticity of the incident light can be determined.

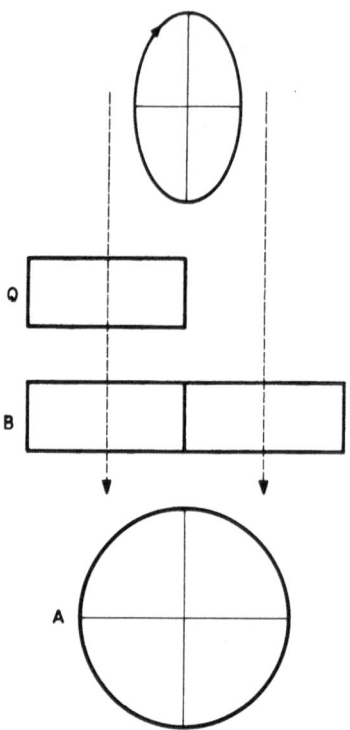

Fig. 5.8. Principle of *Richartz*'s method, which uses a quarter-wave plate Q that covers one-half of the field of view and a rotating biplate B that covers the whole field of view, to determine the azimuth and ellipticity of elliptically polarized light

Both the azimuth and the ellipticity of a light beam can also be determined by placing two crossed polarizers and two quarter-wave plates in the light path and detecting the output field of view. The combinations of crossed polarizers and quarter-wave plates can be separately rotated. This setting, corresponding to two crossed-elliptical analyzers, constitutes the basic arrangement of the photoelastic method of stress analysis and will be studied in detail in Chapter 7.

Compensators

Compensators are optical elements that provide variable optical retardation at the point in question, while their orientation is kept constant. This can be achieved by use of a wedge-like plate, which can be conveniently displaced to yield the desired retardation at a particular point in the optical field. However, in order to have the possibility of introducing variable optical retardations of opposite signs, the compensator is composed of two wedges, whose fast and slow directions are interchanged, or two wedges with their axes parallel but followed by another plate of constant thickness and opposite sign. The first type of compensator is known as the *Babinet compensator* (Fig. 5.9a); the second type is called the *Babinet-Soleil compensator* (Fig. 5.9b).

The only difference between these two types of compensators is that in the Babinet compensator the optical retardation introduced by the compensator

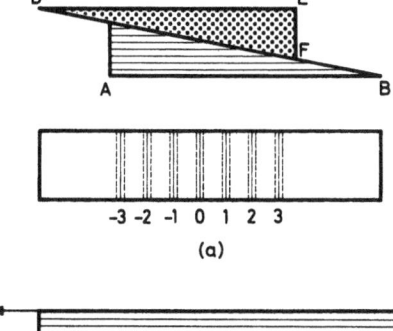

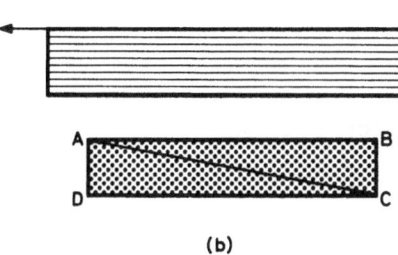

Fig. 5.9 a and b. Elements of the Babinet compensator, with the corresponding field of view (a) and the Babinet-Soleil compensator (b). The field of view of the Babinet compensator is covered by dark and bright equally spaced fringes, whereas in the Babinet-Soleil compensator, the field is uniform

varies linearly over the field of view, whereas in the Babinet-Soleil compensator this optical retardation is constant. In both types, the variable optical retardation is introduced by moving one wedge relative to the other by means of a micrometric screw.

If the Babinet compensator is placed between two crossed polarizers with its principal axes at 45° to the axes of the polarizers, then a series of dark fringes appears in the field of view. Unlike this, the field of view in the Babinet-Soleil compensator is uniformly illuminated.

The retardation at the point under study is measured with either of these compensators by setting the compensator with its principal axes parallel to the principal directions of the light ellipse. By varying the retardation introduced by the compensator, a polarized light field can be extinguished. The retardation measured at this adjustment and the retardation of the compensator have the same absolute value and opposite signs. The retardation in the compensator is varied by moving the wedges of the compensator; this movement is translated to retardation by calibration. The movement of the micrometric screw in the compensator, necessary to produce a retardation equal to 360°, is called the *compensator constant*.

For a more detailed analysis of the problems encountered with the two types of compensators the reader is referred to [5.33–40].

5.5.3 Determination of Handedness

The handedness, that is the sense with which the light vector describes the light ellipse, is the third quantity that, together with the azimuth and the ellipticity, determines completely an elliptically polarized light. As defined in Section 3.3, elliptically polarized light is said to be right handed when an observer looking towards the light source sees the light vector move in the clockwise sense. It was also shown in Section 3.4.1, p. 25, that the phase retardation δ, defined by (3.12), lies in the interval $0 < \delta < 180°$ for right-handed elliptical polarization. Thus, in all methods that measure phase retardation δ, such as compensators, the handedness of elliptically polarized light is determined by the value of δ.

The handedness of polarization can also be determined by the methods of measuring the ellipticity of the polarized light, by use of retardation plates, such as half- or quarter-wave plates. Thus, in methods that use a quarter-wave plate with an analyzer, the light ellipse is right handed when we place the fast axis of the quarter-wave plate parallel to the major axis of the ellipse of the polarized light and the analyzer has to be rotated through an angle γ in the interval $0 < \gamma < 180°$ to produce extinction.

5.6 Determination of the Matrix Elements of an Optical Device

As shown in Chapter 4, an optical device is completely characterized by the corresponding matrix, which, in the Mueller calculus, is a four-by-four real matrix; in the Jones calculus, it is a two-by-two complex matrix. It was also shown that a train of retarders is equivalent to a linear retarder and a pure rotator and that the Mueller or the Jones matrix of the whole system is obtained by multiplying the corresponding matrices of each separate optical element. Thus, either in the case of a single optical device or in the case of a train of devices, the system is characterized by its Mueller or Jones matrix. Knowledge of the matrix of a simple device or of a system of devices enables one to obtain the Stokes or Jones vector of the light beam that emerges from the device, by multiplying the vector of the incident beam by the Mueller or Jones matrix, respectively, that represents the optical system.

As will be shown later, a three-dimensional photoelastic model is equivalent to a system of optical retarders; therefore determination of the matrix of the model is of particular value for photoelastic analysis of the stress system developed in the real object.

The elements of the Mueller or the Jones matrix of an optical device can be determined by letting various forms of light impinge on the device and determining the corresponding Stokes or Jones vectors of the outcoming light. They can be determined by use of one of the various techniques described in preceding sections of this chapter.

In the following, we shall examine separately the determination of the elements of a Mueller or a Jones matrix. We shall assume that the Stokes or the Jones vector of the output light can be experimentally determined.

5.6.1 Determination of a Mueller Matrix

Let M be the Mueller matrix

$$M = \begin{bmatrix} m_{11} & m_{12} & m_{13} & m_{14} \\ m_{21} & m_{22} & m_{23} & m_{24} \\ m_{31} & m_{32} & m_{33} & m_{34} \\ m_{41} & m_{42} & m_{43} & m_{44} \end{bmatrix}, \tag{5.30}$$

whose elements m_{ij} $(i,j=1,2,3,4)$ are to be determined. For this purpose we assume that light beams of different forms are incident on the device and, furthermore, that for each incident form, the Stokes vector of the output form is completely determined. The required input forms are

I) Unpolarized light of unit intensity with Stokes vector S_1 given by (Sec. 3.4.4)

$$S_1 = \begin{bmatrix} 1 \\ 0 \\ 0 \\ 0 \end{bmatrix}, \tag{5.31}$$

The Stokes vector S_1' of the output light is

$$S_1' = M S_1, \tag{5.32}$$

or

$$S_1' = \begin{bmatrix} m_{11} \\ m_{21} \\ m_{31} \\ m_{41} \end{bmatrix}. \tag{5.33}$$

Thus, the elements m_{i1} $(i=1,2,3,4)$ of the matrix M are determined.

II) Linearly polarized light of unit intensity, whose light vector is parallel to the Ox axis. Its Stokes vector S_2 is

$$S_2 = \begin{bmatrix} 1 \\ 1 \\ 0 \\ 0 \end{bmatrix}. \tag{5.34}$$

Then, the Stokes vector S_2' of the outcoming light is

$$S_2' = \begin{bmatrix} m_{11} + m_{12} \\ m_{21} + m_{22} \\ m_{31} + m_{32} \\ m_{41} + m_{42} \end{bmatrix}. \tag{5.35}$$

Relation (5.35) permits evaluation of the elements m_{i2} ($i = 1, 2, 3, 4$), after the elements m_{i1} ($i = 1, 2, 3, 4$) have been determined.

III) Linearly polarized light of unit intensity, whose light vector makes an angle $45°$ with the Ox axis. Its Stokes vector S_3 is

$$S_3 = \begin{bmatrix} 1 \\ 0 \\ 1 \\ 0 \end{bmatrix}. \tag{5.36}$$

The Stokes vector S_3' of the output light is

$$S_3' = \begin{bmatrix} m_{11} + m_{13} \\ m_{21} + m_{23} \\ m_{31} + m_{33} \\ m_{41} + m_{43} \end{bmatrix}. \tag{5.37}$$

Vector S_3' permits determination of the elements m_{i3} ($i = 1, 2, 3, 4$), after the elements m_{i1} ($i = 1, 2, 3, 4$) have been determined.

IV) Right-handed circularly polarized light of unit intensity. Its Stokes vector S_4 is

$$S_4 = \begin{bmatrix} 1 \\ 0 \\ 0 \\ 1 \end{bmatrix}. \tag{5.38}$$

The Stokes vector S_4' of the output light is

$$S_4' = \begin{bmatrix} m_{11} + m_{14} \\ m_{21} + m_{24} \\ m_{31} + m_{34} \\ m_{41} + m_{44} \end{bmatrix}. \tag{5.39}$$

Vector S_4' permits determination of the elements m_{i4} $(i=1,2,3,4)$, after the elements m_{i1} $(i=1,2,3,4)$ have been determined.

5.6.2 Determination of a Jones Matrix

Let J be the Jones matrix

$$J = \begin{bmatrix} j_{11} & j_{12} \\ j_{21} & j_{22} \end{bmatrix}, \tag{5.40}$$

whose elements $j_{\kappa l}$ $(\kappa, l = 1,2)$ are to be determined.

For this ˙purpose, we assume that two linearly polarized light beams impinge on the optical device whose Jones matrix is given by (5.40). One of the two linearly polarized incident beams has its light vector along the Ox axis; the other has its light vector along the Oy axis. The Jones vector a_1 of the beam, whose light vector is along the Ox axis, is

$$a_1 = \begin{bmatrix} 1 \\ 0 \end{bmatrix}; \tag{5.41}$$

the Jones vector a_1' of the beam emerging from the device is

$$a_1' = J\,a_1, \tag{5.42}$$

or

$$a_1' = \begin{bmatrix} j_{11} \\ j_{21} \end{bmatrix}. \tag{5.43}$$

Similarly, when linearly polarized light whose light vector is along the Oy axis, and whose Jones vector a_2 is

$$a_2 = \begin{bmatrix} 0 \\ 1 \end{bmatrix}, \tag{5.44}$$

is incident on the device, the Jones vector of the output beam is

$$a_2' = \begin{bmatrix} j_{12} \\ j_{22} \end{bmatrix}. \tag{5.45}$$

By determination of the output Jones vectors a_1' and a_2', all four elements $j_{\kappa l}$ $(\kappa, l = 1,2)$ of the Jones matrix (5.40) can be determined.

References

5.1 C. V. Kent, J. Lawson: J. Opt. Soc. Am. **27**, 117 (1937)
5.2 J. F. Archard, P. L. Clegg, A. M. Taylor: Proc. Phys. Soc. London B **65**, 758 (1952)
5.3 A. Robert, E. Guillemet: Rev. Fr. Méc. Nos **5-6**, 147 (1963)
5.4 A. Robert, J. L. Vernet: Conf. au troisième Congrès international d'analyse des contraintes (Berlin 1966)
5.5 A. Robert, C. Bourdon, J. L. Le Goer: Rev. Fr. Méc. No 24, 93 (1967)
5.6 A. Robert: Bull. Soc. Fr. Minéral. Cristallogr. **91**, 415 (1968).
5.7 A. Robert, M. Ferre: Bull. ATMA **69**, 1 (1969)
5.8 A. J. Robert: Exp. Mech. **7**, 224 (1967)
5.9 A. Robert: Int. J. Solids Struct. **6**, 423 (1970)
5.10 A. Robert: *Polarimétrie et Photoélasticimétrie* (Serv. Techn. Const. Armes Navales. Paris 1972)
5.11 Z. Sekera: Adv. Geophys. **3**, 43 (1956)
5.12 W. Budde: Appl. Opt. **1**, 201 (1962)
5.13 B. A. Ioshpa, V. N. Obridko: Opt. Spectrosc. USSR **15**, 60 (1963)
5.14 H. Takasaki, N. Okazaki, K. Kida: Appl. Opt. **3**, 833 (1964)
5.15 H. Takasaki, M. Isobe, T. Masaki, A. Konda, T. Agatsuma, Y. Watanabe: Appl. Opt. **3**, 345 (1964)
5.16 M. Richartz, H. Y. Hsü: J. Opt. Soc. Am. **39**, 136 (1949)
5.17 J. H. Jellett: Rpt. Br. Assoc. **30**, 13 (1860)
5.18 M. A. Cornu: Bull. Soc. Chim. **14**, 140 (1870)
5.19 O. Schönrock: „Polarimetrie", in *Elektrische Leistungsphänomene I*, Handbuch der Physik, Hrsg. H. Geiger, K. Scheel, Bd. 19 (Springer, Berlin 1928) pp. 705–776
5.20 F. Lippich: Wien. Ber. **91**, 1059 (1885)
5.21 S. Nakamura: Zentralblatt f. Min. Vol. 267 (1905)
5.22 E. Bertrand: Bull. Soc. Mineral. **1**, 22 (1878)
5.23 J. Strong: Rev. Sci. Instr. **6**, 243 (1935)
5.24 M. Chauvin: Ann. de Toulouse **3**, 30 (1889)
5.25 M. L. Chaumont: Ann. Phys. (Paris) **4**, 175 (1915)
5.26 G. G. Stokes: Math. Phys. Pap. Cambridge **3**, 197 (1901)
5.27 J. MacCullagh: *Collected Works*, Dublin, London (1880) pp. 138, 230
5.28 A. Q. Tool: Phys. Rev. **31**, 1 (1910)
5.29 C. A. Skinner: J. Opt. Soc. Am. **10**, 491 (1925)
5.30 G. Szivessy: Z. Instrumentenkd. **47**, 148 (1927)
5.31 C. Bergholm, Y. Björnstahl: Physik. Zeitschr. **21**, 137 (1920)
5.32 M. Richartz: Z. Instrumentenkd. **60**, 357 (1940)
5.33 G. Szivessy: In *Kristalloptik*, Handbuch der Physik, Hrsg. H. Geiger, K. Scheel, Bd. 20 (Springer, Berlin 1928) pp. 635–904
5.34 H. G. Jerrard: J. Opt. Soc. Am. **38**, 35 (1948)
5.35 H. G. Jerrard: J. Sci. Instrum. **28**, 10 (1951)
5.36 H. G. Jerrard: J. Sci. Instrum. **26**, 353 (1949)
5.37 H. G. Jerrard: J. Sci. Instrum. **27**, 62 (1950)
5.38 H. G. Jerrard: J. Sci. Instrum. **27**, 164 (1950)
5.39 H. G. Jerrard: J. Sci. Instrum. **30**, 65 (1953)
5.40 M. Françon, B. Sergent: Opt. Acta **2**, 182 (1955)

6. The Photoelastic Phenomenon

6.1 Introduction

In Chapter 2, where the electromagnetic theory of light was briefly described, the basic properties of the propagation of a light ray in a crystal were studied. It was shown that when a light ray enters a crystal, it splits into two rays that are linearly polarized at right angles to each other, which propagate with different velocities. The refractive index associated with each light ray in the crystal can be found by use of Fresnel's ellipsoid, which shows, for any point in the crystal, the variation of refractive index with direction of the light ray. This splitting of rays into two rays that have mutually perpendicular polarizations results from a physical property of crystalline materials that is called *optical birefringence* or simply *birefringence*.

Besides crystals, birefringence can also be observed in certain noncrystalline and initially optically isotropic materials, when they are subjected to a stress field. Under stress, these materials behave like crystals. However, the crystalline nature of the material lasts only during the application of the loads; it vanishes when the loads are removed. This phenomenon, called *temporary or artificial birefringence*, was discovered by Sir *David Brewster*, in 1816 [6.1].

Twenty-five years later, the phenomenon of temporary birefringence was reconsidered by *Neumann* [6.2], who formulated the first systematic theory of the phenomenon. *Neumann* gave the complete relationships between the principal refractive indices of the temporarily birefringent body and the cause of the birefringence, which he assumed was the strain.

Eleven years after *Neumann*, *Maxwell* [6.3], probably unaware of *Neumann*'s investigations, studied the phenomenon of temporary birefringence and attributed it to stress. He gave relations completely analogous to those of *Neumann*, but which connected the principal refractive indices with the principal stresses.

However, both *Neumann* and *Maxwell* restricted themselves to the simple case of elastic deformations, in which there is a linear relationship between stress and strain. For this simple case, there is no distinction whether the birefringence is connected with the strain or the stress and *Maxwell*'s relationships can be deduced from *Neumann*'s or vice-versa, by use of *Hooke*'s linear relations, connecting the components of stress and strain. However, when viscoelastic or plastic phenomena are involved in the deformation of the body, the question whether birefringence is related to stress or to strain, or to both, is still open.

The discovery of the phenomenon of *temporary* or *artificial birefringence*, the so-called *photoelastic phenomenon*, opened the route to a new method of experimental stress analysis, namely, to *photoelasticity*. Photoelasticity was rather slow to develop. Although the photoelastic phenomenon was discovered in 1816, not until the turn of the twentieth century was it first used for structural analysis. This delay of development was mainly due to the lack of convenient photoelastic materials, suitable to exhibit the phenomenon of temporary birefringence.

Photoelasticity began to be a powerful method of experimental stress analysis, since the beginning of the twentieth century, as a result of the pioneering work of *Coker* and *Filon* [6.4], mainly because of the discovery of a suitable photoelastic material, bakelite. Since then it developed rapidly and reached in its full maturity around the middle of this century.

Photoelasticity, in its early period of development, was greatly advanced by *Mesnager*, whose researches contributed to the foundation of the new method. Among his contributions were construction of a perfect model of a bridge from glass pieces glued together [6.5], invention and construction of an extremely sensitive lateral extensometer for the derivation of the sum of the principal stresses in a two-dimensional model [6.6], and introduction of the birefringent coating technique for extending photoelasticity to the investigation of stresses in metallic surfaces [6.7].

However, development of photoelasticity as a general experimental method for investigating any type of stress field by use of polarized light has continued until our day. New methods, based on photoelasticity, as well as on interference, holography, and scattering of light, which have recently been introduced, are continuously appearing in the literature. These new tendencies aim to establish simple, easy, and rapid methods for complete evaluation of stress fields of any type by the use of light.

6.2 The Photoelastic Law

6.2.1 Principal Birefringent Directions

As is well known, when a body is subjected to a three-dimensional stress field, the induced stresses can be represented at each point of the body by a symmetric second-order tensor. This stress tensor can therefore be diagonalized; its *principal directions*, which are mutually perpendicular, give the *principal planes* of the state of stress at the point considered. Across these planes the developed stresses are normal to the respective plane. Across any other plane the resultant stress is oblique relative to the plane.

Similarly, the state of strain of a three-dimensional body can be represented by the strain tensor at each point of the body considered, which is also symmetric. Along the *principal strain directions* only normal and non shear deformation is developed.

For an elastically deformed body, both the stress and strain tensors have the same principal directions; therefore, their respective stress and strain ellipsoids, or the so-called Cauchy ellipsoids, are coaxial. The mechanically introduced birefringence in an isotropic material has its origin in the physical deformations induced in the body by the stress or strain. Stress or strain cause intermolecular variations of the structure of the body, which alter its optically isotropic character. In the case of elastic deformations, in which the principal axes of the stress and strain ellipsoids coincide, and stress, strain, and birefringence can each be inferred from the presence of the others, it is reasonable to conclude that the principal birefringent axes coincide with the principal stress or strain axes. Thus, Fresnel's refractive-index ellipsoid and Cauchy's stress and strain ellipsoids are coaxial.

The coaxiality of the three ellipsoids, which results from considerations of symmetry for the case of the elastic deformations constitutes the one part of the stress-optical or photoelastic law. We will use this result to complete the photoelastic law with the relationships that exist between the principal stresses or strains and the principal refractive indices.

However, although in the simple case of an elastically deformed body the stress, strain, and birefringence tensors are coaxial, in the case of a material that exhibits viscoelastic behavior, subjected to any type of loading or displacement, the correspondence between stress, strain, and birefringence tensors is not known in advance to be so simple as for perfectly elastic materials. Even for a rheo-optically simple material, the interrelation of these three tensors is not very simple, especially in the transition region. In a one-dimensional case, the axes of principal stress, strain, and birefringence are obviously aligned, but this alignment is not necessarily the case in a general two- or three-dimensional photoviscoelastic test configuration. It is, therefore, necessary to explore the phase-angle relationships between the principal mechanical axes and principal optical axes in a convenient two- or three-dimensional photoviscoelastic test configuration.

A comprehensive analysis of the problems encountered in the interrelation between the stress, strain, and birefringence tensors as well as a bibliographical review of the subject was provided by *Theocaris* [6.8]. He proved that, even for rheo-optically simple materials, there is a shift in time between the mechanical and optical viscoelastic behavior of polymers at the transition region. *A fortiori*, for nonlinear polymers or for polymers that exhibit peculiarities of optical behavior, this shift is increased and also extended to the glassy or rubbery states.

However, because in the present work only elastic stress states will be studied by photoelasticity, and we will not become involved with plastic, viscoelastic, or viscoplastic states of stress, the stress-optical law will be developed only for the case of the elastically deformed materials.

6.2.2 Principal Refractive Indices

Let us refer both Fresnel's birefringence ellipsoid and Cauchy's stress ellipsoid to their common principal axes. Then Fresnel's ellipsoid described by (2.30) takes the form

$$\frac{x^2}{n_1^2} + \frac{y^2}{n_2^2} + \frac{z^2}{n_3^2} = 1,\tag{6.1}$$

where n_1, n_2, and n_3 are the principal refractive indices along the axes Ox, Oy, and Oz, respectively.

Similarly, Cauchy's stress ellipsoid takes the form

$$\sigma_1 x^2 + \sigma_2 y^2 + \sigma_3 z^2 = 1,\tag{6.2}$$

where σ_1, σ_2, and σ_3 are the principal stresses along the axes Ox, Oy, and Oz, respectively.

As these two ellipsoids are coaxial in the case of elastic stress fields, their planes of circular cross sections must be parallel. These planes coincide with the intersections of each ellipsoid by a concentric sphere; therefore, they may be put into the form

$$\frac{x^2}{n_1^2} + \frac{y^2}{n_2^2} + \frac{z^2}{n_3^2} = A(x^2 + y^2 + z^2)\tag{6.3}$$

for Fresnel's ellipsoid, and into the form

$$\sigma_1 x^2 + \sigma_2 y^2 + \sigma_3 z^2 = B(x^2 + y^2 + z^2)\tag{6.4}$$

for Cauchy's stress ellipsoid.

If the first member of (6.4) and the second one of (6.3) are multiplied by a constant D, then the resulting equations must coincide. By writing these two equations in the form

$$\left(\frac{1}{n_1^2} - AD\right)x^2 + \left(\frac{1}{n_2^2} - AD\right)y^2 + \left(\frac{1}{n_3^2} - AD\right)z^2 = 0$$

$$(D\sigma_1 - B)x^2 + (D\sigma_2 - B)y^2 + (D\sigma_3 - B)z^2 = 0,$$

we get

$$\frac{1}{n_1^2} - AD = D\sigma_1 - B, \quad \frac{1}{n_2^2} - AD = D\sigma_2 - B, \quad \frac{1}{n_3^2} - AD = D\sigma_3 - B,\tag{6.5}$$

or

$$\frac{1}{n_1^2} = D\sigma_1 + C, \quad \frac{1}{n_2^2} = D\sigma_2 + C, \quad \frac{1}{n_3^2} = D\sigma_3 + C, \tag{6.6}$$

where

$$C = AD - B. \tag{6.7}$$

Relations (6.6) interrelate the principal refractive indices n_1, n_2, n_3 with the principal stresses $\sigma_1, \sigma_2, \sigma_3$ through the quantities D and C; they are the consequence of the coaxiality of Fresnel's and Cauchy's ellipsoids. As can be seen from (6.6), D and C must be symmetric with respect to the stresses $\sigma_1, \sigma_2, \sigma_3$.

From Fresnel's ellipsoid, we conclude that the refractive index n_s along a given direction s (s_x, s_y, s_z) is given by

$$\frac{1}{n_s^2} = \frac{s_x^2}{n_1^2} + \frac{s_y^2}{n_2^2} + \frac{s_z^2}{n_3^2}. \tag{6.8}$$

Similarly, from Cauchy's stress ellipsoid the stress σ_s along the same direction s is

$$\sigma_s = \sigma_1 s_x^2 + \sigma_2 s_y^2 + \sigma_3 s_z^2. \tag{6.9}$$

When (6.6) are taken into account, (6.8) can be written in the form

$$\frac{1}{n_s^2} = D(\sigma_1 s_x^2 + \sigma_2 s_y^2 + \sigma_3 s_z^2) + C(s_x^2 + s_y^2 + s_z^2), \tag{6.10}$$

or

$$\frac{1}{n_s^2} = D\sigma_s + C, \tag{6.11}$$

by use of (6.9).

Expressing the constant C in the form

$$C = \frac{1}{n^2} + C_2'(\sigma_s + \sigma_t + \sigma_r), \tag{6.12}$$

where $\sigma_s, \sigma_t, \sigma_r$ are the normal stresses along three mutually perpendicular directions, and n is the refractive index of the unloaded material, we get from (6.11)

$$\frac{1}{n_s^2} = \frac{1}{n^2} + C_2'(\sigma_s + \sigma_t + \sigma_r) + D\sigma_s, \tag{6.13}$$

or

$$\frac{1}{n_s^2} = \frac{1}{n^2} + C_1' \sigma_s + C_2'(\sigma_t + \sigma_r),$$ (6.14)

with

$$C_1' = C_2' + D.$$ (6.15)

Equation (6.14) can be written in the form

$$n^2 - n_s^2 = n^2 n_s^2 \left[C_1' \sigma_s + C_2'(\sigma_t + \sigma_r) \right],$$ (6.16)

or

$$n - n_s = \frac{n^2 n_s^2}{n + n_s} \left[C_1' \sigma_s + C_2'(\sigma_t + \sigma_r) \right],$$ (6.17)

and because the difference $(n - n_s)$ is small, compared with n or n_s, the following approximation is valid

$$\frac{n^2 n_s^2}{n + n_s} \simeq \frac{n^3}{2}.$$

Then, by putting

$$C_1 = -\frac{n^3}{2} C_1', \quad C_2 = -\frac{n^3}{2} C_2',$$ (6.18)

we get from (6.17)

$$n_s - n = C_1 \sigma_s + C_2(\sigma_t + \sigma_r).$$ (6.19)

Similarly, we obtain

$$n_t - n = C_1 \sigma_t + C_2(\sigma_r + \sigma_s)$$
$$n_r - n = C_1 \sigma_r + C_2(\sigma_s + \sigma_t).$$ (6.20)

Equations (6.19) and (6.20) give the variations of the refractive indices n_s, n_t, n_r along the directions s, t, r, as linear functions of the corresponding stresses and the sums of the other two stresses; they constitute the second part of *Maxwell*'s photoelastic law. The proportionality constants C_1 and C_2 are called stress-optical coefficients; they depend on the properties of the material of the birefringent body. These three equations are completely analogous to Hooke's stress-strain relations in isotropic media, where $(n_s - n)$, $(n_t - n)$, and $(n_r - n)$ replace the components of strain, σ_s, σ_t, and σ_r replace the components of stress, and C_1, C_2 replace $(1/E)$ and $(-v/E)$, respectively, v being Poisson's ratio and E the modulus of elasticity.

By introducing Hooke's stress-strain relations, expressed by

$$\varepsilon_1 = \frac{1}{E}[\sigma_1 - v(\sigma_2 + \sigma_3)]$$

$$\varepsilon_2 = \frac{1}{E}[\sigma_2 - v(\sigma_3 + \sigma_1)] \tag{6.21}$$

$$\varepsilon_3 = \frac{1}{E}[\sigma_3 - v(\sigma_1 + \sigma_2)],$$

where $\varepsilon_1, \varepsilon_2, \varepsilon_3$ represent the principal strains and the indices $1, 2, 3$ stand for s, t, r, respectively, into (6.19) and (6.20), we obtain

$$n_1 - n = b_1 \varepsilon_1 + b_2(\varepsilon_2 + \varepsilon_3)$$
$$n_2 - n = b_1 \varepsilon_2 + b_2(\varepsilon_3 + \varepsilon_1) \tag{6.22}$$
$$n_3 - n = b_1 \varepsilon_3 + b_2(\varepsilon_1 + \varepsilon_2),$$

where

$$C_1 = \frac{1}{E}(b_1 - 2vb_2), \quad C_2 = \frac{1}{E}(b_2 - vb_1 - vb_2). \tag{6.23}$$

Equations (6.22) were first set up by *Neumann*, independently of Maxwell's stress-optical equations. Because in the elastic range of the material, both stress and strain principal directions coincide and stresses and strains are linearly related, (6.19), (6.20), and (6.22) are referred to as *Neumann-Maxwell stress-optical equations*.

For the case in which we are interested only in the differences between the refractive indices along the principal stress directions, we obtain from (6.19) and (6.20)

$$n_1 - n_2 = (C_1 - C_2)(\sigma_1 - \sigma_2)$$
$$n_2 - n_3 = (C_1 - C_2)(\sigma_2 - \sigma_3) \tag{6.24}$$
$$n_3 - n_1 = (C_1 - C_2)(\sigma_3 - \sigma_1),$$

and, by putting

$$C_0 = C_1 - C_2, \tag{6.25}$$

we have

$$n_1 - n_2 = C_0(\sigma_1 - \sigma_2)$$
$$n_2 - n_3 = C_0(\sigma_2 - \sigma_3) \tag{6.26}$$
$$n_3 - n_1 = C_0(\sigma_3 - \sigma_1).$$

Similarly, from (6.22), we obtain

$$n_1 - n_2 = b_0(\varepsilon_1 - \varepsilon_2)$$
$$n_2 - n_3 = b_0(\varepsilon_2 - \varepsilon_3) \tag{6.27}$$
$$n_3 - n_1 = b_0(\varepsilon_3 - \varepsilon_1)$$

where

$$b_0 = b_1 - b_2. \tag{6.28}$$

Relations (6.26) and (6.27) were first experimentally discovered by *Wertheim* [6.9]; they are known as the Wertheim stress or strain optical law.

References

6.1 D. Brewster: Phil. Trans. R. Soc. London **105**, 60 (1815); **106**, 156 (1816)
6.2 F. E. Neumann: Berichte Königl. Preuss. Akad. Wissensch. 1840; Abh. der Königl. Akad. der Wissensch. Berlin Part II 50 (1841); Pogg. Ann. **54**, 449 (1841)
6.3 J. C. Maxwell: Trans. R. Soc. Edinburgh **20**, 87 (1853)
6.4 E. G. Coker, L. N. G. Filon: *A Treatise on Photoelasticity* 2nd. ed. (University Press, Cambridge 1957)
6.5 A. Mesnager: Compt. Rend. **155**, 1071 (1912)
6.6 A. Mesnager: Ann. Ponts Chaussées **4**, 129 (1901)
6.7 A. Mesnager: Compt. Rend. **190**, 1249 (1930)
6.8 P. S. Theocaris: "Phenomenological Analysis of Mechanical and Optical Behaviour of Rheo-Optically Simple Materials", in *The Photoelastic Effect and Its Applications*, (IUTAM Symposium) ed. by J. Kestens (Springer, Berlin, Heidelberg, New York 1975) pp. 146–230
6.9 M. G. Wertheim: Compt. Rend. **32**, 289 (1851); **33**, 576 (1851); **35**, 276 (1852); Ann. Chim. Phys. III **40**, 156 (1854)

7. Two-Dimensional Photoelasticity

7.1 Introduction

Two-dimensional photoelasticity deals with the determination of two-dimensional stress fields by use of polarized light. This method is based on the temporary- or artificial-birefringence effect, first discovered by Sir *David Brewster* [7.1], according to which some transparent materials when subjected to a stress system behave like birefringent crystals. The birefringence in the material is retained only during the application of the loads and disappears when they are removed. According to the temporary-birefringence effect, as explained in the previous chapter, the principal-birefringence axes of the model coincide with the principal axes of the induced two-dimensional state of stress, and the optical birefringence δ induced in a thickness d of the model is

$$\delta = C(\sigma_1 - \sigma_2)d, \tag{7.1}$$

where σ_1 and σ_2 are the two principal stresses and C is the stress-optical constant of the material of the model.

Two-dimensional photoelasticity is the oldest method that uses light in the investigation of stress fields; it was established by the pioneering works of *Coker* and *Filon* [7.2] and *Mesnager* [7.3], almost a century after the discovery of the photoelastic effect by *Brewster*. Since then, the method of photoelasticity became very popular in the engineering world and attained its full maturity in the middle of our century. The main merit of the method is its simplicity in the experimentation as well as in the evaluation of the stress field from the information gained in the experiments. An approximate picture of the stresses developed in a structure can be obtained from only one picture of the optical pattern obtained by placing a loaded transparent model of the structure between two crossed polarizers and using an ordinary light source.

The theory of two-dimensional photoelasticity is now well established; it is described in many relevant books [7.2, 4–8]. In all of these references, the basic formulae for the optical arrangements used in photoelasticity, such as the plane and circular polariscopes, and the Tardy and Senarmont compensation methods, were derived by use of the vector description of polarized light.

In the present chapter, a new formulation of the theory of two-dimensional photoelasticity is presented, based on the modern methods of handling polarization-optics problems, by use of the Poincaré sphere and the Mueller and Jones calculi. The governing formulae of all important optical arrangements used in two-dimensional photoelasticity are derived in a simple and unified manner by use of these modern methods. Details of instrumentation of the optical arrangements, model materials, as well as on the thus-obtained optical patterns, and on the use of these patterns for the complete evaluation of the parameters of the induced stress fields are not given, because they are found in all books about two-dimensional photoelasticity.

7.2 The Plane Polariscope

A plane polariscope consists of a pair of linear polarizers that have their optical axes, perpendicular to each other, in the optical field of which the model is inserted (Fig. 7.1). When the model is stressed, it behaves like a birefringent plate whose retardation at a given point in the field is δ and whose principal axes subtend an angle β with the Ox axis. Usually, the first polarizer, following the light source, is called the *polarizer*, whereas the second polarizer, following the specimen, is called the *analyzer*. Let the axis of the polarizer be vertical, then, the plane polariscope consists of

I)   A linear polarizer P_{90}, whose optical axis is vertical.
II) A birefringent plate (specimen) $R_{\beta}(\delta)$ with retardation δ, whose fast axis subtends an angle β with the Ox axis.
III) A linear polarizer P_0, whose optical axis is horizontal.

We will now consider the optical transformations in a plane polariscope by using the Jones and the Mueller calculi.

7.2.1 The Jones Calculus for a Plane Polariscope

The normalized Jones vector a for the linearly polarized light that emerges from the polarizer P_{90}, whose optical axis is vertical, is expressed by (Table 3.2)

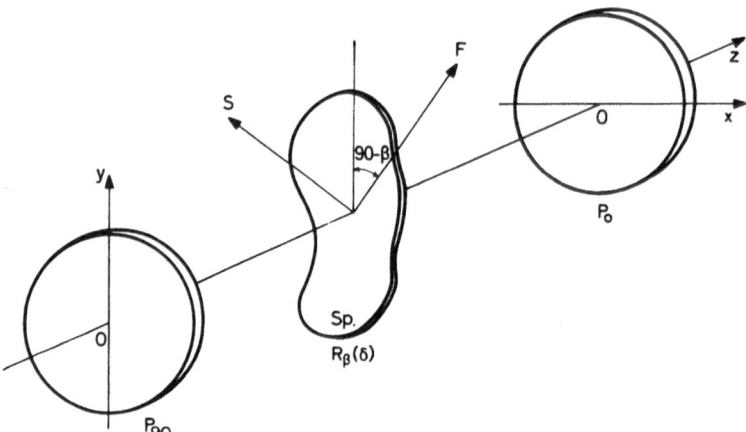

Fig. 7.1. Arrangement of the optical elements in the plane polariscope with dark background

$$a = \begin{bmatrix} 0 \\ 1 \end{bmatrix}. \qquad (7.2)$$

The two-by-two Jones matrix of the birefringent plate $R_\beta(\delta)$ with retardation δ, whose fast axis subtends an angle β with the Ox axis is given by (Table 4.2)

$$R_\beta(\delta) = \begin{bmatrix} e^{i\delta}\cos^2\beta + \sin^2\beta & (e^{i\delta}-1)\sin\beta\cos\beta \\ (e^{i\delta}-1)\sin\beta\cos\beta & e^{i\delta}\sin^2\beta + \cos^2\beta \end{bmatrix}. \qquad (7.3)$$

The Jones matrix of the second linear polarizer (analyzer), whose optical axis is horizontal, is given by (Table 4.1)

$$P_0 = \begin{bmatrix} 1 & 0 \\ 0 & 0 \end{bmatrix}. \qquad (7.4)$$

According to Jones calculus, the Jones vector a' of the light beam that emerges from the analyzer is given by

$$a' = P_0 R_\beta(\delta) a . \qquad (7.5)$$

Introducing in (7.5) the expressions (7.2) to (7.4) we obtain

$$a' = \begin{bmatrix} 1 & 0 \\ 0 & 0 \end{bmatrix} \begin{bmatrix} e^{i\delta}\cos^2\beta + \sin^2\beta & (e^{i\delta}-1)\sin\beta\cos\beta \\ (e^{i\delta}-1)\sin\beta\cos\beta & e^{i\delta}\sin^2\beta + \cos^2\beta \end{bmatrix} \begin{bmatrix} 0 \\ 1 \end{bmatrix}$$

$$= \begin{bmatrix} (e^{i\delta}-1)\sin\beta\cos\beta \\ 0 \end{bmatrix}. \qquad (7.6)$$

The light intensity I of the beam, expressed by (7.6), is

$$I = \tilde{a}' a' = \begin{bmatrix} (e^{-i\delta}-1)\sin\beta\cos\beta & 0 \end{bmatrix} \begin{bmatrix} (e^{i\delta}-1)\sin\beta\cos\beta \\ 0 \end{bmatrix}$$

$$= [2-(e^{i\delta}+e^{-i\delta})]\sin^2\beta\cos^2\beta = \sin^2\frac{\delta}{2}\sin^2 2\beta . \qquad (7.7)$$

From this relation, we obtain $I = 0$ when

$$\beta = 0, \frac{\pi}{2}, \ldots, n\frac{\pi}{2} \quad \text{or} \quad \delta = 0, 2\pi, \ldots, n2\pi . \qquad (7.8)$$

Thus, the loci of points on the specimen for which $\beta = \text{constant}$, that is the loci where the principal stresses have the same inclination, form dark bands, which are called the *isoclinics*, whereas the loci of points for which $\delta = 2\pi n$ form dark bands, which are called the *isochromatics*, because they become colored when white instead of monochromatic light is used. It is, therefore, concluded that a plane polariscope yields the families of isochromatics and isoclinics superposed on each other when a stressed specimen made of bire-fringent material is inserted in the optical field of the instrument. This result is the keystone of photoelastic method of stress analysis; it is well established in all classical books about photoelasticity [7.2, 4–8].

7.2.2 The Mueller Calculus for a Plane Polariscope

The normalized Stokes vector S of the unpolarized light beam that emerges from the monochromatic light source, placed behind the polarizer is (p. 43)

$$S = \begin{bmatrix} 1 \\ 0 \\ 0 \\ 0 \end{bmatrix}. \tag{7.9}$$

The Mueller matrices P_{90}, $R_\beta(\delta)$, and P_0 of the polarizer, the birefringent plate, and the analyzer of the plane polariscope are (Tables 4.3, 4.4)

$$P_{90} = \frac{1}{2} \begin{bmatrix} 1 & -1 & 0 & 0 \\ -1 & 1 & 0 & 0 \\ 0 & 0 & 0 & 0 \\ 0 & 0 & 0 & 0 \end{bmatrix}, \tag{7.10}$$

$$R_\beta(\delta) = \begin{bmatrix} 1 & 0 & 0 & 0 \\ 0 & \cos^2 2\beta + \sin^2 2\beta \cos\delta & (1-\cos\delta)\sin 2\beta \cos 2\beta & -\sin 2\beta \sin\delta \\ 0 & (1-\cos\delta)\sin 2\beta \cos 2\beta & \sin^2 2\beta + \cos^2 2\beta \cos\delta & \cos 2\beta \sin\delta \\ 0 & \sin 2\beta \sin\delta & -\cos 2\beta \sin\delta & \cos\delta \end{bmatrix}, \tag{7.11}$$

$$P_0 = \frac{1}{2} \begin{bmatrix} 1 & 1 & 0 & 0 \\ 1 & 1 & 0 & 0 \\ 0 & 0 & 0 & 0 \\ 0 & 0 & 0 & 0 \end{bmatrix}. \tag{7.12}$$

According to the Mueller calculus, the Stokes vector S' of the light that emerges from the analyzer is

$$S' = P_0 R_\beta(\delta) P_{90} S,$$ (7.13)

or

$$
\begin{bmatrix} s'_0 \\ s'_1 \\ s'_2 \\ s'_3 \end{bmatrix} = \frac{1}{2}
\begin{bmatrix}
1 & 1 & 0 & 0 \\
1 & 1 & 0 & 0 \\
0 & 0 & 0 & 0 \\
0 & 0 & 0 & 0
\end{bmatrix}
$$

$$
\begin{bmatrix}
1 & 0 & 0 & 0 \\
0 & \cos^2 2\beta + \sin^2 2\beta \cos\delta & (1-\cos\delta)\sin 2\beta \cos 2\beta & -\sin 2\beta \sin\delta \\
0 & (1-\cos\delta)\sin 2\beta \cos 2\beta & \sin^2 2\beta + \cos^2 2\beta \cos\delta & \cos 2\beta \sin\delta \\
0 & \sin 2\beta \sin\delta & -\cos 2\beta \sin\delta & \cos\delta
\end{bmatrix}
$$

$$
\frac{1}{2}
\begin{bmatrix}
1 & -1 & 0 & 0 \\
-1 & 1 & 0 & 0 \\
0 & 0 & 0 & 0 \\
0 & 0 & 0 & 0
\end{bmatrix}
\begin{bmatrix} 1 \\ 0 \\ 0 \\ 0 \end{bmatrix}
= \frac{1}{2}
\begin{bmatrix}
\sin^2 2\beta \sin^2(\delta/2) \\
\sin^2 2\beta \sin^2(\delta/2) \\
0 \\
0
\end{bmatrix}.
$$ (7.14)

The intensity I of the Stokes vector, expressed by (7.14) is

$$I = \frac{1}{2}\sin^2 2\beta \sin^2 \frac{\delta}{2},$$ (7.15)

which was previously obtained by using the Jones calculus [see (7.7)].

The difference between the multiplying factors $(1/2)$ in (7.7) and (7.15) is due to the fact that in (7.15) the intensity of the unpolarized light used was equal to unity whereas in (7.7) it was assumed that the intensity of the plane-polarized light that emerges from the polarizer is equal to unity. This last intensity is double the intensity of the unpolarized used in (7.15).

7.3 The Circular Polariscope

The circular polariscope is obtained from the plane polariscope by introducing two quarter-wave plates, the first after the polarizer and the second before the analyzer, whose fast axes make angles $\pm 45°$ with the axes of polarizer and analyzer, respectively. When the fast axes of the two quarter-wave plates

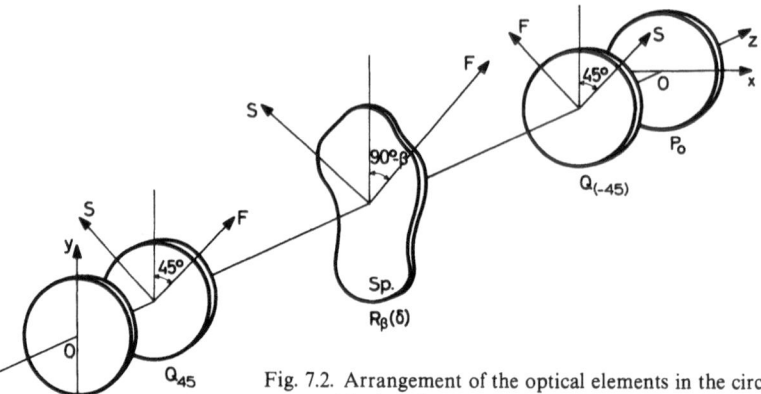

Fig. 7.2. Arrangement of the optical elements in the circular polariscope with dark background (crossed quarter-wave plates as well
as polarizer and analyzer)

are crossed, no light emerges from the analyzer, so that the *dark-field circular
polariscope* is obtained, whereas when the fast axes of the two quarter-wave
plates are parallel, the *bright-field circular polariscope* is obtained.

Thus, the circular polariscope consists of the optical elements (Fig. 7.2):

I)   A linear polarizer P_{90}, whose optical axis is vertical.
II) A quarter-wave plate Q_{45}, whose fast axis makes an angle $45°$ with
     the Ox axis.
III) A birefringent plate (specimen) $R_\beta(\delta)$ with retardation δ, whose fast
     axis makes an angle β with the Ox axis.
IV) A quarter-wave plate $Q_{\pm 45}$ whose fast axis makes an angle of $-45°$
     with the Ox axis (for the dark-field circular polariscope) or $+45°$
     with the Ox axis (for the light-field circular polariscope).
V)   A linear polarizer P_0, whose optical axis is horizontal.

We shall now consider the optical transformations in a circular polariscope
by use of the Jones calculus, the Mueller calculus, and the Poincaré sphere.

7.3.1 The Jones Calculus for a Circular Polariscope

The Jones matrices for the quarter-wave plate, whose axis makes an angle
$\pm 45°$ with the Ox axis, are obtained from (7.3) by putting $\delta = 90°$ and
$\beta = \pm 45°$. Thus, we obtain the matrices Q_{45} or Q_{-45} for the cases when
$\beta = 45°$ or $\beta = -45°$,

$$Q_{45} = \frac{1}{2}\begin{bmatrix} i+1 & i-1 \\ i-1 & i+1 \end{bmatrix} = \frac{i+1}{2}\begin{bmatrix} 1 & i \\ i & 1 \end{bmatrix},$$

$$Q_{-45} = \frac{1}{2}\begin{bmatrix} i+1 & -(i-1) \\ -(i-1) & i+1 \end{bmatrix} = \frac{i+1}{2}\begin{bmatrix} 1 & -i \\ -i & 1 \end{bmatrix}.$$

(7.16)

According to the Jones calculus, the Jones vector a' of the light that emerges from the analyzer, for the case of the dark-field circular polariscope, is

$$a' = P_0 Q_{-45} R_\beta(\delta) Q_{45} a . \tag{7.17}$$

By taking into account (7.2) to (7.4) and (7.16), we obtain

$$a' = \begin{bmatrix} 1 & 0 \\ 0 & 0 \end{bmatrix} \frac{i+1}{2} \begin{bmatrix} 1 & -i \\ -i & 1 \end{bmatrix} \begin{bmatrix} e^{i\delta}\cos^2\beta + \sin^2\beta & (e^{i\delta}-1)\sin\beta\cos\beta \\ (e^{i\delta}-1)\sin\beta\cos\beta & e^{i\delta}\sin^2\beta + \cos^2\beta \end{bmatrix}$$
$$\frac{i+1}{2} \begin{bmatrix} 1 & i \\ i & 1 \end{bmatrix} \begin{bmatrix} 0 \\ 1 \end{bmatrix} . \tag{7.18}$$

This vector, when the matrix multiplications are performed, can be reduced to

$$a' = \frac{1}{2} \begin{bmatrix} e^{i\delta}-1 \\ 0 \end{bmatrix} . \tag{7.19}$$

The intensity of the beam expressed by (7.19) is

$$I = \tilde{a}'a' = \frac{1}{4}\begin{bmatrix} e^{-i\delta}-1 & 0 \end{bmatrix}\begin{bmatrix} e^{i\delta}-1 \\ 0 \end{bmatrix} = \sin^2\frac{\delta}{2} . \tag{7.20}$$

Relation (7.20) indicates that, in the dark-field circular polariscope, only integer-order isochromatics are obtained.

For the case of the bright-field circular polariscope, we obtain for the Jones vector a' of the emerging light

$$a' = P_0 Q_{45} R_\beta(\delta) Q_{45} a , \tag{7.21}$$

or

$$a' = \begin{bmatrix} 1 & 0 \\ 0 & 0 \end{bmatrix} \frac{i+1}{2} \begin{bmatrix} 1 & -i \\ -i & 1 \end{bmatrix} \begin{bmatrix} e^{i\delta}\cos^2\beta + \sin^2\beta & (e^{i\delta}-1)\sin\beta\cos\beta \\ (e^{i\delta}-1)\sin\beta\cos\beta & e^{i\delta}\sin^2\beta + \cos^2\beta \end{bmatrix}$$
$$\frac{i+1}{2} \begin{bmatrix} 1 & i \\ i & 1 \end{bmatrix} \begin{bmatrix} 0 \\ 1 \end{bmatrix} . \tag{7.22}$$

Vector a', when the matrix multiplications are performed, becomes

$$a' = \frac{1}{2} \begin{bmatrix} e^{i\delta}+1 \\ 0 \end{bmatrix} . \tag{7.23}$$

The intensity of the light, with Jones vector expressed by (7.23), is

$$I = \tilde{a}' a' = \frac{1}{4}[e^{-i\delta}+1 \quad 0]\begin{bmatrix} e^{i\delta}+1 \\ 0 \end{bmatrix} = \cos^2 \frac{\delta}{2}. \tag{7.24}$$

Relation (7.24) shows that, in the bright-field circular polariscope, only half-integer-order isochromatics are obtained.

7.3.2 The Mueller Calculus for a Circular Polariscope

The Mueller matrices for the quarter-wave plate whose fast axis makes an angle of $\pm 45°$ with the Ox axis, are obtained from (7.11), by putting $\delta = 90°$ and $\beta = \pm 45°$, respectively. Thus, we obtain the matrices $\boldsymbol{Q}_{45}$ or $\boldsymbol{Q}_{-45}$ for the cases when $\beta = 45°$ or $\beta = -45°$, respectively,

$$\boldsymbol{Q}_{45} = \begin{bmatrix} 1 & 0 & 0 & 0 \\ 0 & 0 & 0 & -1 \\ 0 & 0 & 1 & 0 \\ 0 & 1 & 0 & 0 \end{bmatrix}, \quad \boldsymbol{Q}_{-45} = \begin{bmatrix} 1 & 0 & 0 & 0 \\ 0 & 0 & 0 & 1 \\ 0 & 0 & 1 & 0 \\ 0 & -1 & 0 & 0 \end{bmatrix}. \tag{7.25}$$

According to the Mueller calculus, the Stokes vector S' of the light that emerges from the analyzer, for the case of the dark-field circular polariscope, is

$$S' = P_0 \boldsymbol{Q}_{-45} \boldsymbol{R}_\beta(\delta) \boldsymbol{Q}_{45} P_{90} S. \tag{7.26}$$

By substituting the values of S, P_{90}, $\boldsymbol{R}_\beta(\delta)$, P_0, $\boldsymbol{Q}_{\pm 45}$ from (7.9) to (7.12) and (7.25), respectively, we obtain

$$S' = \frac{1}{4}\begin{bmatrix} 1 & 1 & 0 & 0 \\ 1 & 1 & 0 & 0 \\ 0 & 0 & 0 & 0 \\ 0 & 0 & 0 & 0 \end{bmatrix}\begin{bmatrix} 1 & 0 & 0 & 0 \\ 0 & 0 & 0 & 1 \\ 0 & 0 & 1 & 0 \\ 0 & -1 & 0 & 0 \end{bmatrix}$$

$$\begin{bmatrix} 1 & 0 & 0 & 0 \\ 0 & \cos^2 2\beta + \sin^2 2\beta \cos\delta & (1-\cos\delta)\sin 2\beta \cos 2\beta & -\sin 2\beta \sin\delta \\ 0 & (1-\cos\delta)\sin 2\beta \cos 2\beta & \sin^2 2\beta + \cos^2 2\beta \cos\delta & \cos 2\beta \sin\delta \\ 0 & \sin 2\beta \sin\delta & -\cos 2\beta \sin\delta & \cos\delta \end{bmatrix}$$

$$\begin{bmatrix} 1 & 0 & 0 & 0 \\ 0 & 0 & 0 & -1 \\ 0 & 0 & 1 & 0 \\ 0 & 1 & 0 & 0 \end{bmatrix}\begin{bmatrix} 1 & -1 & 0 & 0 \\ -1 & 1 & 0 & 0 \\ 0 & 0 & 0 & 0 \\ 0 & 0 & 0 & 0 \end{bmatrix}\begin{bmatrix} 1 \\ 0 \\ 0 \\ 0 \end{bmatrix}. \tag{7.27}$$

The Stokes vector S' given by (7.27), when the indicated matrix multiplications are performed, takes the form

$$S' = \frac{1}{4} \begin{bmatrix} 1 - \cos \delta \\ 1 - \cos \delta \\ 0 \\ 0 \end{bmatrix}. \tag{7.28}$$

The intensity I of the light, expressed by (7.28), is

$$I = \frac{1}{4}(1 - \cos \delta) = \frac{1}{2} \sin^2 \frac{\delta}{2}. \tag{7.29}$$

Equation (7.29), which expresses the intensity of the light that emerges from the dark-field circular polariscope, is half the intensity found in (7.20), which was obtained by use of the Jones calculus. This is due to the fact that the intensity of the unpolarized light used in (7.29) was equal to unity, whereas in (7.20) the intensity of the polarized light emerging from the polarizer is equal to unity.

For the case of the bright-field circular polariscope, the Stokes vector S' of the light that emerges from the analyzer is:

$$S' = P_0 Q_{45} R_\beta(\delta) Q_{45} P_{90} S, \tag{7.30}$$

or

$$S' = \frac{1}{4} \begin{bmatrix} 1 & 1 & 0 & 0 \\ 1 & 1 & 0 & 0 \\ 0 & 0 & 0 & 0 \\ 0 & 0 & 0 & 0 \end{bmatrix} \begin{bmatrix} 1 & 0 & 0 & 0 \\ 0 & 0 & 0 & -1 \\ 0 & 0 & 1 & 0 \\ 0 & 1 & 0 & 0 \end{bmatrix}$$

$$\begin{bmatrix} 1 & 0 & 0 & 0 \\ 0 & \cos^2 2\beta + \sin^2 2\beta \cos \delta & (1 - \cos \delta) \sin 2\beta \cos 2\beta & -\sin 2\beta \sin \delta \\ 0 & (1 - \cos \delta) \sin 2\beta \cos 2\beta & \sin^2 2\beta + \cos^2 2\beta \cos \delta & \cos 2\beta \sin \delta \\ 0 & \sin 2\beta \sin \delta & -\cos 2\beta \sin \delta & \cos \delta \end{bmatrix}$$

$$\begin{bmatrix} 1 & 0 & 0 & 0 \\ 0 & 0 & 0 & -1 \\ 0 & 0 & 1 & 0 \\ 0 & 1 & 0 & 0 \end{bmatrix} \begin{bmatrix} 1 & -1 & 0 & 0 \\ -1 & 1 & 0 & 0 \\ 0 & 0 & 0 & 0 \\ 0 & 0 & 0 & 0 \end{bmatrix} \begin{bmatrix} 1 \\ 0 \\ 0 \\ 0 \end{bmatrix}. \tag{7.31}$$

Relation (7.31), when the matrix multiplications indicated in this relation are performed, gives for the emerging light

$$
S' = \frac{1}{4}
\begin{bmatrix}
1 + \cos \delta \\
1 + \cos \delta \\
0 \\
0
\end{bmatrix}.
\tag{7.32}
$$

The intensity of the light whose Stokes vector is given by (7.32), is

$$
I = \frac{1}{4}(1 + \cos \delta) = \frac{1}{2}\cos^2 \frac{\delta}{2},
\tag{7.33}
$$

which is half the intensity of light in (7.24), which was found by use of the Jones calculus, due to the fact that the intensity of unpolarized light used in (7.33) is equal to unity, whereas, in (7.24), the intensity of the polarized light emerging from the polarizer is equal to unity.

7.3.3 The Poincaré Sphere in the Circular Polariscope

We shall now consider the optical transformations that take place in the circular polariscope, by using the Poincaré sphere. The light that emerges from the linear polarizer whose axis is parallel to the Oy direction is represented by the point A on the equatorial plane. In order to find the point that represents the light that emerges from the first quarter-wave plate, whose fast axis is at an angle $45°$ with the Ox axis, the Poincaré sphere must be rotated $90°$, that is, the angle that represents the retardation of the quarter-wave plate, around the Oy axis, which axis represents the eigenvector of the fast axis of the plate. Thus, the right-circularly polarized light that emerges

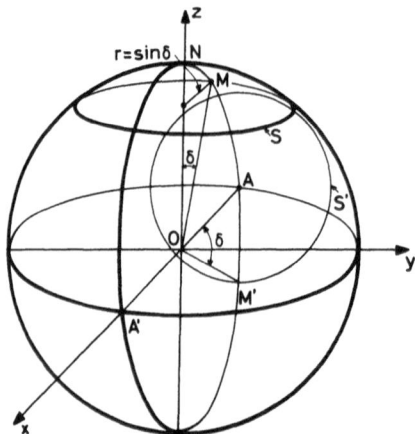

Fig. 7.3. Representation on the Poincaré sphere of the optical transformations that occur in the circular polariscope

from the first quarter-wave plate, which is represented by the north pole N of the sphere, is obtained.

Let us now consider all points of the specimen that have the same retardation δ, but different principal axes. The specimen at each such point behaves like a linear retarder that has retardation δ and arbitrary principal directions. In order to find the positions M of the point N that represent the circularly polarized light that emerges from the quarter-wave plate, after passage through all retarders that have the same retardation δ but arbitrary principal directions, the Poincaré sphere must be rotated through an angle δ around an arbitrary axis that lies in its equatorial plane. Thus, we obtain the circle S of radius $r = \sin \delta$ parallel to the equator, which represents the polarization states of all beams that emerge from the specimen with the same retardation δ.

Now, if we consider the case of a dark-field polariscope, the fast axis of the second quarter-wave plate is represented in the Poincaré sphere with the negative Oy axis. The optical effect of the second quarter-wave plate on the light that emerges from the specimen, which is represented by the circle S, is obtained by rotating this circle about the Oy axis through an angle of $90°$. Thus, all of the polarization forms of the light that emerges from the second quarter-wave plate are represented by the points M' of the circle S'.

The principal axis of the analyzer, whose fast axis coincides with the Ox axis, is represented by point A' on the equatorial plane. According to the property of the Poincaré sphere, established in Section 4.3.5, the intensity of the light that emerges from the analyzer is

$$I = \cos^2 \frac{\widehat{A'M'}}{2} = \sin^2 \frac{\widehat{AM'}}{2} = \sin^2 \frac{\delta}{2}. \tag{7.34}$$

Similarly working for the case of the bright-field circular polariscope we obtain

$$I = \cos^2 \frac{\delta}{2}. \tag{7.35}$$

Expressions (7.34) and (7.35) for the intensity of the light that emerges from either the dark- or the bright-field circular polariscope coincide with those found previously.

7.4 The Senarmont Compensation Method

In the Senarmont compensation method the optical elements of a circular polariscope, suitably arranged, are used. In this method, the first quarter-wave plate is removed, the axis of polarizer is put at $45°$ with the principal stress directions at the point considered on the specimen, and one axis of the second quarter-wave plate is arranged to be parallel to the axis of polarizer.

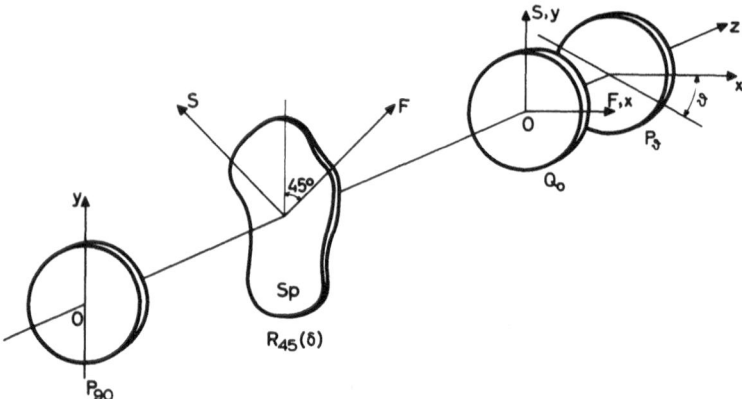

Fig. 7.4. Arrangement of the optical elements in the Senarmont compensation method

The arrangement of the optical elements and the specimen for the Senarmont compensation method is shown in Fig. 7.4.

A light ray emitted from a monochromatic source of light, placed behind the polarizer, encounters the following optical elements, successively:

I)   A linear polarizer P_{90}, whose optical axis is vertical.

II) A birefringent plate (specimen) $R_{45}(\delta)$ with retardation δ, whose fast axis makes an angle of $45°$ with the Ox axis.

III) A quarter-wave plate, whose fast axis is horizontal.

IV) A linear polarizer P_{θ}, whose optical axis makes an angle θ with the Ox axis.

We shall now consider the optical transformations that occur when a light ray passes successively through the optical elements of the Senarmont compensation method, by using the Jones calculus, the Mueller calculus, and the Poincaré sphere.

7.4.1 The Jones Calculus in the Senarmont Method

The Jones vector a' of the light that emerges from the quarter-wave plate is, according to Jones calculus,

$$a' = Q_0 R_{45}(\delta)a , \tag{7.36}$$

where Q_0 and $R_{45}(\delta)$ are the Jones matrices of the quarter-wave plate whose fast axis is horizontal and the birefringent plate whose fast axis makes an angle of $45°$ with the Ox axis, and a is the Jones vector of the linearly polarized light that emerges from the polarizer with the vertical plane of vibration.

From (7.3), which expresses the Jones matrix $R_\beta(\delta)$ for a linear retarder with retardation δ whose fast axis makes an angle β with the Ox axis, we obtain, for $\beta=0$, $\delta=90°$ (for Q_0) and $\beta=45°$ [for $R_{45}(\delta)$],

$$Q_0 = \begin{bmatrix} i & 0 \\ 0 & 1 \end{bmatrix}, \quad R_{45}(\delta) = \frac{1}{2}\begin{bmatrix} e^{i\delta}+1 & e^{i\delta}-1 \\ e^{i\delta}-1 & e^{i\delta}+1 \end{bmatrix}. \tag{7.37}$$

Thus, we have from (7.36)

$$\begin{aligned} a' &= \begin{bmatrix} i & 0 \\ 0 & 1 \end{bmatrix}\frac{1}{2}\begin{bmatrix} e^{i\delta}+1 & e^{i\delta}-1 \\ e^{i\delta}-1 & e^{i\delta}+1 \end{bmatrix}\begin{bmatrix} 0 \\ 1 \end{bmatrix} \\ &= e^{i\delta/2}\begin{bmatrix} -\sin\dfrac{\delta}{2} \\ \cos\dfrac{\delta}{2} \end{bmatrix} = e^{i\delta/2}\begin{bmatrix} \cos\left(\dfrac{\pi}{2}+\dfrac{\delta}{2}\right) \\ \sin\left(\dfrac{\pi}{2}+\dfrac{\delta}{2}\right) \end{bmatrix}. \end{aligned} \tag{7.38}$$

That is, the light that emerges from the quarter-wave plate is linearly polarized; its plane of polarization makes an angle $\delta/2$ with the vertical axis.

If the analyzer is rotated anticlockwise through an angle $\theta=(\delta/2)$, its principal axis becomes perpendicular to the linearly polarized light that emerges from the quarter-wave plate; therefore, extinction occurs at the point considered.

Thus, the relative retardation N at the point considered is

$$N = k + \frac{\theta}{\pi} \quad (k=1,2,\ldots,n). \tag{7.39}$$

7.4.2 The Mueller Calculus in the Senarmont Method

The Stokes vector S' of the light that emerges from the quarter-wave plate, in the Senarmont method of compensation, is given according to Mueller calculus by

$$S' = Q_0 R_{45}(\delta) P_{90} S, \tag{7.40}$$

where Q_0 and $R_{45}(\delta)$ are the Mueller matrices of the quarter-wave plate and the specimen, respectively, P_{90} is the matrix that represents the polarizer, and S is the Stokes vector of the unpolarized light that is incident on the polarizer.

Matrices Q_0 and $R_{45}(\delta)$, obtained from the general matrix $R_\beta(\delta)$ of the birefringent plate, given by (7.11) for $\beta=0$, $\delta=90°$ (for Q_0), and $\beta=45°$ [for $R_{45}(\delta)$], are

$$Q_0 = \begin{bmatrix} 1 & 0 & 0 & 0 \\ 0 & 1 & 0 & 0 \\ 0 & 0 & 0 & 1 \\ 0 & 0 & -1 & 0 \end{bmatrix}, \tag{7.41}$$

and

$$
R_{45}(\delta) = \begin{bmatrix} 1 & 0 & 0 & 0 \\ 0 & \cos\delta & 0 & -\sin\delta \\ 0 & 0 & 1 & 0 \\ 0 & \sin\delta & 0 & \cos\delta \end{bmatrix}.
$$

(7.42)

Thus, (7.40) takes the form

$$
S' = \begin{bmatrix} 1 & 0 & 0 & 0 \\ 0 & 1 & 0 & 0 \\ 0 & 0 & 0 & 1 \\ 0 & 0 & -1 & 0 \end{bmatrix} \begin{bmatrix} 1 & 0 & 0 & 0 \\ 0 & \cos\delta & 0 & -\sin\delta \\ 0 & 0 & 1 & 0 \\ 0 & \sin\delta & 0 & \cos\delta \end{bmatrix} \frac{1}{2} \begin{bmatrix} 1 & -1 & 0 & 0 \\ -1 & 1 & 0 & 0 \\ 0 & 0 & 0 & 0 \\ 0 & 0 & 0 & 0 \end{bmatrix} \begin{bmatrix} 1 \\ 0 \\ 0 \\ 0 \end{bmatrix}.
$$

(7.43)

When the matrix multiplications are performed,

$$
S' = \frac{1}{2} \begin{bmatrix} 1 \\ -\cos\delta \\ -\sin\delta \\ 0 \end{bmatrix} = \frac{1}{2} \begin{bmatrix} 1 \\ \cos(\pi+\delta) \\ \sin(\pi+\delta) \\ 0 \end{bmatrix}.
$$

(7.44)

That is, the light that emerges from the quarter-wave plane is linearly polarized and its plane of polarization makes an angle $(\delta/2)$ with the vertical axis. This result is the same as found by use of the Jones calculus.

7.4.3 The Poincaré Sphere in the Senarmont Method

The linearly polarized light that emerges from the polarizer, whose principal axis is vertical, is presented on the Poincaré sphere by point A on the negative Ox axis (Fig. 7.5). The fast and slow axes of the birefringent plate, being at an angle of 45° with the Ox axis, are represented on the sphere by points Q and Q'. Thus, the light that emerges from the specimen is determined by the new position A' of point A, which is found by rotating the sphere about the axis QQ' through an angle δ.

The light that leaves the specimen encounters the quarter-wave plate, whose fast and slow axes are represented by points B and A on the positive and negative Ox axes, respectively. Thus, the position of the light that emerges from the quarter-wave plate is found by rotating the sphere about the Ox axis, through an angle of 90°. The new position of point A' is A'' on the equatorial plane, such that $\widehat{AA''}=\delta$, that is the light that emerges from the quarter-wave plate is linearly polarized and its plane of polarization makes an angle $(180 + \delta)/2$ with the Ox axis. This is the same result that was found by use of the Jones and Mueller calculi.

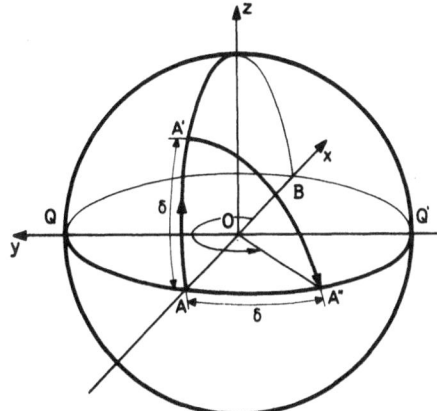

Fig. 7.5. Representation on the Poincaré sphere of the optical transformations that occur in the Senarmont compensation method

7.5 The Tardy Compensation Method

In the Tardy compensation method none of the elements of the circular polariscope is removed; the whole system of polarizer, analyzer, and quarter-wave plates is rotated until the axes of the polarizer and analyzer become parallel to the principal axes of the point considered in the specimen (Fig. 7.6). The determination of the isochromatic-fringe order at the point considered is found by rotating the analyzer until the light is extinguished at this point.

A light ray emitted from a monochromatic light source, placed behind the polarizer encounters the following optical elements, in succession (Fig. 7.6).

I) A linear polarizer P_{90}, whose optical axis is vertical.

II) A quarter-wave plate Q_{45}, whose fast axis makes an angle 45° with the Ox axis.

III) A birefringent plate (specimen) $R_0(\delta)$ with retardation δ, whose fast axis is parallel to the Ox axis.

IV) A quarter-wave plate Q_{-45}, whose fast axis makes an angle $-45°$ with the Ox axis.

V) A linear polarizer P_θ, whose pass axis makes an angle θ with the Ox axis.

We shall now consider the optical transformations that take place when a light ray passes through the optical elements of the Tardy method of compensation, by using the Jones calculus, the Mueller calculus, and the Poincaré sphere, respectively.

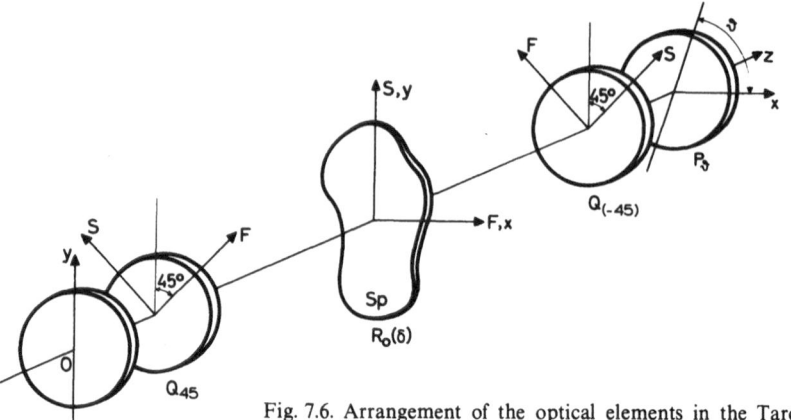

Fig. 7.6. Arrangement of the optical elements in the Tardy compensation method

7.5.1 The Jones Calculus in the Tardy Method

The Jones vector a' of the light that emerges from the second quarter-wave plate is

$$a' = Q_{-45}\, R_0(\delta)\, Q_{45}\, a \,, \tag{7.45}$$

where $R_0(\delta)$ and $Q_{\pm45}$ are the Jones matrices of the specimen and the quarter-wave plate whose fast axis makes an angle of $\pm45°$ with the Ox axis.

Matrices $R_0(\delta)$ and $Q_{\pm45}$, obtained from (7.3), by putting $\beta=0$ [for $R_0(\delta)$], and $\beta=\pm45°$, $\delta=90°$ (for $Q_{\pm45}$), respectively, are

$$R_0(\delta) = \begin{bmatrix} e^{i\delta} & 0 \\ 0 & 1 \end{bmatrix}, \quad Q_{45} = \frac{i+1}{2}\begin{bmatrix} 1 & i \\ i & 1 \end{bmatrix}, \quad Q_{-45} = \frac{i+1}{2}\begin{bmatrix} 1 & -i \\ -i & 1 \end{bmatrix}. \tag{7.46}$$

Then, (7.45) takes the form

$$a' = \left(\frac{i+1}{2}\right)^2 \begin{bmatrix} 1 & -i \\ -i & 1 \end{bmatrix}\begin{bmatrix} e^{i\delta} & 0 \\ 0 & 1 \end{bmatrix}\begin{bmatrix} 1 & i \\ i & 1 \end{bmatrix}\begin{bmatrix} 0 \\ 1 \end{bmatrix}$$

$$= \frac{(i+1)^2}{2}\, e^{i\delta/2} \begin{bmatrix} \cos\left(\dfrac{\pi}{2}+\dfrac{\delta}{2}\right) \\[2mm] \sin\left(\dfrac{\pi}{2}+\dfrac{\delta}{2}\right) \end{bmatrix}. \tag{7.47}$$

That is, the light that emerges from the second quarter-wave plate is linearly polarized at an angle $\delta/2$ with the Oy axis.

Therefore, when the analyzer is rotated counterclockwise through an angle $\theta = (\delta/2)$, extinction occurs at the point considered; the fringe order N at this point is

$$N = k + \frac{\theta}{\pi}, \quad (k=1,2,...,n).$$ (7.48)

7.5.2 The Mueller Calculus in the Tardy Method

The Stokes vector S' of the light that emerges from the second quarter-wave plate is

$$S' = Q_{-45} R_0(\delta) Q_{45} P_{90} S,$$ (7.49)

where $R_0(\delta)$ and $Q_{\pm 45}$ are the Mueller matrices of the specimen and the quarter-wave plate whose fast axis makes an angle of $\pm 45°$ with the Ox axis.

Matrices $R_0(\delta)$ and $Q_{\pm 45}$, obtained from (7.11) by putting $\beta = 0$ and $\beta = \pm 45°$, $\delta = 90°$, are

$$R_0(\delta) = \begin{bmatrix} 1 & 0 & 0 & 0 \\ 0 & 1 & 0 & 0 \\ 0 & 0 & \cos\delta & \sin\delta \\ 0 & 0 & -\sin\delta & \cos\delta \end{bmatrix},$$ (7.50)

$$Q_{45} = \begin{bmatrix} 1 & 0 & 0 & 0 \\ 0 & 0 & 0 & -1 \\ 0 & 0 & 1 & 0 \\ 0 & 1 & 0 & 0 \end{bmatrix}, \quad Q_{-45} = \begin{bmatrix} 1 & 0 & 0 & 0 \\ 0 & 0 & 0 & 1 \\ 0 & 0 & 1 & 0 \\ 0 & -1 & 0 & 0 \end{bmatrix}.$$ (7.51)

Then, (7.49) takes the form

$$S' = \begin{bmatrix} 1 & 0 & 0 & 0 \\ 0 & 0 & 0 & 1 \\ 0 & 0 & 1 & 0 \\ 0 & -1 & 0 & 0 \end{bmatrix} \begin{bmatrix} 1 & 0 & 0 & 0 \\ 0 & 1 & 0 & 0 \\ 0 & 0 & \cos\delta & \sin\delta \\ 0 & 0 & -\sin\delta & \cos\delta \end{bmatrix} \begin{bmatrix} 1 & 0 & 0 & 0 \\ 0 & 0 & 0 & -1 \\ 0 & 0 & 1 & 0 \\ 0 & 1 & 0 & 0 \end{bmatrix}$$
$$\frac{1}{2} \begin{bmatrix} 1 & -1 & 0 & 0 \\ -1 & 1 & 0 & 0 \\ 0 & 0 & 0 & 0 \\ 0 & 0 & 0 & 0 \end{bmatrix} \begin{bmatrix} 1 \\ 0 \\ 0 \\ 0 \end{bmatrix}.$$ (7.52)

When the matrix multiplications are performed

$$
S' = \frac{1}{2}\begin{bmatrix} 1 \\ -\cos\delta \\ -\sin\delta \\ 0 \end{bmatrix} = \frac{1}{2}\begin{bmatrix} 1 \\ \cos(\pi+\delta) \\ \sin(\pi+\delta) \\ 0 \end{bmatrix}. \tag{7.53}
$$

That is, the light that emerges from the second quarter-wave plate is linearly polarized with its plane of polarization at an angle $\delta/2$ with the Oy axis, which is the same result as was found by use of the Jones calculus.

7.5.3 The Poincaré Sphere in the Tardy Method

The linearly polarized light that emerges from the polarizer, whose principal axis is vertical, is represented on the Poincaré sphere by point A on the negative Ox axis (Fig. 7.7). In order to find the point that represents the light that emerges from the first quarter-wave plate, the sphere must be rotated around the axis $QQ' \equiv Oy$ through an angle of 90°, so that the new position of point A coincides with the north pole N of the sphere. The fast and slow axes of the birefringent plate (specimen) are represented by points A''' and A on the positive and negative Ox axis, so that, in order to find the point that represents the light that emerges from the specimen, the Poincaré sphere must be rotated around the Ox axis through an angle of δ equal to the retardation introduced by the specimen when loaded. Thus, point N takes its new position A', where $\widehat{AN'} = \delta$.

The effect of the second quarter-wave plate is to rotate the sphere through an angle 90° about the $Q'Q$ axis. Therefore the final position of point A' is A'' on the equatorial plane, such that $\widehat{AA''} = \delta$.

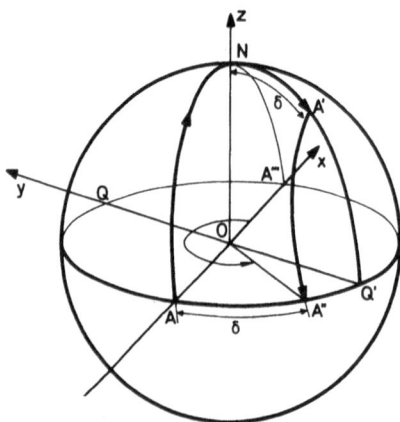

Fig. 7.7. Representation on the Poincaré sphere of the optical transformations that occur in the Tardy compensation method

Thus, the light that emerges from the second quarter-wave plate is linearly polarized at an angle of $(90° + \delta/2)$ with the Ox axis. This result is the same as that found by use of the Jones and Mueller calculi.

References

7.1   D. Brewster: Phil. Trans. R. Soc. London, **105**, 60 (1815); **106**, 156 (1816)
7.2   E. G. Coker, L. N. G. Filon: *A Treatise on Photoelasticity* (University Press, Cambridge 1957)
7.3   A. Mesnager: Compt. Rend. **155**, 1071 (1912)
7.4   M. M. Frocht: *Photoelasticity*, Vol. I, Vol. II (Wiley and Sons, New York 1941, 1948)
7.5   A. J. Durelli, W. F. Riley: *Introduction to Photomechanics* (Prentice-Hall, Englewood Cliffs, NJ 1965)
7.6   M. Hetényi: *Handbook of Experimental Stress Analysis* (Wiley and Sons, New York 1950)
7.7   R. B. Heywood: *Photoelasticity for Designers* (Pergamon Press, Oxford 1969)
7.8   A. Kuske, G. Robertson: *Photoelastic Stress Analysis* (Wiley and Sons, New York 1974)

8. Three-Dimensional Photoelasticity

8.1 Introduction

Three-dimensional photoelasticity provides one of the commonest and most widely used experimental methods for determination of three-dimensional states of stress. This method, like two-dimensional photoelasticity, is based upon the phenomenon of temporary or artificial birefringence, according to which when a transparent material is loaded its optical properties change; measurement of these changes provides data adequate for quantitative determination of the corresponding state of stress. Although the method of two-dimensional photoelasticity is simple and straightforward and the optical patterns obtained are directly related to the stresses developed in the material, many difficulties arise in the evaluation of a triaxial stress field, when the method of three-dimensional photoelasticity is applied. This is mainly due to variation of stress distribution from point to point along the light path; this causes rotation of the principal-stress directions from layer to layer through a three-dimensional photoelastic model.

Use of the conventional vectorial representation of polarized light for the interpretation of the optical patterns in three-dimensional photoelasticity requires analysis of the incident light relative to the principal axes of each separate layer in the three-dimensional model, which coincide with the principal axes of a birefringent plate made of the same substance and corresponding to this layer. If the thickness of each layer is small, so that the principal-stress directions do not vary significantly through the thickness of the layer, then to predict the polarization state of the emerging light, the corresponding stress-optical retardation introduced by the layer has to be added to either of the linearly polarized beams at the entrance of the layer. Even if the stress distribution along the light path were known, a large amount of labor would be required to predict the polarization form of the emerging light. Solution of the inverse problem, that is, determination of the stress distribution along the light path from the observed optical pattern, is even more difficult and laborious. These difficulties of interpretation of stress-induced optical patterns in three-dimensional photoelasticity led to the development of two separate techniques for photoelastic determination of three-dimensional stress states, namely the *frozen-stress* and the *scattered-light* methods.

The frozen-stress method is based on a behavior that is exhibited by polymeric substances when they are heated. When a high polymer is heated to a temperature slightly above its softening point, or its critical temperature (glass-transition temperature), its rigidity, expressed for example by the modulus of elasticity, is greatly reduced, although the material retains its elastic behavior. When, at this temperature, a model is loaded and the applied loads are retained while the model is cooled to ambient temperature, a state of stress is locked in the model. This state of stress is not disturbed even though the model is cut into slices; it remains unaltered for a long period of time. By use of this property of polymeric materials a model may be made of the three-dimensional body under study and subjected to the procedure described. When the model is cooled, it is cut into slices of a suitable thickness, such that the variation of the principal stress directions through the thickness of each slice may be considered negligible; each separate slice is examined in the circular and plane polariscope. By cutting duplicate models, into slices at three orthogonal directions, the principal stress differences in each direction may be determined. Thus, five of the six quantities needed for the characterization of a three-dimensional state of stress are determined. The sixth quantity is calculated by applying either the equilibrium or the compatibility equations.

This mechanical behavior of polymers may be explained as follows. Most polymeric substances used in photoelasticity are of the cross-linked type and are rigid at ambient temperature. The basic structure of a polymer molecule is that of a flexible linear chain composed of a high number of links of various lengths hinged together so that, by thermal agitation, the molecular chains may adopt a great number of conformations. For a cross-linked polymer, besides these main-chain links, there are also crosslinks that transversely join neighboring chains.

At ambient temperature, all cross-linked polymers used in photoelasticity are at a solid state, that is, they behave like linearly elastic glasses. When a cross-linked polymer, being initially at its glassy state, is progressively heated, only the cross-links are agitated; the main backbone chains remain unaltered. Therefore, the elastic modulus is very high, the deformations of the substance are very small and elastic, and no significant viscoelastic phenomena appear. When the temperature is continuously raised, thermal agitations of the cross links increase rapidly and some insignificant backbone movements appear. As a result of these movements, a time-dependent (viscoelastic) variation of deformations of the body appears, although the applied stress may be constant (creep phenomenon). Similarly, if the deformation of the body is kept constant, the applied stress necessary to keep this deformation constant is continuously reduced (relaxation phenomenon). The characteristic temperature above which these phenomena are perceptible is called the glass-transition temperature (T_g). Above this temperature the viscoelastic phenomena become progressively more and more pronounced, owing to the progressively increasing relative movements between the backbone chains. This process continues until another temperature limit is reached, at which all backbone-chain movements are exhausted and only small, elastic cross-link movements still persist. Above this temperature limit the material passes into its rubbery state where it again behaves elastically. In the rubbery plateau, the various moduli remain almost constant, although they are considerably reduced from their values in the glassy state.

According to the Boltzmann principle, and the time-temperature super-position principle, which are both valid throughout the whole viscoelastic spectrum of linear viscoelastic substances, increase of temperature corresponds to logarithmic increase of the duration of the phenomena. Thus, heating of the specimen in the frozen-stress procedure to a temperature above T_g has the same effect as loading the specimen for the corresponding time duration in the respective creep or relaxation curve of the polymer. Because recovery phenomena in linear viscoelasticity require time spans generally longer than the corresponding loading periods, the practice in the frozen-stress method, in which the specimen is loaded at high temperature, close to the rubbery plateau, corresponds to a very long period of static loading at ambient temperature, which in turn assures a much longer recovery period.

During this long period of recovery of the loaded specimen, any type of measurement may be executed. Consequently, there is ample time during which to obtain all of the information needed for a three-dimensional photo-elastic analysis. For example, for a cold-setting pure-epoxy polymer of the type of Araldite B, heating at 130 °C and loading for the frozen-stress procedure corresponds to the creep during approximately 300 years of static loading at ambient temperature. The even longer recovery time is more than enough for any photoelastic measurement.

The frozen-stress method for photoelastic investigation of three-dimensional stress states has become popular among experimenters since its first days of development. A great many papers dealing with the development of the method and its various applications for the solution of engineering problems have appeared in the literature. For a comprehensive review of these works the reader is referred to [8.1–5]. Studies dealing with the development of suitable polymeric substances for three-dimensional photoelastic applications of the frozen-stress method are included in the above references.

The second approach for investigating three-dimensional stress states photoelastically, the so-called scattered-light method, is based on the phe-nomenon of scattering of light when it passes through the model. This method, which has not received great attention in the literature, owing to inherent difficulties in the experimentation, will be described in Chapter 9.

However, both the frozen-stress and the scattered-light methods have serious limitations and disadvantages, which restrict their applicability. Thus, the frozen-stress method necessitates the preparation of three similar models, from which slices in three different orthogonal directions must be cut. It is restricted to static loadings only. The scattered-light method presents many difficulties in its application.

On the other hand, interpretation of optical patterns obtained from the intact three-dimensional model presents many difficulties, owing to the continuous rotation of the principal-stress directions along the light path. Solution of the problem by the well-known methods that use conventional vectorial representation of polarized light is very difficult. For this reason, nondestructive methods for evaluation of stress distributions in three-dimensional photoelastic problems have not made much progress, and the frozen-stress method continues to offer definite advantages for solving problems of three-dimensional photoelasticity.

The influence of the rotation of the principal-stress directions on the optical patterns obtained from three-dimensional models was considered in earlier studies of photoelasticity. Thus, *Neumann* [8.6], using purely kinematical considerations, came to the conclusion that the existence of rotation in the principal-stress directions influences the resulting retardation of the birefringent element. He established the corresponding equation that expresses the total birefringence of each volume element as the sum of two terms, one of which expresses the birefringence in the absence of rotation and the other gives the part of birefringence due to the rotation of the principal stresses.

The same phenomenon was subsequently studied by *Coker* and *Filon* [8.7], who also derived *Neumann's* equations from kinematic considerations, however, different than those used by *Neumann*. *Mindlin* and *Goodman* [8.8] derived the optical equations of two- and three-dimensional photoelasticity by using *Maxwell's* electromagnetic equations and again obtained *Neumann's* equations. *Drucker* and *Mindlin* [8.9] formulated the equations of the three-dimensional photoelastic problem by using the ether theory of light and considered the case of a constant value of the ratio C of the rotation of the secondary principal axes to the corresponding retardation. They found that the resultant retardation δ is equal to $\delta = \delta_0 \sqrt{1 + 4C^2}$, where δ_0 is the retardation in the absence of rotation of the principal axes.

All of these pioneering works in the nondestructive or so-called integrated photoelasticity were carried out by use of the concept of the light vector, which is defined in the electromagnetic theory of light as the electric or the magnetic field vector, whereas in the ether theory of light it is the vector that defines the position of the ether particle at each moment during its oscillation. The solution of these problems, by use of the concept of the light vector, necessitated lengthy, tedious, cumbersome and error-prone calculations. This was the reason why nondestructive photoelasticity, using the conventional concept of the light vector, did not make much progress.

Modern methods of description of polarized light, based on the concepts of the Poincaré sphere and the Mueller and Jones calculi, revived interest in nondestructive photoelasticity. The elegance and powerfulness of these methods made the formulation and solution of three-dimensional stress problems easier. The equivalence theorem, proved in Section 4.5, by use of the Jones calculus, greatly facilitated solution of the three-dimensional photoelastic problem. According to this theorem, an optical system containing any number of birefringent plates is optically equivalent to a system that contains a birefringent plate and a rotator. The term "optically equivalent" is used in the sense that for the same incident light form the emerging light from the initial system and its optical equivalent are the same. This result may be applied to any three-dimensional photoelastic model, which is equivalent to a number of birefringent plates. Thus, the optical effect of the three-dimensional model is defined by three independent quantities, i. e., the orientation and the birefringence of the equivalent birefringent plate, and the rotation of the equivalent rotator. From this equivalence theorem, it is deduced that, for a three-dimensional photoelastic model, there always exist two directions for which, when the incident light is linearly polarized, the emerging light is also linearly polarized. These two directions are defined, for the incident light, by the principal axes of the birefringent plate and, for the emerging light, by the directions derived from these axes by adding the rotation introduced by the rotator.

The equivalence theorem was used by *Srinath* and *Sarma* [8.10] for formulation and solution of the three-dimensional photoelastic problem. They also established [8.11, 12] differential equations that the Stokes parameters of polarized light must satisfy when it passes through a three-dimensional photoelastic model, and applied these equations for solution of the Drucker-Mindlin problem of constant rotation of the principal-stress directions along the light path. *Srinath* and *Bhave* [8.13], by use of Jones calculus, also obtained differential equations that should be satisfied by the elements of the Jones matrix along the axis of propagation of a polarized light beam in a three-dimensional photoelastic model. In order to obtain additional optical data, they used different wavelengths of incident light. *Theocaris* and *Gdoutos* [8.14] in a recent paper developed a new nondestructive method of three-dimensional photoelasticity based on the differential equations satisfied by the Stokes and Jones vectors along the axis of propagation of a polarized light beam in a three-dimensional photoelastic model. They solved these equations in a matrix-series form by use of the Peano-Baker method and proved that they lead immediately to the *Neumann* equations. As an application of the Peano-Baker solution they reconsidered and solved the Drucker-Mindlin problem. *Mylonas* and *Brown* [8.15], in an interesting report, presented a brief review of the method of the Poincaré sphere and the Mueller and Jones calculi, and they used the Jones calculus to study the rotational effect in birefringent coatings. The experimental data corroborated satisfactorily with the theoretical predictions.

However, nondestructive three-dimensional photoelasticity has been greatly advanced by the pioneering works of the Russian researcher *Aben*, during the last two decades. In a series of papers [8.16–18] *Aben* stated the foundations of nondestructive photoelasticity, explained its difficulties, and gave relevant solutions for the problem. By using *Ginsburg's* (*Maxwell's*) electromagnetic equations, he derived the most-general equations that govern the propagation of polarized light in a three-dimensional photoelastic model, and deduced as special cases the Neumann, the Drucker-Mindlin and the Wertheim photoelastic equations. He proved, independently of the Jones calculus, the equivalence theorem in a three-dimensional model and expressed it in terms of the so-called characteristic directions, that is the directions for which, when linearly polarized light is incident on the model with its plane of polarization parallel to the characteristic directions, the emerging light is also linearly polarized. He showed that two such characteristic directions exist in a three-dimensional model and that they are mutually perpendicular. Experimental determination of the characteristic directions, as well as the associated retardation of the photoelastic model, gives for every light path three quantities, which are related to the stress field across the light path. Thus, for the determination of the stress components along the light path, additional data are needed. To obtain additional data, *Aben* introduced the idea of using various wavelengths and developed a general algorithm, based on the measured experimental data, for determination of the state of stress along the light path. He applied this algorithm to the solution of shell problems.

However, for antisymmetrical stress distributions, developed for example in flexed plates, the integrated optical effect is zero, so that the method of non-destructive photoelasticity failed for the solution of such problems. To overcome these difficulties in such problems and to obtain additional data for the general three-dimensional photoelastic problem, *Aben* introduced the idea of using the Faraday and Kerr effects. In the Faraday effect, when a photoelastic model is inserted in a magnetic field, the plane of polarization is rotated, whereas in the Kerr effect, when the model is placed in an electric field, additional birefringence is introduced. By use of these effects, *Aben* formulated two new methods of nondestructive photoelasticity, called magnetophotoelasticity [8.19–23] and electrophotoelasticity [8.23] based on the Faraday and Kerr effects, respectively. However, owing to the need for large electric fields to obtain appreciable additional birefringence, electrophotoelasticity has not made much progress. On the other hand, magnetophotoelasticity has been used successfully by *Aben* in the solution of many problems in which the principal directions rotate.

In the present chapter, the general problem of the nondestructive three-dimensional photoelasticity will be formulated and solved by use of modern methods of description of polarized light, that is, the Poincaré sphere and the Mueller and Jones calculi. Finally, the approach developed by *Aben* for the solution of this problem will be presented in detail.

8.2 The Equivalence Theorem

Let us consider a three-dimensional photoelastic model referred to a cartesian system of coordinates $Oxyz$. If a light ray, parallel to the Oz axis, impinges on the specimen, then this ray, according to the photoelastic law, is split into two plane-polarized beams along the principal-stress directions at the point of entrance. Travelling through the model an infinitesimal distance dz, the two plane-polarized rays are retarded, one relative to the other, by a phase angle δ, which is proportional to the difference of the principal stresses and the thickness dz, and which depends also on the material of the model. This dependence is specified by a proportionality factor C, which is called the *stress-optical constant*. Thus, the layer of the three-dimensional photoelastic model with thickness dz behaves like a birefringent plate (linear retarder), whose principal axes coincide with the principal-stress directions of the layer and whose retardation δ is given by (7.1).

As the state of stress in the three-dimensional model varies along the Oz axis, that is, the principal stress directions and the difference of the principal stresses vary with the z coordinate, the three-dimensional photoelastic model can be considered as optically equivalent to an infinite number of birefringent plates with different orientations of their principal axes and different values of their optical retardations δ. The expression "optically equivalent" is used in the sense that, for a given light incident on the photoelastic model and on the optically equivalent model, which consists of a train of retarders, the polarization form of the emerging light is the same for both.

The train of retarders that is optically equivalent to the photoelastic model can be further simplified by making use of the theorem developed in Section 4.5. According to this theorem, an optical system that contains any number of retarders is optically equivalent to a system that contains one retarder and one rotator. Thus the theorem may be derived: *Any three-dimensional photoelastic model is optically equivalent to a system consisting of a retarder (birefringent plate) and a rotator (active plate).*

This theorem constitutes the basis of three-dimensional photoelasticity. It will be referred in the sequence as the *equivalence theorem* of three-dimensional photoelasticity.

8.3 Basic Equations of Three-Dimensional Photoelasticity

When a beam of polarized light passes through a three-dimensional photoelastic model in which the principal stress directions rotate along the light path, its characteristic quantities, defined either by the azimuth, the ellipticity, and the handedness of the corresponding light ellipse, or by the elements of the Stokes and Jones vectors, vary from point to point along the light path. We shall now establish the differential equations satisfied by the Stokes and Jones vectors when a polarized beam propagates through a three-dimensional photoelastic model.

8.3.1 The Stokes Vector

Let us consider an infinitesimal layer of the model of thickness dz, retardation $d\delta$, and whose principal stress or birefringent axes subtend an angle φ with the Ox axis (Fig. 8.1). It is assumed that a light beam propagating along the Oz axis traverses the layer normally and that the stress components in the

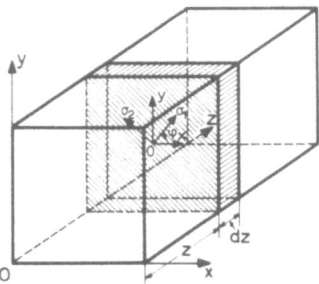

Fig. 8.1. Geometry of an infinitesimal element of a three-dimensional photoelastic model normal to a polarized light beam

plane of the layer are σ_x, σ_y, τ_{xy}. According to Wertheim's photoelastic law, the retardation $d\delta$ is given by

$$d\delta = C_\sigma [\sqrt{(\sigma_x - \sigma_y)^2 + 4\tau_{xy}^2}] \, dz , \tag{8.1}$$

where C_σ is the photoelastic constant of the material of the model; C_σ depends, among others, on the wavelength of the light used.

The angle φ that defines the principal birefringent axes of the state of stress developed in this element is defined by

$$\tan 2\varphi = \frac{2\tau_{xy}}{\sigma_x - \sigma_y} . \tag{8.2}$$

If $d\varphi$ is the variation of angle φ along the thickness dz of the layer, then the effect of this infinitesimal layer on the incident light beam can be obtained by considering the layer to consist of a birefringent plate with retardation $d\delta$ and a rotator with rotation $d\varphi$. If the Stokes vector of the light incident upon the layer is referred to the principal directions of the layer, then the Mueller matrices of the infinitesimal birefringent plate M and the infinitesimal rotator M' are obtained from (4.44) with $\delta \to (-d\delta)$, $\theta = 0$ for M and from (4.47) with $\theta = \varphi \to d\varphi$ for M'

$$M = \begin{bmatrix} 1 & 0 & 0 & 0 \\ 0 & 1 & 0 & 0 \\ 0 & 0 & 1 & -d\delta \\ 0 & 0 & d\delta & 1 \end{bmatrix}, \tag{8.3}$$

$$M' = \begin{bmatrix} 1 & 0 & 0 & 0 \\ 0 & 1 & 2d\varphi & 0 \\ 0 & -2d\varphi & 1 & 0 \\ 0 & 0 & 0 & 1 \end{bmatrix}. \tag{8.4}$$

Thus, the Stokes vector S' of the emerging light is

$$S' = M'MS,$$ (8.5)

where S is the Stokes vector of the incident light

$$S = \begin{bmatrix} s_0 \\ s_1 \\ s_2 \\ s_3 \end{bmatrix}.$$

By eliminating second-order differential quantities in the matrix S', we obtain

$$S' = \begin{bmatrix} 1 & 0 & 0 & 0 \\ 0 & 1 & 2d\varphi & 0 \\ 0 & -2d\varphi & 1 & -d\delta \\ 0 & 0 & d\delta & 1 \end{bmatrix} S,$$ (8.6)

and since

$$S' = S + dS,$$ (8.7)

we obtain from (8.6) and (8.7) that

$$d\begin{bmatrix} s_1 \\ s_2 \\ s_3 \end{bmatrix} = \begin{bmatrix} 0 & 2d\varphi & 0 \\ -2d\varphi & 0 & -d\delta \\ 0 & d\delta & 0 \end{bmatrix} \begin{bmatrix} s_1 \\ s_2 \\ s_3 \end{bmatrix}.$$ (8.8)

Equation (8.8) defines the differential equations that the Stokes parameters s_1, s_2, s_3 must satisfy when a light beam propagates through a three-dimensional photoelastic model. Equation (8.8) is equivalent to a system of three linear differential equations of first order. For the Stokes parameter s_0, which defines the intensity of the incident light beam,

$$ds_0 = 0,$$

because the medium is considered nonabsorbing.

8.3.2 The Jones Vector

Working similarly with the Jones calculus, we obtain from (4.9) and (4.15), with $\delta \rightarrow d\delta$, $\theta = 0$, and $\theta = \varphi \rightarrow d\varphi$, respectively, the Jones matrices J and J' of the infinitesimal birefringent plate and the infinitesimal rotator,

$$J = \begin{bmatrix} e^{id(\delta/2)} & 0 \\ 0 & e^{-id(\delta/2)} \end{bmatrix}, \quad J' = \begin{bmatrix} 1 & d\varphi \\ -d\varphi & 1 \end{bmatrix}. \tag{8.9}$$

The Jones vector a' of the emerging light is

$$a' = J'J a, \tag{8.10}$$

where a is the Jones vector of the incident light,

$$a = \begin{bmatrix} a_x \\ a_y \end{bmatrix},$$

or

$$a' = \begin{bmatrix} a_x e^{id(\delta/2)} + a_y e^{-id(\delta/2)} d\varphi \\ -a_x e^{id(\delta/2)} d\varphi + a_y e^{-id(\delta/2)} \end{bmatrix}. \tag{8.11}$$

Since

$$a' = a + da, \tag{8.12}$$

we obtain from (8.11), by eliminating second-order differential quantities,

$$d\begin{bmatrix} a_x \\ a_y \end{bmatrix} = \begin{bmatrix} id(\delta/2) & d\varphi \\ -d\varphi & -id(\delta/2) \end{bmatrix} \begin{bmatrix} a_x \\ a_y \end{bmatrix}. \tag{8.13}$$

Equation (8.13) defines the differential equation that is satisfied by the Jones vector when light passes through a three-dimensional photoelastic model. This equation is equivalent to a system of two linear differential equations of first order.

Equations (8.8) and (8.13) connect the values of the Stokes and Jones vectors at each position along the light path with the corresponding values of the stress-optical retardation $d\delta$ and the rotation $d\varphi$ of the principal stress directions caused when the light beam passes through an infinitesimal element of thickness dz. Both of these quantities are directly connected with the difference $(\sigma_x - \sigma_y)$ of the normal stresses σ_x, σ_y and the shear stress τ_{xy} in the plane normal to the light path, through (8.1) and (8.2). Thus, by integrating either

(8.8) or (8.13), the values of either the Stokes or the Jones vectors at each position can be expressed in terms of the quantities $(\sigma_x - \sigma_y)$ and τ_{xy} and the initial values of either vector, which are assumed to be known. It is therefore concluded that, by measuring the values of the Stokes or Jones vectors at each position along the light path, the corresponding values of the quantities $(\sigma_x - \sigma_y)$ and τ_{xy} can be determined.

Equation (8.13), which is satisfied by the Jones vector parameters a_x, a_y, is much simpler than (8.8), which is satisfied by the Stokes parameters s_1, s_2, s_3, because it is equivalent to two scalar differential equations, whereas (8.8) is equivalent to three such equations. However, (8.13) deals with complex quantities, whereas (8.8) deals with real ones.

8.4 Neumann's Equations

Equation (8.8) gives three differential equations for the Stokes parameters s_1, s_2, s_3

$$
\begin{aligned}
ds_1 &= 2s_2\,d\varphi \\
ds_2 &= -2s_1\,d\varphi - s_3\,d\delta \\
ds_3 &= s_2\,d\delta,
\end{aligned}
\tag{8.14}
$$

where $d\delta$ is the birefringence of the infinitesimal layer of thickness dz and $d\varphi$ is the rotation of the principal axes through dz.

If ψ is the phase difference between the two linearly polarized light components at the entrance of the layer, then we obtain, from (3.34), that

$$
\begin{aligned}
ds_1 &= 2A_x\,dA_x - 2A_y\,dA_y \\
ds_2 &= 2A_x \cos\psi\,dA_y + 2A_y \cos\psi\,dA_x - 2A_x A_y \sin\psi\,d\psi \\
ds_3 &= 2A_x \sin\psi\,dA_y + 2A_y \sin\psi\,dA_x + 2A_x A_y \cos\psi\,d\psi.
\end{aligned}
\tag{8.15}
$$

From the first equations of (8.14) and (8.15), we obtain

$$
A_x\,dA_x - A_y\,dA_y = 2A_x A_y \cos\psi\,d\varphi,
\tag{8.16}
$$

and by putting

$$
\begin{aligned}
A_x &= k \cos\gamma \\
A_y &= k \sin\gamma,
\end{aligned}
\tag{8.17}
$$

we obtain from (8.16),

$$dy = -\cos\psi \, d\varphi.$$ (8.18)

Equation (8.18) is the first Neumann equation (see [8.7]).

Similarly, from the two last equations of relations (8.14) and (8.15), we obtain

$$\cos\psi \left(\frac{dA_x}{A_x} + \frac{dA_y}{A_y}\right) - \sin\psi \, d\psi = -\frac{A_x^2 - A_y^2}{A_x A_y} d\varphi - \sin\psi \, d\delta,$$

$$\sin\psi \left(\frac{dA_x}{A_x} + \frac{dA_y}{A_y}\right) + \cos\psi \, d\psi = \cos\psi \, d\delta,$$ (8.19)

and by eliminating the factor $(dA_x/A_x + dA_y/A_y)$, we get

$$d\psi = d\delta + \left(\frac{A_x}{A_y} - \frac{A_y}{A_x}\right) \sin\psi \, d\varphi.$$ (8.20)

Equation (8.20) indicates that, in the presence of rotation of the principal-stress directions, the resulting retardation $d\psi$ is the sum of the retardation $d\delta$, without rotation of the principal stresses, and the second term in the right-hand side of (8.20).

When (8.17) are taken into account, (8.20) becomes

$$d\psi = d\delta + 2\cot 2\gamma \sin\psi \, d\varphi.$$ (8.21)

Equation (8.20) or (8.21) constitutes the second Neumann equation (see [8.7]).

Neumann's equations, (8.18) and (8.21), can also be derived from the differential equation (8.13), satisfied by the Jones vector $\boldsymbol{a}$.

When (8.17) are taken into account, the normalized Jones vector $\boldsymbol{a}$ at the entrance of the infinitesimal layer can be expressed by

$$\boldsymbol{a} = \begin{bmatrix} a_x \\ a_y \end{bmatrix} = \begin{bmatrix} \cos\gamma \; e^{i\psi} \\ \sin\gamma \end{bmatrix}.$$ (8.22)

By combining (8.13) and (8.22), we obtain

$$-\sin\gamma \, e^{i\psi} dy + i\cos\gamma \, e^{i\psi} d\psi = i\cos\gamma \, e^{i\psi} d(\delta/2) + \sin\gamma \, d\varphi$$ (8.23)

$$\cos\gamma \, dy = -\cos\gamma \, e^{i\psi} d\varphi - i\sin\gamma \, d(\delta/2).$$

From the second of (8.23) it follows that

$$dy = -\cos\psi \, d\varphi,$$ (8.24)

$$d\delta = -2\cot\gamma \sin\psi \, d\varphi.$$ (8.25)

Introducing (8.24) and (8.25) into the first of (8.23), we obtain again (8.21).

8.5 The Drucker-Mindlin Problem

As an application of the differential equation (8.13), we will consider the case in which the ratio C of rotation $d\varphi$ of the principal axes to the corresponding retardation $d\delta$ is constant, that is

$$C = \frac{d\varphi}{d\delta}. \tag{8.26}$$

This problem was first considered by *Drucker* and *Mindlin* and is well known as the Drucker-Mindlin problem.

For this case, (8.13) gives two separate equations,

$$\frac{da_x}{d(\delta/2)} = ia_x + 2Ca_y,$$

$$\frac{da_y}{d(\delta/2)} = -2Ca_x - ia_y. \tag{8.27}$$

Since the phase angle related to a_x is $d(\delta/2)$, we differentiate the first of these equations with respect to $d(\delta/2)$ and, by taking into account the second of (8.27), we obtain

$$\frac{d^2 a_x}{d(\delta/2)^2} = -(4C^2 + 1)a_x. \tag{8.28}$$

Similarly, we obtain

$$\frac{d^2 a_y}{d(\delta/2)^2} = -(4C^2 + 1)a_y. \tag{8.29}$$

The solutions of both (8.28) and (8.29) have the form

$$a_x, a_y = A\cos(4C^2 + 1)^{1/2}\delta, \tag{8.30}$$

where A is a constant.

Equation (8.30) indicates that, in the presence of optical rotation, the resulting retardation Δ is equal to that without rotation δ, multiplied by the factor $(4C^2 + 1)^{1/2}$. This result was first obtained by *Drucker* and *Mindlin* [8.9].

8.6 Solution of the Problem by Use of the Equivalence Theorem

8.6.1 Formulation of the Problem

According to the equivalence theorem, of three-dimensional photoelasticity developed in Section 8.2, the Jones matrix J of a three-dimensional photoelastic model can be written in the form

$$J = R(\theta)\,R_\varphi(\delta),$$ (8.31)

where the matrix $R_\varphi(\delta)$ of the retarder is given according to (4.16) by

$$R_\varphi(\delta) = \begin{bmatrix} \cos\varphi & -\sin\varphi \\ \sin\varphi & \cos\varphi \end{bmatrix} \begin{bmatrix} e^{i\delta} & 0 \\ 0 & 1 \end{bmatrix} \begin{bmatrix} \cos\varphi & \sin\varphi \\ -\sin\varphi & \cos\varphi \end{bmatrix}.$$ (8.32)

The matrix $R(\theta)$ of the rotator is given by

$$R(\theta) = \begin{bmatrix} \cos\theta & -\sin\theta \\ \sin\theta & \cos\theta \end{bmatrix}.$$ (8.33)

Thus, the matrix J that represents the photoelastic model can be expressed by

$$J = \begin{bmatrix} \cos\theta & -\sin\theta \\ \sin\theta & \cos\theta \end{bmatrix} \begin{bmatrix} \cos\varphi & -\sin\varphi \\ \sin\varphi & \cos\varphi \end{bmatrix} \begin{bmatrix} e^{i\delta} & 0 \\ 0 & 1 \end{bmatrix} \begin{bmatrix} \cos\varphi & \sin\varphi \\ -\sin\varphi & \cos\varphi \end{bmatrix},$$ (8.34)

or

$$J = \begin{bmatrix} \cos(\varphi+\theta) & -\sin(\varphi+\theta) \\ \sin(\varphi+\theta) & \cos(\varphi+\theta) \end{bmatrix} \begin{bmatrix} e^{i\delta} & 0 \\ 0 & 1 \end{bmatrix} \begin{bmatrix} \cos\varphi & \sin\varphi \\ -\sin\varphi & \cos\varphi \end{bmatrix}.$$ (8.35)

Equation (8.35) indicates that any three-dimensional photoelastic model can be represented in terms of three quantities, which will be called *characteristic parameters* of the photoelastic model; they are the azimuth φ of the fast axis of the equivalent retarder, the retardation δ of the equivalent retarder, and the characteristic angle θ of the equivalent rotator.

From (8.35) it also follows that when a linearly polarized light at an angle φ or $(\varphi+90°)$ enters the photoelastic model, then the emerging light is also linearly polarized at an angle $(\varphi+\theta)$ or $[(\varphi+\theta)+90°]$ respectively. This can be expressed by the theorem:

For any three-dimensional photoelastic model there are always two perpendicular directions along which, when a plane-polarized light beam is incident on the model, the emerging light beam is also plane-polarized.

The direction defined by the angle φ was called by *Aben* the *primary char-acteristic direction* [8.16]; that defined by the angle $(\varphi+\theta)$ was called the *secondary characteristic direction*.

According to the Jones calculus, the Jones vectors of the incident light a and the emerging light a' from the photoelastic model are

$$
a = \begin{bmatrix} A_x\,e^{i\varphi_x} \\ A_y\,e^{i\varphi_y} \end{bmatrix}, \qquad a' = \begin{bmatrix} A'_x\,e^{i\varphi'_x} \\ A'_y\,e^{i\varphi'_y} \end{bmatrix}. \tag{8.36}
$$

These vectors are connected by

$$
\begin{bmatrix} A'_x\,e^{i\varphi'_x} \\ A'_y\,e^{i\varphi'_y} \end{bmatrix} = \begin{bmatrix} \cos(\varphi+\theta) & -\sin(\varphi+\theta) \\ \sin(\varphi+\theta) & \cos(\varphi+\theta) \end{bmatrix}
$$
$$
\begin{bmatrix} e^{i\delta} & 0 \\ 0 & 1 \end{bmatrix} \begin{bmatrix} \cos\varphi & \sin\varphi \\ -\sin\varphi & \cos\varphi \end{bmatrix} \begin{bmatrix} A_x\,e^{i\varphi_x} \\ A_y\,e^{i\varphi_y} \end{bmatrix}, \tag{8.37}
$$

or

$$
a' = J\,a, \tag{8.38}
$$

with

$$
J = \begin{bmatrix} J_{11} & J_{12} \\ J_{21} & J_{22} \end{bmatrix}, \tag{8.39}
$$

given by (8.35).

8.6.2 Solution of the Problem

As was previously proved, any three-dimensional photoelastic model can be optically specified in terms of the three characteristic quantities, that is the angles φ and $(\varphi+\theta)$ and the retardation δ. Equation (8.37) suggests that these three quantities can be measured by letting a known polarized light form a impinge on the specimen and by detecting the emergent light, specified by the vector a', by use of the methods of Chapter 5. As was shown for the special case, when the incident light is linearly polarized along the direction φ, the emerging light is also linearly polarized along the direction $(\varphi+\theta)$. These two directions can be determined by use of two rotating polarizers.

On the other hand, the three characteristic quantities can be specified in terms of the principal stress directions and the differences of the principal stresses of all points along the ray that traverses the specimen. Indeed, by considering the three-dimensional model as a train of n birefringent plates, the Jones matrix J that specifies its optical behavior can be represented by

$$
J = \begin{bmatrix} \cos\varphi_1 & -\sin\varphi_1 \\ \sin\varphi_1 & \cos\varphi_1 \end{bmatrix} \begin{bmatrix} e^{i\delta_1} & 0 \\ 0 & 1 \end{bmatrix} \begin{bmatrix} \cos\varphi_1 & \sin\varphi_1 \\ -\sin\varphi_1 & \cos\varphi_1 \end{bmatrix} \cdots
$$

$$
\begin{bmatrix} \cos\varphi_i & -\sin\varphi_i \\ \sin\varphi_i & \cos\varphi_i \end{bmatrix} \begin{bmatrix} e^{i\delta_i} & 0 \\ 0 & 1 \end{bmatrix} \begin{bmatrix} \cos\varphi_i & \sin\varphi_i \\ -\sin\varphi_i & \cos\varphi_i \end{bmatrix} \cdots
$$

$$
\begin{bmatrix} \cos\varphi_n & -\sin\varphi_n \\ \sin\varphi_n & \cos\varphi_n \end{bmatrix} \begin{bmatrix} e^{i\delta_n} & 0 \\ 0 & 1 \end{bmatrix} \begin{bmatrix} \cos\varphi_n & \sin\varphi_n \\ -\sin\varphi_n & \cos\varphi_n \end{bmatrix}, \tag{8.40}
$$

where φ_i and δ_i represent the inclination of the principal stress directions and the birefringence, respectively, for the i layer of the model. These quantities are expressed in terms of the stress components $\sigma_x^i, \sigma_y^i, \tau_{xy}^i$ of the i layer by the relations

$$
\tan 2\varphi_i = \frac{2\tau_{xy}^i}{\sigma_x^i - \sigma_y^i} \quad \delta_i = C_{\sigma i}(\sigma_1^i - \sigma_2^i)\Delta d_i = C_{\sigma i}\sqrt{(\sigma_x^i - \sigma_y^i)^2 + (2\tau_{xy}^i)^2}\,\Delta d_i. \tag{8.41}
$$

Thus, by considering the three-dimensional model as a train of n layers, we can express the three characteristic quantities of the model in terms of the $2n$ quantities $\varphi_1, \varphi_2, ..., \varphi_n$ and $\delta_1, \delta_2, ..., \delta_n$. For calculation of the latter $2n$ quantities, related to the stress components $\sigma_x^i, \sigma_y^i, \tau_{xy}^i$ along the light ray through (8.41), we must use $2n$ characteristic quantities experimentally measured. There are two methods of solution of the problem, that is either by using different wavelengths of the incident light and measuring for each the emergent polarization form, or by using the Faraday and Kerr effects, first introduced in photoelasticity by *Aben* [8.19–23].

Indeed, because of the dependence of the stress-optical constant $C_{\sigma i}$ of the i layer on the wavelength of the light that passes through the layer, for each wavelength, a separate set of three characteristic directions for each value of the wavelength of light chosen is formed. Thus, by use of $(2n/3)$ different wavelengths, the $2n$ unknown quantities φ_i and δ_i $(i=1,2,...,n)$ can be determined.

The use of the Faraday and Kerr effects for the determination of the $2n$ unknown quantities φ_i and δ_i will be explained.

The Faraday Effect

Because of the Faraday effect, when a model that is interposed in a light beam is placed in a magnetic field that is parallel to the direction of the light beam, the plane of polarization of the incident light is rotated. The angle of rotation ψ is given by

$$
\psi = KH, \tag{8.42}
$$

where K is the *Verdet constant* and H is the intensity of the magnetic field.

Thus, in the presence of a magnetic field, the Jones matrix J of the i layer of the three-dimensional photoelastic model is

$$J_i = \begin{bmatrix} \cos\varphi_i & -\sin\varphi_i \\ \sin\varphi_i & \cos\varphi_i \end{bmatrix} \begin{bmatrix} e^{i\delta_i} & 0 \\ 0 & 1 \end{bmatrix} \begin{bmatrix} \cos\varphi_i & \sin\varphi_i \\ -\sin\varphi_i & \cos\varphi_i \end{bmatrix} \begin{bmatrix} \cos\psi & \sin\psi \\ -\sin\psi & \cos\psi \end{bmatrix},$$

or

$$J_i = R_{\varphi_i} R_0(\delta_i) R_{(-\varphi_i)} R_\psi, \tag{8.43}$$

and the Jones vector J of the whole three-dimensional model is

$$J = R_{\varphi_1} R_0(\delta_1) R_{(-\varphi_1)} R_\psi \cdots R_{\varphi_i} R_0(\delta_i) R_{(-\varphi_i)} R_\psi \cdots R_{\varphi_n} R_0(\delta_n) R_{(-\varphi_n)} R_\psi. \tag{8.44}$$

From (8.44) it follows that to each intensity H of the magnetic field correspond three different characteristic directions of the photoelastic model. Thus, by use of m different magnetic intensities H_i $(i=1,2,...,m)$ to the model, the $2n$ unknown quantities φ_i and δ_i $(i=1,2,...,n)$ can be determined.

The Kerr Effect

Because of the Kerr effect, when an optically isotropic medium is placed in an electric field, it becomes birefringent and its optical axis coincides with the direction of the electric field. The induced birefringence δ is given by

$$\delta = \frac{360}{\lambda} B d E^2, \tag{8.45}$$

where B is the *Kerr constant*, E is the intensity of the electric field, d is the thickness of the model, and λ is the wavelength of the light used.

Thus, in the presence of an electric field, the Jones matrix J_i of the i layer of the three-dimensional photoelastic model is

$$J_i = \begin{bmatrix} \cos\varphi_i & -\sin\varphi_i \\ \sin\varphi_i & \cos\varphi_i \end{bmatrix} \begin{bmatrix} e^{i\delta_i} & 0 \\ 0 & 1 \end{bmatrix} \begin{bmatrix} \cos\varphi_i & \sin\varphi_i \\ -\sin\varphi_i & \cos\varphi_i \end{bmatrix} \begin{bmatrix} e^{i\delta} & 0 \\ 0 & 1 \end{bmatrix}, \tag{8.46}$$

or

$$J_i = R_{\varphi_i} R_0(\delta_i) R_{(-\varphi_i)} R_0(\delta), \tag{8.47}$$

and the Jones matrix J of the whole three-dimensional model is

$$J = R_{\varphi_1} R_0(\delta_1) R_{(-\varphi_1)} R_0(\delta) \cdots R_{\varphi_i} R_0(\delta_i) R_{(-\varphi_i)} R_0(\delta) \cdots$$
$$R_{\varphi_n} R_0(\delta_n) R_{(-\varphi_n)} R_0(\delta). \tag{8.48}$$

From (8.48) it follows that to each intensity E of the electric field three different characteristic directions of the photoelastic model correspond. Thus, by use of m electric-field intensities E_i $(i=1,2,...,m)$, the $2n$ unknown quantities φ_i and δ_i $(i=1,2,...,n)$ can be determined.

However, in order to achieve an appreciable amount of birefringence δ, we must apply extremely high voltages; this makes the Kerr effect of only academic interest for three-dimensional photoelasticity.

8.7 Solution of the Problem by Integrating the Equations of Three-Dimensional Photoelasticity

In Section 8.3, the equations of three-dimensional photoelasticity were established in terms of the Stokes and Jones vectors. Both of these equations, expressed by (8.8) and (8.13), can be put in the form

$$\frac{d\omega}{dz} = u\omega, \tag{8.49}$$

where

$$\omega = \begin{bmatrix} s_1 \\ s_2 \\ s_3 \end{bmatrix} \quad \text{and} \quad u = \begin{bmatrix} 0 & 2d\varphi/dz & 0 \\ -2d\varphi/dz & 0 & -d\delta/dz \\ 0 & d\delta/dz & 0 \end{bmatrix}, \tag{8.50}$$

for (8.8), and

$$\omega = \begin{bmatrix} a_x \\ a_y \end{bmatrix} \quad \text{and} \quad u = \begin{bmatrix} (\mathrm{i}/2)d\delta/dz & d\varphi/dz \\ -d\varphi/dz & (-\mathrm{i}/2)\,d\delta/dz \end{bmatrix}, \tag{8.51}$$

for (8.13). For a light ray that travels through the model parallel to the Oz axis, the matrices u, whose elements are defined by (8.50) and (8.51) for the two cases considered, are functions of the z coordinate.

The solution of (8.49) by the Peano-Baker method [8.24], can be represented in the form

$$\omega = \Omega(u)\omega_0, \tag{8.52}$$

where the matrix $\Omega(u)$ is given by the series of matrices

$$\Omega(u) = i_0 + Qu + QuQu + QuQuQu + ... \tag{8.53}$$

In (8.52) and (8.53), ω is the matrix at the position z, ω_0 is the same matrix at the position z_0, i_0 is the unit matrix, Qu represents the matrix obtained by integrating every element of the matrix u from z_0 to z along the light path; uQu denotes the product of the matrices u and Qu; $QuQu$ denotes the matrix $Q(uQu)$, and so on.

For a given three-dimensional photoelastic model, the elements of the matrix Ω, for both the Mueller and Jones calculi, can be determined experimentally by use of the methods of Chapter 5. On the other hand, the elements of the same matrix Ω can be determined from (8.50), (8.51), and (8.53) by assuming that the functions $\delta = \delta(z)$ and $\varphi = \varphi(z)$ along the light path are known or can be approximated by simple functions. We can, for example, assume that these functions can be expressed by series expansions of the form

$$\varphi = \varphi(z) = \sum_{k=1}^{n} \alpha_k z^k, \qquad \delta = \delta(z) = C_\sigma \sum_{k=1}^{n} b_k z^k.$$

Then, the matrix Ω is expressed in terms of the coefficients α_k and b_k. Any desired number of the series coefficients α_k and b_k can be obtained by using different wavelengths λ and determining the elements of the matrix Ω for each wavelength. It is obvious that these elements of the matrix Ω are functions of the wavelength λ used, because the stress-optical constant C_σ of (8.41), connecting the maximum shear stress with the retardation δ, depends on the wavelength λ.

8.7.1 Application of the Peano-Baker Solution to the Drucker-Mindlin Problem

The above-described Peano-Baker solution of the differential equation (8.49) with matrices ω and u, defined from (8.50) and (8.51) for the Mueller and Jones calculi, respectively, will now be applied to the Drucker-Mindlin problem.

For the Mueller calculus, we obtain from (8.50) and (8.26) for the matrix u (normalized to the quantity $dz/d\delta$)

$$u = \begin{bmatrix} 0 & 2C & 0 \\ -2C & 0 & -1 \\ 0 & 1 & 0 \end{bmatrix} \quad \text{and}$$

$$u^2 = u \cdot u = \begin{bmatrix} -4C^2 & 0 & -2C \\ 0 & -(4C^2+1) & 0 \\ -2C & 0 & -1 \end{bmatrix}.$$

(8.54)

Further, we have

$$Qu = \delta u$$

$$QuQu = \frac{\delta^2}{2!}u^2$$

$$QuQuQu = -\frac{\delta^3}{3!}(4C^2+1)u$$

$$QuQuQuQu = -\frac{\delta^4}{4!}(4C^2+1)u^2$$

$$QuQuQuQuQu = \frac{\delta^5}{5!}(4C^2+1)^2 u$$

$$QuQuQuQuQuQu = \frac{\delta^6}{6!}(4C^2+1)^2 u^2$$

$$QuQuQuQuQuQuQu = -\frac{\delta^7}{7!}(4C^2+1)^3 u$$

$$QuQuQuQuQuQuQuQu = -\frac{\delta^8}{8!}(4C^2+1)^3 u^2$$

(8.55)

and so on.

Applying (8.53), we obtain for the matrix Ω

$$\Omega = i_0 + \left[\delta - \frac{\delta^3}{3!}(4C^2+1) + \frac{\delta^5}{5!}(4C^2+1)^2 - \frac{\delta^7}{7!}(4C^2+1)^3 + ...\right]u$$

$$+ \left[\frac{\delta^2}{2!} - \frac{\delta^4}{4!}(4C^2+1) + \frac{\delta^6}{6!}(4C^2+1)^2 - \frac{\delta^2}{8!}(4C^2+1)^3 + ...\right]u^2,$$

(8.56)

or

$$\Omega = i_0 + \frac{1}{(4C^2+1)^{1/2}}\sin(4C^2+1)^{1/2}\delta u + \frac{2}{4C^2+1}\sin^2\left(\frac{(4C^2+1)^{1/2}}{2}\delta\right)u^2.$$

(8.57)

From (8.57) it follows that the induced optical retardation δ, when the ratio of rotation to retardation is constant, is increased by the factor $(4C^2+1)^{1/2}$. Similarly, for the Jones calculus, we have

$$u = \begin{bmatrix} i & 2C \\ -2C & -i \end{bmatrix}, \quad u^2 = -(4C^2+1)i_0, \quad i_0 = \begin{bmatrix} 1 & 0 \\ 0 & 1 \end{bmatrix}.$$

(8.58)

We further obtain

$$Qu = \quad \delta u$$

$$QuQu = -\frac{\delta^2}{2!}(4C^2+1)i_0$$

$$QuQuQu = -\frac{\delta^3}{3!}(4C^2+1)u$$

$$QuQuQuQu = \frac{\delta^4}{4!}(4C^2+1)^2 i_0$$

$$QuQuQuQuQu = \frac{\delta^5}{5!}(4C^2+1)^2 u$$

$$QuQuQuQuQuQu = -\frac{\delta^6}{6!}(4C^2+1)^3 i_0$$

$$QuQuQuQuQuQuQu = -\frac{\delta^7}{7!}(4C^2+1)^3 u$$

$$QuQuQuQuQuQuQuQu = \frac{\delta^8}{8!}(4C^2+1)^4 i_0$$

$$(8.59)$$

and so on.

From (8.53), we obtain for the matrix Ω

$$\Omega = i_0 + \left[\delta - \frac{\delta^3}{3!}(4C^2+1) + \frac{\delta^5}{5!}(4C^2+1)^2 - \frac{\delta^7}{7!}(4C^2+1)^3 + \ldots\right]u$$

$$+ \left[-\frac{\delta^2}{2!}(4C^2+1) + \frac{\delta^4}{4!}(4C^2+1)^2 - \frac{\delta^6}{6!}(4C^2+1)^3 + \frac{\delta^8}{8!}(4C^2+1)^4 + \ldots\right]i_0.$$

$$(8.60)$$

or

$$\Omega = \frac{1}{(4C^2+1)^{1/2}}\sin(4C^2+1)^{1/2}\delta u + \cos(4C^2+1)^{1/2}\delta i_0.$$

$$(8.61)$$

From (8.61), we obtain again that the retardation in the case of the Drucker-Mindlin problem is multiplied by the factor $(4C^2+1)^{1/2}$.

8.8 Another Approach to the Problem by Using the Characteristic Matrix

In Section 8.7, the three-dimensional photoelastic problem was solved by integrating the differential equations satisfied by the Stokes or Jones vectors along the light path in the three-dimensional model. In the present section the problem will be solved by formulating and solving the differential equations that are satisfied by the elements of the Jones characteristic matrix of the photoelastic model.

Let us consider an infinitesimal layer of thickness Δz normal to the light ray in the photoelastic model. The Jones matrix $\boldsymbol{J}_{\Delta z}$ of this element, which may be considered as a birefringent plate with retardation $\Delta\delta$ and principal axes at azimuth φ with respect to the fixed axes $Ox - Oy$ can be expressed by the matrix

$$\boldsymbol{J}_{\Delta z} = \begin{bmatrix} \cos\varphi & -\sin\varphi \\ \sin\varphi & \cos\varphi \end{bmatrix} \begin{bmatrix} e^{i\Delta\delta/2} & 0 \\ 0 & e^{-i\Delta\delta/2} \end{bmatrix} \begin{bmatrix} \cos\varphi & \sin\varphi \\ -\sin\varphi & \cos\varphi \end{bmatrix}. \tag{8.62}$$

By introduction of the values of $\Delta\delta$ and φ from (8.1) and (8.2), (8.62) takes the form

$$\boldsymbol{J}_{\Delta z} = \begin{bmatrix} 1 & 0 \\ 0 & 1 \end{bmatrix} + \frac{iC_\sigma}{2} \begin{bmatrix} (\sigma_x - \sigma_y) & 2\tau_{xy} \\ 2\tau_{xy} & -(\sigma_x - \sigma_y) \end{bmatrix} \Delta z, \tag{8.63}$$

which, because,

$$\boldsymbol{J}_{\Delta z} = \boldsymbol{J}_{z+\Delta z} \boldsymbol{J}_z^{-1},$$

takes the form

$$\frac{\boldsymbol{J}_{z+\Delta z} - \boldsymbol{J}_z}{\Delta z} \boldsymbol{J}_z^{-1} = \frac{iC_\sigma}{2} \begin{bmatrix} (\sigma_x - \sigma_y) & 2\tau_{xy} \\ 2\tau_{xy} & -(\sigma_x - \sigma_y) \end{bmatrix}. \tag{8.64}$$

When $\Delta z \to 0$, (8.64) becomes

$$\frac{d\boldsymbol{J}}{dz} \boldsymbol{J}^{-1} = \frac{iC_\sigma}{2} \begin{bmatrix} (\sigma_x - \sigma_y) & 2\tau_{xy} \\ 2\tau_{xy} & -(\sigma_x - \sigma_y) \end{bmatrix}, \tag{8.65}$$

or

$$\begin{bmatrix} \dfrac{dJ_{11}}{dz} & \dfrac{dJ_{12}}{dz} \\ \dfrac{dJ_{21}}{dz} & \dfrac{dJ_{22}}{dz} \end{bmatrix} \begin{bmatrix} J_{22} & -J_{12} \\ -J_{21} & J_{11} \end{bmatrix} = \frac{iC_\sigma}{2} \begin{bmatrix} (\sigma_x - \sigma_y) & 2\tau_{xy} \\ 2\tau_{xy} & -(\sigma_x - \sigma_y) \end{bmatrix}. \tag{8.66}$$

Equation (8.65) or (8.66) is the differential equation satisfied by the Jones matrix $\boldsymbol{J}$ when a light beam is propagated through the three-dimensional photoelastic model. This equation is equivalent to

$$\frac{d}{dz}\begin{bmatrix} J_{11} \\ J_{12} \\ J_{21} \\ J_{22} \end{bmatrix} = \frac{iC_\sigma}{2}\begin{bmatrix} (\sigma_x-\sigma_y) & 0 & 2\tau_{xy} & 0 \\ 0 & (\sigma_x-\sigma_y) & 0 & 2\tau_{xy} \\ 2\tau_{xy} & 0 & -(\sigma_x-\sigma_y) & 0 \\ 0 & 2\tau_{xy} & 0 & -(\sigma_x-\sigma_y) \end{bmatrix}\begin{bmatrix} J_{11} \\ J_{12} \\ J_{21} \\ J_{22} \end{bmatrix}. \quad (8.67)$$

The solution of (8.67) by use of the Peano-Baker method can be expressed in the form,

$$\begin{bmatrix} J_{11} \\ J_{12} \\ J_{21} \\ J_{22} \end{bmatrix}_z = \Omega \begin{bmatrix} J_{11} \\ J_{12} \\ J_{21} \\ J_{22} \end{bmatrix}_{z_0}, \quad (8.68)$$

where

$$\Omega = i_0 + Qu + QuQu + QuQuQu + \dots, \quad (8.69)$$

with

$$i_0 = \begin{bmatrix} 1 & 0 & 0 & 0 \\ 0 & 1 & 0 & 0 \\ 0 & 0 & 1 & 0 \\ 0 & 0 & 0 & 1 \end{bmatrix}, \quad (8.70)$$

and

$$u = \frac{iC_\sigma}{2}\begin{bmatrix} (\sigma_x-\sigma_y) & 0 & 2\tau_{xy} & 0 \\ 0 & (\sigma_x-\sigma_y) & 0 & 2\tau_{xy} \\ 2\tau_{xy} & 0 & -(\sigma_x-\sigma_y) & 0 \\ 0 & 2\tau_{xy} & 0 & -(\sigma_x-\sigma_y) \end{bmatrix}. \quad (8.71)$$

In (8.69), Qu represents the matrix obtained by integrating every element of the matrix u from z_0 to z along the light path; uQu denotes the matrix product of the matrices u and Qu; $QuQu$ denotes the matrix $Q(uQu)$, and so on.

8.9 The Poincaré Sphere in Three-Dimensional Photoelasticity

The problem of three-dimensional photoelasticity can also be solved by use of the Poincaré sphere and suitable manipulations on it. Let us first consider the case of an infinitesimal layer of thickness dz and establish the change of the state of polarization of the incident light induced by this layer.

The action of the infinitesimal layer on the incident light can be analyzed into the action of a linear retarder that produces a retardation $d\delta$, and a pure rotator that rotates the light ellipse through an angle $d\varphi$. If the fast axis of the linear retarder subtends an angle θ with the Ox axis, then the effect of passage of the light through the retarder is found by rotating the Poincaré sphere through an angle $d\delta$ about an axis AA' that makes an angle 2θ with the Ox axis. The effect of the pure rotator is found by rotating the sphere about its polar axis NS through an angle $2d\varphi$ (Fig. 8.2).

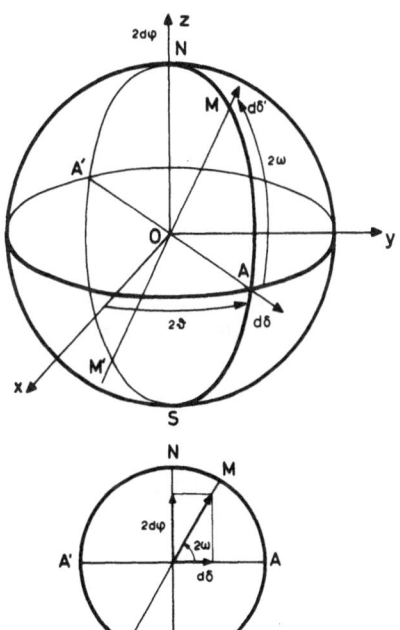

Fig. 8.2. Geometrical representation on the Poincaré sphere of the action of an infinitesimal element of a three-dimensional photoelastic model. The action of the element consists of a retarder with eigenvectors A and A' and retardation $d\delta$, which results in a rotation of the Poincaré sphere about the axis AA', through an angle $d\delta$ and a pure rotator with rotation $d\varphi$, which results in a rotation of the Poincaré sphere about the axis NS through an angle $2d\varphi$. These rotations have their axes coplanar; therefore, they combine to a single rotation to the axis MM', through an angle $d\delta'$

Thus, the effect of the infinitesimal layer on the incident light corresponds to two successive rotations of the sphere, about the axes AA' and NS, through angles $d\delta$ and $2d\varphi$, respectively. As is well-known, these two rotations, whose axes are coplanar, may be combined into a single rotation about an axis that lies in the plane of the axes AA' and NS and subtending an angle 2ω with the AA' axis, such that

$$\tan 2\omega = \frac{2\,d\varphi}{d\delta}\,; \tag{8.72}$$

the rotation about that axis should by through an angle $d\delta'$ equal to

$$d\delta' = (d\delta^2 + 4\,d\varphi^2)^{1/2} = d\delta(1 + 4C^2)^{1/2}\,, \tag{8.73}$$

where

$$C = \frac{d\varphi}{d\delta}\,;$$

that is, the rotation of the principal axes of the retarder increases the resultant retardation by the factor equal to $(1 + 4C^2)^{1/2}$.

For the case in which the ratio C of the rotation $d\varphi$ to the retardation $d\delta$ is constant along the light path in the three-dimensional model, (8.73) takes the form

$$\delta' = \delta(1 + 4C^2)^{1/2}\,. \tag{8.74}$$

Therefore, the result of *Drucker* and *Mindlin*, found previously by use of the Mueller and Jones calculis, is reestablished by use of the Poincaré representation of a polarization state.

After establishing the effect of an infinitesimal layer on an incident polarized light, we proceed to the general case of a three-dimensional model. This model can be considered as a train of infinitesimal layers, the effect of each being to rotate the Poincaré sphere about an axis by an infinitesimal angle. Thus, the combined effect of the train of the infinitesimal layers is a single rotation of the sphere about a definite axis, through a definite angle, both determined by the combination of the successive rotations of the sphere that correspond to each infinitesimal layer.

The single rotation of the Poincaré sphere can be analyzed into two rotations about perpendicular axes, one of which is the polar axis NS and the other is a particular axis in the equatorial plane. Therefore: *The three-dimensional photoelastic model is optically equivalent to a linear retarder and a pure rotator*, that is, we again obtain the equivalence theorem of three-dimensional photoelasticity, by using the Poincaré sphere.

8.10 The Electromagnetic Approach to the Three-Dimensional Photoelastic Problem

The basic equations that govern the propagation of light in a three-dimensional photoelastic medium were derived by *Aben* [8.16–23], who used Maxwell's electromagnetic equations (2.1) and (2.2). By introducing (2.8) which connects the magnetic induction B with the magnetic vector H through the magnetic permeability coefficient μ in either an isotropic or an anisotropic medium into (2.2), and by eliminating the magnetic vector H, we obtain

$$\operatorname{curl} \operatorname{curl} E = -\mu \frac{\partial^2 D}{\partial t^2} . \tag{8.75}$$

For the usual case of a plane harmonic wave, we seek a solution of this equation in the form

$$E = E' e^{i(\omega \sqrt{K} t)}, \qquad D = D' e^{i(\omega \sqrt{K} t)}, \tag{8.76}$$

where K is the dielectric coefficient in vacuum, and ω is the circular frequency of the light wave. Introducing (8.76) into (8.75), we obtain

$$\operatorname{curl} \operatorname{curl} E' = \frac{\omega^2}{c^2} D' . \tag{8.77}$$

In this relation, c is the velocity of light in vacuum, given by $c^2 = (1/K\mu)$ (see p. 9).

Considering the case of a plane anisotropic medium, referred to a system $Oxyz$ of cartesian coordinates with the Oz axis coinciding with the direction of the light propagation, we obtain from (8.77) the system of equations

$$\frac{d^2 E_x}{dz^2} + \frac{\omega^2}{c^2} D_x = 0$$

$$\frac{d^2 E_y}{dz^2} + \frac{\omega^2}{c^2} D_y = 0, \tag{8.78}$$

where E_x, E_y and D_x, D_y are the components of the vectors E' and D' along the Ox and Oy axes, respectively.

If we introduce the material equations, expressed by (2.17) with $K_{ij} = K_{ji} \, (i,j = x, y)$ into (8.78), we obtain

$$\frac{d^2 E_x}{dz^2} + \frac{\omega^2}{c^2} (K_{xx} E_x + K_{xy} E_y) = 0$$

$$\frac{d^2 E_y}{dz^2} + \frac{\omega^2}{c^2} (K_{yx} E_x + K_{yy} E_y) = 0. \tag{8.79}$$

Equations (8.79) are the differential equations that govern the propagation of light in a two-dimensional anisotropic medium. These equations are referred to a fixed frame $Oxyz$ in space.

As was stated in Section 2.4, the dielectric tensor K_{ij} is symmetric; therefore it can be diagonalized when it is referred to its principal directions. If K_x and K_y are the principal dielectric coefficients and φ is the angle between the direction of K_x and the Ox axis, then we have the well-known tensorial relation

$$
\begin{pmatrix} K_{xx} & K_{xy} \\ K_{yx} & K_{yy} \end{pmatrix} = \begin{pmatrix} \cos\varphi & -\sin\varphi \\ \sin\varphi & \cos\varphi \end{pmatrix} \begin{pmatrix} K_x & 0 \\ 0 & K_y \end{pmatrix} \begin{pmatrix} \cos\varphi & \sin\varphi \\ -\sin\varphi & \cos\varphi \end{pmatrix}. \tag{8.80}
$$

Furthermore, the components E_x^P, E_y^P of the electric vector E' along the principal dielectric directions are

$$
\begin{pmatrix} E_x^P \\ E_y^P \end{pmatrix} = \begin{pmatrix} \cos\varphi & \sin\varphi \\ -\sin\varphi & \cos\varphi \end{pmatrix} \begin{pmatrix} E_x \\ E_y \end{pmatrix}. \tag{8.81}
$$

Substituting the values of the principal dielectric coefficients K_x and K_y and the values of the electric vector E' along the principal dielectric directions E_x^P, E_y^P given by (8.80) and (8.81), into (8.79), we obtain the differential equations that are satisfied by the electric vector, referred to the principal dielectric directions,

$$
\frac{d^2 E_x^P}{dz^2} + \left[\frac{\omega^2}{c^2} K_x - \left(\frac{d\varphi}{dz} \right)^2 \right] E_x^P - 2\frac{d\varphi}{dz}\frac{dE_y^P}{dz} - \frac{d^2\varphi}{dz^2} E_y^P = 0 ,
$$
$$
\frac{d^2 E_y^P}{dz^2} + \left[\frac{\omega^2}{c^2} K_y - \left(\frac{d\varphi}{dz} \right)^2 \right] E_y^P + 2\frac{d\varphi}{dz}\frac{dE_x^P}{dz} + \frac{d^2\varphi}{dz^2} E_x^P = 0 . \tag{8.82}
$$

We seek solutions of the system (8.79), of the form

$$
E_i = A_i(z)e^{-ikz} \quad (i = x,y) , \tag{8.83}
$$
$$
k = \frac{\omega}{c}\sqrt{K} ,
$$

where K is the dielectric coefficient in vacuum.

With the transformation defined by (8.83), the phase difference and the ratio of the components of the electric vector remain unchanged. Therefore, the light ellipse remains the same if, instead of the components of the electric vector E_i, the quantities $A_i(z)$ are introduced. Introducing the values of E_x and E_y from (8.83) into (8.79), we obtain the two differential equations that are satisfied by the quantities A_x and A_y,

$$
\frac{d^2 A_x}{dz^2} - 2ik\frac{dA_x}{dz} + \frac{\omega^2}{c^2}(K_{xx} - K)A_x + \frac{\omega^2}{c^2}K_{xy}A_y = 0 ,
$$
$$
\frac{d^2 A_y}{dz^2} - 2ik\frac{dA_y}{dz} + \frac{\omega^2}{c^2}K_{yx}A_x + \frac{\omega^2}{c^2}(K_{yy} - K)A_y = 0 . \tag{8.84}
$$

For the case of the usual photoelastic materials, the value of the optical anisotropy due to loading is of the order of $(K_{ii}-K)\sim 10^{-3}\div 10^{-4}$, so that the coefficients of the system (8.84) are of the order of $k\sim 10^5$, $(\omega^2/c^2)(K_{ii}-K)\sim 10^6\div 10^7$, and $(\omega^2/c^2)K_{ij}\sim 10^6\div 10^7$. Thus, the first terms (d^2A_i/dz^2) in (8.84) are insignificant when compared to the remaining terms; therefore, they can be omitted. Then, (8.84) take the form

$$\frac{dA_x}{dz}=-iC(K_{xx}-K)A_x-iCK_{xy}A_y,$$

$$\frac{dA_y}{dz}=-iCK_{yx}A_x-iC(K_{yy}-K)A_y,$$

(8.85)

with

$$C=\frac{\omega}{2c\sqrt{K}}.$$

If we refer (8.85) to the principal directions of the vector (A_x, A_y), whose components A_x^P, A_y^P along the principal directions are

$$\begin{pmatrix}A_x^P\\A_y^P\end{pmatrix}=\begin{pmatrix}\cos\varphi & \sin\varphi\\-\sin\varphi & \cos\varphi\end{pmatrix}\begin{pmatrix}A_x\\A_y\end{pmatrix},$$

(8.86)

we obtain the differential equations,

$$\frac{dA_x^P}{dz}\cos\varphi-A_x^P\frac{d\varphi}{dz}\sin\varphi-\frac{dA_y^P}{dz}\sin\varphi-A_y^P\frac{d\varphi}{dz}\cos\varphi$$

$$+iC(K_{xx}-K)(A_x^P\cos\varphi-A_y^P\sin\varphi)+iCK_{xy}(A_x^P\sin\varphi+A_y^P\cos\varphi)=0,$$

(8.87)

$$\frac{dA_x^P}{dz}\sin\varphi+A_x^P\frac{d\varphi}{dz}\cos\varphi+\frac{dA_y^P}{dz}\cos\varphi-A_y^P\frac{d\varphi}{dz}\sin\varphi$$

$$+iCK_{yx}(A_x^P\cos\varphi-A_y^P\sin\varphi)+iC(K_{yy}-K)(A_x^P\sin\varphi+A_y^P\cos\varphi)=0.$$

Multiplying the first of (8.87) by $\sin\varphi$ and the second equation by $\cos\varphi$ and subtracting the resulting two equations, and then multiplying the first of (8.87) by $\cos\varphi$ and the second equation by $\sin\varphi$ and adding the resulting two equations, we obtain, by taking into account (8.80), the two differential equations

$$\frac{dA_x^P}{dz}=-iC(K_x-K)A_x^P+\frac{d\varphi}{dz}A_y^P,$$

$$\frac{dA_y^P}{dz}=-\frac{d\varphi}{dz}A_x^P-iC(K_y-K)A_y^P.$$

(8.88)

If now we introduce the transformation

$$A_i = B_i(z) e^{-if(z)}, \qquad (i = x, y)$$

$$f(z) = \frac{1}{2} C \int (K_{xx} + K_{yy} - 2K) dz,$$

(8.89)

which alters neither the phase difference between the two components of the vector (A_x, A_y) nor their ratio, into (8.85), we obtain

$$\frac{dB_x}{dz} = -\frac{iC}{2}(K_{xx} - K_{yy})B_x - iCK_{xy}B_y,$$

$$\frac{dB_y}{dz} = -iCK_{yx}B_x + \frac{iC}{2}(K_{xx} - K_{yy})B_y.$$

(8.90)

Similarly, we obtain from (8.88), when the quantities B_x and B_y are referred to the principal directions,

$$\frac{dB_x^P}{dz} = -\frac{iC}{2}(K_x - K_y)B_x^P + \frac{d\varphi}{dz}B_y^P,$$

$$\frac{dB_y^P}{dz} = -\frac{d\varphi}{dz}B_x^P + \frac{iC}{2}(K_x - K_y)B_y^P.$$

(8.91)

Equations (8.90) and (8.91) are the differential equations of light propagation in an optically anisotropic plane medium. Although (8.90) are referred to a fixed cartesian frame, (8.91) are referred to a frame varying with the principal dielectric directions. When the medium into which the light is propagated is elastic, the dielectric coefficients K_{ij}, or their principal values K_i, and the accidental birefringence, introduced in the medium by the applied stress or strain, are linearly related to the corresponding stresses or strains. We now introduce into (8.90) and (8.91) the linearity condition, expressed by the equation,

$$K_{ij} = K\delta_{ij} + 2C_1\sqrt{K}\,\sigma_{ij} + 2C_2\sqrt{K}\sum_{k=1}^{3}\sigma_{kk}\delta_{ij},$$

(8.92)

where σ_{ij} is the stress tensor, δ_{ij} is the Kronecker symbol, and C_1 and C_2 are the photoelastic constants of the medium considered and we obtain from (8.90),

$$\frac{dB_x}{dz} = -\frac{iC_0}{2}(\sigma_{xx} - \sigma_{yy})B_x - iC_0\sigma_{xy}B_y,$$

$$\frac{dB_y}{dz} = -iC_0\sigma_{yx}B_x + \frac{iC_0}{2}(\sigma_{xx} - \sigma_{yy})B_y.$$

(8.93)

Similarly, from (8.91),

$$\frac{dB_x^P}{dz} = -i\frac{C_0}{2}(\sigma_1 - \sigma_2)B_x^P + \frac{d\varphi}{dz}B_y^P,$$

$$\frac{dB_y^P}{dz} = -\frac{d\varphi}{dz}B_x^P + i\frac{C_0}{2}(\sigma_1 - \sigma_2)B_y^P,$$

(8.94)

where $C_0 = 2CC_1\sqrt{K}$. Equations (8.93) and (8.94) are the differential equations of three-dimensional photoelasticity. They are referred either to a fixed cartesian system (8.93) or to a varying system with the principal-stress directions (8.94). These directions coincide with the principal dielectric directions, owing to the linearity conditions (8.92) between the dielectric and the stress tensors. Then, (8.94) is the same as (8.13).

We conclude, therefore, that the solution of the three-dimensional photoelastic problem developed by *Aben* by use of the electromagnetic equations of the light propagation in a plane anisotropic medium leads to exactly the same results as the Jones calculus. This was expected, because the Jones calculus is based on the results of the electromagnetic theory of light. Therefore both approaches lead to the same results.

Equations (8.93) and (8.94) are the most-general forms of the three-dimensional photoelastic equations. All other well-known photoelastic equations can be deduced from these equations as special cases. Thus, in Section 8.4, (8.94) were shown to lead to Neumann's equations of three-dimensional photoelasticity, and for the case of a constant ratio of the rotation of the principal directions to the corresponding retardation, we have derived in Section 8.5 the Drucker-Mindlin equations. For the case of constant principal directions $(d\varphi/dz = 0)$, we obtain, from (8.94),

$$\frac{dB_x^P}{dz} = -i\frac{C_0}{2}(\sigma_1 - \sigma_2)B_x^P,$$

$$\frac{dB_y^P}{dz} = i\frac{C_0}{2}(\sigma_1 - \sigma_2)B_y^P,$$

(8.95)

which by integration give

$$B_{x,y}^P = \Gamma e^{\mp i\frac{C_0}{2}\int(\sigma_1 - \sigma_2)dz},$$

(8.96)

with Γ the constant of integration.

From (8.96) it follows that the phase retardation δ is given by

$$\delta = C_0 \int(\sigma_1 - \sigma_2)dz,$$

(8.97)

and for the case of a constant difference of the principal stresses $(\sigma_1 - \sigma_2)$ along the light ray, we obtain

$$d\delta = C_0(\sigma_1 - \sigma_2)dz .\qquad(8.98)$$

Equation (8.98) is the well-known Wertheim photoelastic law.

The solution of the system of (8.93) or (8.94) can be represented in the form

$$\begin{pmatrix} B_x \\ B_y \end{pmatrix} = \begin{pmatrix} u_{11} & u_{12} \\ u_{21} & u_{22} \end{pmatrix} \begin{pmatrix} B_x \\ B_y \end{pmatrix}_0 ,\qquad(8.99)$$

where $u_{ij}(i,j = 1,2)$ represents a unitary matrix.

Equation (8.99) leads to the result that *any three-dimensional photoelastic model can be represented by a two-by-two complex matrix.* This result, for any optical element and for the most-general case of a three-dimensional photoelastic model, was proved by use of the Jones calculus.

It is, therefore, concluded that the equations of three-dimensional photoelasticity can be derived by two distinct procedures, that is, one that uses the electromagnetic equations of light propagation and an other that uses the Jones calculus. If we begin with the electromagnetic equations, we are led to the result that polarization-optics problems can be handled by the Jones calculus, whereas by starting with the Jones calculus we get the equations of three-dimensional photoelasticity. *Aben* [8.16, 23] used the electromagnetic equations to obtain the equations of three-dimensional photoelasticity, whereas use of the Jones calculus to obtain the same equations was developed in Section 8.3 of this chapter.

When the photoelastic model is placed in a magnetic field, the Faraday effect rotates the principal stress directions through an angle ψ. In this case, the total rotation of the principal stress directions is $(\varphi - \psi)$; therefore (8.94) take the form

$$\frac{dB_x^P}{dz} = -i\frac{C_0}{2}(\sigma_1 - \sigma_2)B_x^P + \frac{d(\varphi - \psi)}{dz}B_y^P ,$$
$$\frac{dB_y^P}{dz} = -\frac{d(\varphi - \psi)}{dz}B_x^P + i\frac{C_0}{2}(\sigma_1 - \sigma_2)B_y^P .\qquad(8.100)$$

Equations (8.100) are the general equations of magnetophotoelasticity; they were derived for the first time and used by *Aben* in a series of publications [8.19–23].

References

8.1 M. M. Frocht: *Photoelasticity* (Wiley and Sons, New York) Vol. I, pp. 334–341 (1941) Vol. II, pp. 364–465 (1948)

8.2 A. J. Durelli, W. F. Riley: *Introduction to Photomechanics* (Prentice-Hall, Englewood Cliffs, N. J. 1965) pp. 173–179, 254–291

8.3 M. Hetényi: *Handbook of Experimental Stress Analysis* (Wiley and Sons, New York 1950) pp. 933–959

8.4 R. B. Heywood: *Photoelasticity for Designers* (Pergamon Press, Oxford 1969) pp. 209–276

8.5 A. Kuske, G. Robertson: *Photoelastic Stress Analysis* (Wiley and Sons, New York 1974) pp. 329–390

8.6 F. E. Neumann: Abh. der Königl. Akad. der Wissensch. Berlin, Part II, 1 (1841)

8.7 E. G. Coker, L. N. G. Filon: *A Treatise on Photoelasticity* (Cambridge Univ. Press, London 1957) pp. 253–258

8.8 R. D. Mindlin, L. E. Goodman: J. Appl. Phys. **20**, 89 (1949)

8.9 D. C. Drucker, R. D. Mindlin: J. Appl. Phys. **11**, 724 (1940)

8.10 L. S. Srinath, A. V. S. S. S. R. Sarma: Exp. Mech. **14**, 118 (1974)

8.11 A. V. S. S. S. R. Sarma, L. S. Srinath: J. Aeron. Soc. Ind. **24**, 300 (1972)

8.12 L. S. Srinath, A. V. S. S. S. R. Sarma: J. Phys. D: Appl. Phys. **5**, 883 (1972)

8.13 L. S. Srinath, S. K. Bhave: Exp. Mech. **14**, 367 (1974)

8.14 P. S. Theocaris, E. E. Gdoutos: Proc. 6th Int. Conf. Exp. Stress Anal. (München, Germany 1978) pp. 599–605

8.15 C. Mylonas, G. M. Brown: Brown Univ. Rpt. NSF G–20259/2, ARPA E (1967)

8.16 H. K. Aben: Exp. Mech. **6**, 13 (1966)

8.17 H. K. Aben: Exp. Mech. **9**, 25 (1969)

8.18 H. K. Aben: J. Text. Inst. **59**, 523 (1968)

8.19 H. K. Aben: Exp. Mech. **10**, 97 (1970)

8.20 H. K. Aben: Proc. 4th Int. Conf. Exp. Stress Anal., Cambridge, 1970, ed. by M. L. Meyer (The Institution of Mechanical Engineers, London 1971) pp. 175–182

8.21 H. K. Aben, S. Idnurm: Proc. 5th Int. Conf. Exp. Stress Anal., Udine, 1974, ed. by G. Bartolozzi (Tecnoprint-Pitagora, Bologna 1974) pap. 37, pp. 4.5–4.10

8.22 H. K. Aben: Exp. Mech. **10**, 97 (1970)

8.23 H. K. Aben, S. J. Idnurm, E. I. Klabunovskii, M. M. Uffert: Exp. Mech. **14**, 361 (1974)

8.24 E. L. Ince: *Ordinary Differential Equations* (Dover, New York 1956)

9. Scattered-Light Photoelasticity

9.1 Introduction

By the method of scattered-light photoelasticity, two- and three-dimensional states of stress are investigated by using the phenomenon of scattering of light from a photoelastic medium, when it is subjected to a system of applied loads. The phenomenon of scattering of light was first established for the case of cloudy media, where it is more pronounced, by the pioneering works of Lord Rayleigh in the turn of the twentieth century [9.1]. Later on, it was observed that light is scattered from every transparent and homogeneous medium, that is, when a light beam passes through such a medium it is dispersed from every point along its path and in a direction perpendicular to the direction of the light propagation.

The phenomenon of scattering of light is attributed to the absorption of the light energy by the particles of the medium, which subsequently are put to a vibrating motion and re-emit the incident light. When the incident light is unpolarized, that is, when it does not have any directional property, the intensity of the scattered light is the same for all directions of observations in the plane normal to the incident light. However, when some order is introduced in the light vector and the light becomes polarized, the intensity of the scattered light varies with the direction of observation. In this case, the scattered-light intensity along a given direction of observation is proportional to the square of the component of the amplitude of the light vector in the direction normal to the direction of observation.

Let us now consider the case when a linearly polarized light beam is incident on a photoelastic model subjected to a system of applied loads. Then, according to the photoelastic phenomenon described in Chapter 6, the incident light at each point of the model is resolved along two principal-stress directions of the state of stress in the plane normal to the direction of light propagation. Between these two linearly polarized light components along the principal-stress directions, a phase retardation proportional to the difference of the principal stresses of the photoelastic medium is introduced, which varies from point to point along the light path. When a given direction of observation is chosen in the plane normal to the incident light, the observed light intensity is proportional to the square of the sum of the components of the two linearly polarized light vectors that lie in a transverse plane in the direction perpendicular to the direction of observation.

The phase difference between the components of the original linearly-polarized light rays along the direction normal to the direction of observation is equal to the phase difference between the original linearly polarized light rays along the principal-stress directions. Because this phase difference varies from point to point along the light path, an interference pattern will be observed when the photoelastic medium is viewed from a direction of observation that is in the plane normal to the incident light.

Therefore, the application of the phenomenon of scattering of light to the case of a photoelastic medium gives the same result as the conventional plane polariscope when the pass axis of the analyzer coincides with the direction normal to the direction of observation in the plane transverse to the incident light beam. Thus, scattering of light may serve as a substitute for the analyzer, when linearly polarized light is used. In the same manner, when the incident light beam is unpolarized and scattered light from a photoelastic medium is observed through an analyzer, the result is similar to that obtained with a plane polariscope, with scattering substituted for the polarizer of the plane polariscope.

The scattered light obtained when ordinary light passes through a transparent isotropic medium is not completely polarized. Some degree of depolarization is always introduced by scattering; thus the light is always partially polarized. The unpolarized light results in reduction of contrast of the observed optical patterns. This and the fact that the intensity of the scattered light is only a small fraction of the intensity of the incident light necessitate use of high-power light sources in scattered-light photoelasticity.

The idea of using light scattering in photoelastic investigation of two- and three-dimensional problems was first suggested by *Weller* [9.2] in 1939. *Weller* [9.3, 4] also developed the theory and the experimental details of the method and applied it to the elementary cases of bars subjected to tension combined with torsion and bending. By this method *Weller* was able to obtain the principal-stress directions and the difference of the principal stresses in a two- or three-dimensional photoelastic model. *Menges* [9.5] used the method in tensile tests and proposed a graphical method for determining the relevant photoelastic quantities by light scattering. *Drucker* and *Frocht* [9.6] studied the case of torsion by scattered light and proved that when the incident light is linearly polarized and normal to the axis of the shaft, the scattered pattern, as seen by observation along the direction of the axis, is equivalent to the contour pattern of the corresponding membrane analogy. Other early works in scattered-light photoelasticity are listed in [9.7–9].

After the above-mentioned pioneering works, the method of scattered light photoelasticity was further developed, elaborated and applied to the solution of three-dimensional problems by *Frocht* and *Srinath* [9.10–12]. Thus, in [9.10, 12] the scattering of light was used to provide the necessary data for the application of the shear-difference method in determining the values of the individual stress components in three-dimensional photoelasticity. The scattered-light method has the advantage over the well-known frozen-stress and slicing technique that it is nondestructive and, therefore, all of the disadvantages that result from locking-in stresses in the three-dimensional photoelastic model are removed. Thus, in addition to the fact that the model does not need to be cut, the laws of transition from model to prototype are more appropriately applied in the scattered-light method, in which the model is tested at room temperature. Therefore, Poisson's ratio is approximately equal to that of the prototype, whereas in the frozen-stress method, because of the heating of the model, the value of Poisson's ratio approaches 0.5, that is the value for all rubbery polymers, which therefore differs greatly from Poisson's ratio for the prototype. *Srinath* and *Frocht* [9.11] described in detail the design and construction of a special polariscope for application of the scattered light method to photoelasticity. In [9.11, 12] they presented the general principles of the method, as well as the methods for determination of the principal-stress directions and the difference of the principal stresses. They applied the method to the case of a sphere subjected to diametral compression; their results were close to the theoretical predictions.

Cheng, in a series of publications [9.13–15], proposed some new techniques for the determination of the principal-stress directions and the difference of principal stresses in scattered-light photoelasticity. In [9.13], he suggested that the principal-stress directions can be determined by setting the pass axis of the analyzer parallel to the direction of observation and rotating the model about the incident light ray, until the scattered intensity becomes minimum. In the same paper, *Cheng* applied the idea of the Senarmont compensating method for determining the fringe order in scattered-light photoelasticity. He also presented, in [9.14], a method for determining the relevant photoelastic parameters in scattered light by simultaneously recording the scattered-light intensities along two directions at an angle of 45 degrees from each other in the plane normal to the incident beam. This method was automated in [9.15]. *Srinath* [9.16] discussed the limitations and inaccuracies involved in the various existing methods for determining the directions and the difference of principal stresses when the principal directions rotate along the light path. The effect of the rotation of the principal-stress directions in scattered-light techniques was also considered by *Aderholdt* et al. [9.17].

The feasibility of the scattered-light method for studying two-dimensional stress states, after the above-mentioned pioneering investigations, was demonstrated by *Shelson* and *Smith* [9.18]. Also *Hemann* and *Becherer* [9.19] presented a method for determining the state of stress in two-dimensional problems from one full-field photograph of the optical pattern, obtained by scattered light.

This brief review of the most pertinent works in scattered-light photoelasticity shows that, although this method is well established in the case when the directions of the principal stresses along the light path are constant, many difficulties arise in the case when the principal axes rotate. In this case, the general stress-optic law, incorporating the rotation of the principal axes developed in Section 8.4, must be used. This makes difficult the analysis and interpretation of scattered-light patterns. This and the special instrumentation needed in the experimental application of the method were the reasons why scattered-light photoelasticity has not progressed much up to now.

However, formulation and interpretation of the optical patterns in scattered-light photoelasticity when the principal axes rotate along the light path are greatly assisted by introduction of modern methods of handling polarization optics, based on the Poincaré sphere and the Mueller and Jones calculi. *Robert* [9.20–23] used the Poincaré-sphere representation and the equivalence theorem of three-dimensional photoelasticity to determine the relevant photoelastic parameters in scattered light. He used polarization by scattering instead of a polarizer and described a method for determining the directions and the difference of the principal stresses in an elementary strip at the interior of a three-dimensional model, by illuminating the ends of the strip by two unpolarized light beams and determining the state of polarization of the corresponding outcoming scattered light. A mathematical formulation of this idea, based on Mueller calculus, for the case when polarization by scattering is used instead of an analyzer, was provided by *Gross-Petersen* [9.24]. *Cernosek* [9.25] solved the general problem of scattered-light photoelasticity by using the equivalence theorem and the matrix formulation of the three-dimensional photoelastic problem, based on the electromagnetic analysis developed by *Aben* (Sec. 8.10).

In the present chapter, the problem of scattered-light photoelasticity will be studied by use of two different approaches: the conventional approach, based on the vectorial representation of the state of polarization for the case when the principal axes do not rotate along the light path, and the new approach, based on modern methods for handling polarization optics for the more general case, when the principal axes rotate along the light path.

9.2 Polarization of Light by Scattering

When a light beam passes through a transparent isotropic medium it is scattered from every particle along its path. When the dimensions of the scattering particle in the medium are of the same order of magnitude as the wavelength of the light used, the scattered light observed from any given direction in the plane normal to the incident light is plane polarized. When the incident light is unpolarized, that is, when it does not present different behavior in any transverse direction, then the light scattered exhibits the same properties in all directions. Therefore, when the incident light is unpolarized, the intensity of the scattered light is the same in all directions of observation in the plane normal to the primary light beam. However, when the incident light is polarized, that is when it exhibits properties dependent on direction, the intensity of the scattered light will not be the same in all directions of observation. The intensity of the scattered light will be the same as that passed through a linear analyzer whose pass axis is perpendicular to the direction of observation. If the light incident at any point in the transparent medium is elliptically polarized, its Stokes vector S, is expressed (as in Table 3.1) by,

$$
S = \begin{bmatrix} A_x^2 + A_y^2 \\ A_x^2 - A_y^2 \\ 2A_x A_y \cos \delta \\ 2A_x A_y \sin \delta \end{bmatrix},
\tag{9.1}
$$

where A_x and A_y are the amplitudes and δ the phase difference of the two linearly polarized light beams along two perpendicular directions into which the elliptically polarized light is resolved.

When the observation direction makes an angle φ with the Ox axis, polarization by scattering acts as a linear analyzer with its pass axis inclined at an angle $(\varphi + 90°)$ to the Ox axis. The Mueller matrix $P_{(\varphi + 90°)}$ of this analyzer is given (as in Table 4.3) by

$$
P_{(\varphi + 90°)} = \frac{1}{2} \begin{bmatrix} 1 & -\cos 2\varphi & -\sin 2\varphi & 0 \\ -\cos 2\varphi & \cos^2 2\varphi & \sin 2\varphi \cos 2\varphi & 0 \\ -\sin 2\varphi & \sin 2\varphi \cos 2\varphi & \sin^2 2\varphi & 0 \\ 0 & 0 & 0 & 0 \end{bmatrix}.
\tag{9.2}
$$

Therefore, the Stokes vector S' of the light beam that passes through the analyzer is

$$
S' = P_{(\varphi + 90°)} S.
\tag{9.3}
$$

Substituting the values of S and $P_{(\varphi + 90°)}$ from (9.1) and (9.2) into (9.3) we obtain for the intensity I'_φ of the light scattered along the observation direction that subtends angle φ with the Ox axis, which is represented by the first element of the Stokes vector S',

$$
I'_\varphi = (A_x^2 + A_y^2) - (A_x^2 - A_y^2)\cos 2\varphi - 2A_x A_y \cos \delta \sin 2\varphi.
\tag{9.4}
$$

If we introduce the quantity θ into (9.4), such that

$$
\tan \theta = \frac{A_y}{A_x},
$$

we have

$$
I'_\varphi = I_0(1 - \cos 2\theta \cos 2\varphi - \sin 2\theta \sin 2\varphi \cos \delta),
\tag{9.5}
$$

where I_0 is the intensity of the incident beam,

$$I_0 = A_x^2 + A_y^2 \ .$$

To obtain the scattered-light intensity I_φ, (9.5) must be multiplied by a proportionality factor k that depends upon the scattering coefficient of the medium and the distance travelled by the scattered light from the scattering point to the observer. Thus,

$$I_\varphi = k I_0 (1 - \cos 2\theta \cos 2\varphi - \sin 2\theta \sin 2\varphi \cos \delta) \ . \tag{9.6}$$

9.3 Polarization by Scattering in a Photoelastic Medium

Because of the photoelastic phenomenon, when a beam impinges on a stressed photoelastic medium the light is resolved into two components linearly polarized along the principal-stress directions. As the beam travels through the medium, a phase retardation is introduced between the two linearly polarized components, so that the light at every point in the photoelastic medium is elliptically polarized. Consider the case when a linearly polarized beam, with a direction of polarization that subtends an angle θ with the Ox axis, is incident on the photoelastic medium and that the phase difference introduced by the medium is δ. Then, the light at each point through the thickness of the medium is elliptically polarized and the intensity scattered in any direction inclined at angle φ with respect to the Ox axis can be expressed by (9.6).

Thus, in the case of a photoelastic medium whose principal axes are along the Ox and Oy axes, when a linearly polarized beam is incident on the medium at an angle θ with respect to the Ox axis, the intensity of the light scattered in a direction that subtends angle φ with the Ox axis, at a point in the medium where the phase difference is δ, is given by (9.6). From this it is concluded that for the general case of a three-dimensional photoelastic medium whose principal-stress directions and phase difference vary from point to point along the light path and whose scattered-light pattern is observed at an angle φ with respect to one of the varying principal-stress directions, the quantities θ, φ, and δ vary. Therefore, the light intensity I will vary along the light path. Thus, an interference-fringe pattern will be formed in any given observation direction.

Let us now consider the case when a right-circularly polarized light beam with Stokes vector S, expressed (as in Table 3.1) by

$$S = \begin{bmatrix} A^2 \\ 0 \\ 0 \\ A^2 \end{bmatrix}, \tag{9.7}$$

is incident on the specimen. When the principal-stress axes of the specimen are along the Ox and Oy axes, the Mueller matrix $R_0(\delta)$ of the specimen will be given (as in Table 4.4) by

$$R_0(\delta) = \begin{bmatrix} 1 & 0 & 0 & 0 \\ 0 & 1 & 0 & 0 \\ 0 & 0 & \cos\delta & \sin\delta \\ 0 & 0 & -\sin\delta & \cos\delta \end{bmatrix}, \tag{9.8}$$

where δ is the phase difference introduced by passage of light through a finite thickness of the specimen.

If the principal-stress directions are constant throughout a finite layer of the specimen, the Stokes vector S' at the exit surface of this layer will be

$$S' = R_0(\delta)S$$

or

$$S' = A^2 \begin{bmatrix} 1 \\ 0 \\ \sin\delta \\ \cos\delta \end{bmatrix}. \tag{9.9}$$

If the observation direction is at an angle φ with the Ox axis, polarization by scattering in the photoelastic medium acts as an analyzer whose pass axis is at $(\varphi + 90°)$ with respect to the Ox axis, whose Mueller matrix is given by (9.2). Thus, we obtain for the Stokes vector S'' of the scattered light

$$S'' = P_{(\varphi + 90°)}S'. \tag{9.10}$$

From (9.2, 9), and (9.10) it follows that the scattered-light intensity I will be

$$I = kA^2(1 - \sin 2\varphi \sin\delta). \tag{9.11}$$

If left-circularly polarized light is used, (9.11) takes the form

$$I = kA^2(1 + \sin 2\varphi \sin\delta). \tag{9.12}$$

Either (9.11) or (9.12) expresses the intensity of the light scattered from a photoelastic medium that introduces a phase retardation δ, when it is illuminated by right- or left-circularly polarized light, respectively, and the observation direction makes an angle φ with respect to the Ox axis. From either of these relations it follows that as δ varies along the light path, I also varies. Therefore, a scattered-light fringe pattern is formed.

9.4 Interpretation of Scattered-Light Fringe Patterns

9.4.1 General Considerations

Equations (9.6) and (9.11) or (9.12) represent the intensity of the light scattered from a point in the photoelastic medium in which a phase retardation δ is produced between the two components, linearly polarized along the principal-stress directions, that subtend an angle φ with the observation direction when the incident light is either linearly or circularly polarized. Because of the variations of the phase retardation δ, of the inclination φ of the principal stresses with respect to the observation direction, as well as of the angle θ between the principal-stress directions and the direction of polarization of the incident polarized light along the light path, the intensity I varies. Therefore a fringe pattern is formed by the scattered light. This scattered-light fringe pattern contains enough information to permit determination of the orientation of the principal stresses and the value of the stress-optical retardation δ.

If we consider the more-general case of a three-dimensional photoelastic model, in which the principal-stress directions vary from point to point along the light path, the stress-optic law, as developed in Section 8.4, is expressed by the two Maxwell-Neumann equations:

$$d\delta = C(\sigma_1 - \sigma_2)ds + 2\cot 2\theta \sin \delta \, d\varphi$$
$$d\vartheta = -\cos \delta \, d\varphi,$$
(9.13)

where the stress-optical retardation $d\delta$ that corresponds to a distance ds along the light path is partly due to the principal-stress difference $(\sigma_1 - \sigma_2)$ and partly to the rotation $d\varphi$ of the principal-stress directions. In (9.13) θ is equal to $\tan^{-1}(A_y/A_x)$ with A_y and A_x the components of the incident light vector along two orthogonal directions.

To obtain the orientations and the difference of the principal stresses from the scattered-light pattern, the system of differential equations (9.13) must be solved along the light path. However, many difficulties are encountered in the solution of this system. This makes difficult the interpretation of the fringe patterns in scattered light for the general case in which the principal axes rotate along the light path. For this reason, the case of variable stress directions will be treated by the modern methods of handling polarization optics that are based on the Poincaré sphere and the Stokes and Jones calculi, which greatly simplify the solution of the problem.

9.4.2 Constant Principal-Stress Directions

Consider the case in which the principal-stress directions remain constant or vary only slightly along the light path. From the first of (9.13) with $d\varphi=0$, we obtained

$$(\sigma_1 - \sigma_2) = \frac{1}{C}\frac{d\delta}{ds}. \tag{9.14}$$

Equation (9.14) indicates that the difference of the principal stresses $(\sigma_1 - \sigma_2)$ in the scattered-light fringe pattern is proportional to the gradient of the fringes along the direction of the incident light ray. Therefore, in the scattered-light pattern, the orders of the fringes are insignificant; only their spacing enters in the calculation of $(\sigma_1 - \sigma_2)$.

For this case of constant principal-stress directions along the light path, if the model is illuminated by linearly polarized light, the angles θ and φ, entering in (9.6) for the scattered-light intensity I_φ, are constant. Therefore, I_φ varies only because of variation of the phase retardation δ along the light path. Relation (9.6) shows that I_φ becomes maximum at points where

$$\delta = (2n-1)\pi \quad (n=0,1,2,\dots),$$

and the corresponding value of I_φ is

$$I_\varphi^{max} = 2kI_0 \sin^2(\theta + \varphi). \tag{9.15}$$

Similarly, I_φ becomes minimum when

$$\delta = 2n\pi \quad (n=0,1,2,\dots);$$

its value is then

$$I_\varphi^{min} = 2kI_0 \sin^2(\theta - \varphi). \tag{9.16}$$

The contrast of fringes in the fringe pattern can then be expressed by the difference $(I_\varphi^{max} - I_\varphi^{min})$ of the maximum and minimum values of the scattered-light intensities. From (9.15) and (9.16) we obtain

$$(I_\varphi^{max} - I_\varphi^{min}) = 2kI_0 \sin 2\theta \sin 2\varphi. \tag{9.17}$$

Equation (9.17) shows that the maximum value of contrast of the scattered-light fringe pattern occurs when $\sin 2\theta = \sin 2\varphi = 1$, that is, when both the directions of polarization and observation are inclined at 45° with the principal-stress directions.

From (9.15) and (9.16) it follows that the maximum scattered intensity is equal to $(2kI_0)$, whereas the minimum intensity is zero. When $\theta = \varphi = \pm 45°$, the minimum intensity is observed with $\delta = 2n\pi$, whereas when $\theta = -\varphi = \pm 45°$ the minimum intensity is observed at points where $\delta = (2n-1)\pi$. Therefore, when the directions of polarization and observation coincide, the entrance point is dark (zero fringe order), whereas when these two directions are perpendicular, the entrance point is white (half fringe order). From (9.17) it also follows that, when either of the directions of polarization or observation is constant, the contrast of the pattern is maximum when the other direction is inclined at 45° with respect to the principal-stress directions.

When left-circularly polarized light is used, the intensity I_φ is independent of ϑ; it is maximum when $\varphi = 45°$ and $\delta = (4n+1)\pi/2$ and minimum (zero) when $\varphi = 45°$ and $\delta = (4n+3)\pi/2$. When right-circularly polarized light is used, the above conditions, for maximum or minimum scattered light intensity are reversed.

For the case when the principal-stress directions rotate along the light path, the fringe pattern will be interpreted and the corresponding photoelastic parameters will be determined in a subsequent section of this chapter by applying modern methods of representing the state of polarization of an elliptically polarized light beam.

9.5 Determination of the Principal-Stress Directions

When the principal-stress directions do not vary along the path of the primary light, their inclination can be determined by using the scattered-light intensity formulas (9.6) or (9.11) and (9.12) for the cases when the incident light is either linearly or circularly polarized, respectively. Indeed, the angles formed either by the directions of the principal stresses and the direction of observation, for the case of the circularly polarized light, or by the directions of principal stresses and of both the incident light and observation, for the case of the linearly polarized light, appear in (9.6, 11, 12). Therefore, the scattered-light-intensity relations make possible the determination of the principal-stress directions from measurements of the intensity of the scattered light. Based on the scattered-light-intensity formulas, the following methods have been developed for determination of the principal-stress directions.

I) *The constant-intensity method.* This method is based on the fact that, when the observation direction coincides with the direction of one of the principal stresses, for the cases of either linearly- or circularly-polarized incident light, the scattered-light intensity is independent of the phase retardation δ and is constant. Indeed, (9.6), for the case of linearly polarized incident light, whose direction of polarization subtends an angle θ with the direction of one of the principal stresses, yields for $\varphi = 0°$ or 90° that

$$I = kI_0(1 \mp \cos 2\theta).   \tag{9.18}$$

The minus sign corresponds to $\varphi = 0°$ and the plus sign to $\varphi = 90°$. Further-more, from either one of (9.11) and (9.12) for the case of circularly polarized incident light it is valid that

$$I = kI_0.$$ (9.19)

From (9.18) it follows that when the polarization and observation direc-tions coincide with the direction of either of principal stresses ($\varphi = \theta = 0°$ or $90°$) then the constant light intensity becomes minimum ($I = 0$).

From the foregoing facts the following procedures are suggested for de-termining the principal stress directions:

a) Determine the scattered-light intensities for various directions of ob-servation and specify the observation direction for which the intensity of the scattered light is constant. This observation direction coincides with the direction of one of the principal stresses. Great accuracy of determination of the principal-stress directions with this method can be obtained by varying the phase retardation δ by means of a compensator for various observation directions and detecting by a photomultiplier the intensity of the scattered light. The direction of observation coincides with the direction of one of the principal stresses when the detected light intensity is independent of the phase retardation.

b) Make the observation direction to coincide with the direction of po-larization of the incident linearly polarized light and detect the scattered-light intensity by rotating the model about the axis of the incident light. Then, when the scattered-light intensity becomes minimum, the identical directions of polarization and observation coincide with the direction of one of the principal stresses.

II) *The zero-intensity or line-ellipse method.* This method is based on the fact that when, at a given point, the stress-optical retardation δ is equal to $2n\pi$ or $(2n-1)\pi$ the state of polarization at that point is linear and its direc-tion is inclined at angles θ and $(-\theta)$, respectively, with the direction of one of the principal stresses (coinciding with the Ox axis). Indeed, as deduced from (9.1), the Stokes vector S of the light beam at a point of the photoelastic model where $\delta = 2n\pi$ or $(2n-1)\pi$ is given by

$$S = \begin{bmatrix} 1 \\ \cos 2\theta \\ \pm \sin 2\theta \\ 0 \end{bmatrix},$$ (9.20)

which represents linearly polarized light with directions of polarization in-clined at θ and $(-\theta)$ with respect to the Ox axis. Therefore, when an arbitrary direction of observation is chosen and the phase retardation at the point in question is varied by means of a compensator, the scattered-light intensity attains its minimum or maximum values when $\delta = 2n\pi$ or $(2n-1)\pi$. Furthermore, when in such a situation, the observation direction coincides with the direction of one of the principal stresses, the scattered-light intensity becomes minimum. Thus, for determination of the principal-stress directions by this method:

a) We choose an arbitrary observation direction at the point considered and vary the optical retardation by means of a compensator until the scattered-light intensity passes from its maximum to its minimum value.

b) At these two positions of the compensator we vary the observation direction until the scattered-light intensity again becomes minimum. These two observation directions are equally inclined with the direction of one of the principal stresses, which, therefore, is coincident with their bisector.

III) *The intensity-variation method.* This method is based on the measurement of the variations of the scattered-light intensities along the primary beam for two observation directions that make an angle of 45°. From (9.6), when the principal-stress directions are constant along the path of the primary beam (θ, φ constant), we obtain, by differentiating with respect to the variable s along the primary beam,

$$\frac{dI_1}{ds} = kI_0 \sin 2\theta \sin 2\varphi \sin \delta \frac{d\delta}{ds}. \tag{9.21}$$

For another observation direction at $(\varphi + 45°)$ we obtain from this relation

$$\frac{dI_2}{ds} = kI_0 \sin 2\theta \cos 2\varphi \sin \delta \frac{d\delta}{ds}. \tag{9.22}$$

From (9.21) and (9.22), we obtain

$$\tan 2\varphi = \frac{(dI_1/ds)}{(dI_2/ds)} = \frac{(I_1)_{s1} - (I_1)_{s2}}{(I_2)_{s1} - (I_2)_{s2}}. \tag{9.23}$$

From (9.23) the angle φ between the observation direction and the direction of one of the two principal stresses is determined by measuring the scattered-light intensities $(I_1)_{s1}$, $(I_1)_{s2}$, $(I_2)_{s1}$, and $(I_2)_{s2}$ along two observation directions that are at 45° with respect to each other, for two positions along the light path.

IV) *The dual-observation method.* This method is based on measurement of the variations of the scattered-light intensities due to the loading of the photoelastic model along two observation directions that are at 45° to each other; it uses circularly polarized light. Thus, for the case of left-circularly polarized light, when the observation directions form angles equal to φ and

$(\varphi + 45°)$, respectively with the direction of one of the principal stresses, (9.12) yields

$$I_1 = kI_0(1 + \sin 2\varphi \sin \delta)$$
$$I_2 = kI_0(1 + \cos 2\varphi \sin \delta).$$

(9.24)

From these relations, we obtain

$$\tan 2\varphi = \frac{I_1 - kI_0}{I_2 - kI_0}$$

(9.25)

and

$$\sin \delta = \frac{1}{(kI_0)}[(I_1 - kI_0)^2 + (I_2 - kI_0)^2]^{1/2}.$$

(9.26)

From (9.25), the angle φ between the observation direction and the direction of one of the principal stresses can be determined by measuring the quantities I_1, I_2, and kI_0.

All of the foregoing methods for measuring the principal-stress directions of a photoelastic model are strictly valid only when the principal-stress directions do not rotate along the path of the primary beam. However, when the rotation of the principal stresses is small, these methods are still valid for determining the principal-stress directions. The errors introduced when the foregoing methods are used for determining the principal directions in the case when the principal stresses rotate along the light path were determined and compared by *Srinath* [9.16]. He concluded that the best method, when the principal stresses rotate, is the zero-intensity or line-ellipse method.

9.6 Determination of the Stress-Optical Retardation

The stress-optical retardation at the point where the incident light enters a photoelastic model is zero. Therefore, as can be shown from (9.6), the light intensity at this point is minimum when the value of the angle φ, which defines the observation direction, is less than 90°. When this condition is satisfied, the point of entrance is dark, although not necessarily black. The light intensity is zero when the direction of observation coincides with the pass axis of the polarizer. For the case in which the angle φ is greater than 90° the point of entrance is bright. The maximum brightness is obtained when the direction of observation is perpendicular to the polarization direction of the polarizer. Therefore, when the angle φ is less than 90°, the dark fringes in the scattered-light pattern represent points where the stress-optical retardation δ contains an integral number of wavelengths; when φ is greater than 90°, the dark fringes in the pattern represent points where δ contains an odd multiple of a half wavelength.

Therefore, when the polarizer and the observation direction are in two perpendicular positions, the points where the stress-optical retardation is either an integral multiple of a wavelength or an odd multiple of a half wavelength are obtained. When the value of the stress-optical retardation at an arbitrary point is to be determined, the following methods may be used:

I) *The compensator method.* The principle of this method is the same as is used for measuring the retardation of a model by transmission light, developed in "Compensators" (see p. 98). The compensator is placed in the path of the primary beam with its axes parallel to the principal-stress directions of the photoelastic model and is adjusted so as to make the nearest fringe at the point in question pass through this point. The amount of the additional retardation, introduced by the compensator, gives the fractional fringe order at the given point. Both Babinet and Babinet-Soleil compensators can be used in the same manner, as developed in "Compensators" for measuring the retardation of a two-dimensional photoelastic model.

II) *The Senarmont method.* The principle of this method is the same as was used in two-dimensional photoelasticity developed in Section 7.4. This principle may be summarized, by using the reversibility principle of optics, as follows: a linearly polarized beam, with a direction of polarization that makes an angle $\theta = (\delta/2)$ with the fast axis of a quarter-wave plate, is incident on this plate. This light ray impinges successively on the given point of the model arranged with its principal axes at 45° with respect to the axes of the quarter-wave plate. Then the emerging light is linearly polarized with its direction of polarization parallel to the fast axis of the quarter-wave plate. Therefore, when the observation direction coincides with the direction of the emerging light, the scattered-light intensity will be minimum.

In accordance with this principle, the axis of polarization of the incident light is rotated until the scattered light intensity at a direction of observation parallel to the fast axis of the quarter-wave plate becomes minimum. Then, the stress-optical retardation at the point considered is equal to half of the angle between the axis of the polarizer and the fast axis of the quarter-wave plate. The sign of the fractional retardation is determined by ascertaining whether it is the fringe with the lower, or the higher order that moves toward the point considered, when the polarizer is rotated.

III) *The dual-observation method.* In accordance with the dual-observation method developed in the previous section, the variations of the scattered-light intensities due to loading the photoelastic model in two observation directions that are at 45° with respect to each other are measured. Then the stress-optical retardation is determined from (9.26).

The errors of the foregoing methods, when they are used in general three-dimensional stress fields, where the principal stress directions rotate along the primary beam, were analyzed by *Srinath* [9.16]. He concluded that the compensation and Senarmont methods are more accurate than the dual-observation method, when there is appreciable rotation of the principal-stress directions.

9.7 The General Three-Dimensional Photoelastic Medium

9.7.1 General Concepts

For the more-general case of a three-dimensional photoelastic medium in which the rotation of the secondary principal stresses, defined as the principal-stresses in the plane normal to the light path, is not negligible. In this case the previously outlined procedures cannot be used for determination of the amount of birefringence and the directions of the principal birefringence axes, and consequently the state of stress in the photoelastic medium. This case can be handled by using the powerful methods of treating polarization optics that are based on the Poincaré sphere and the Mueller and Jones calculi. Two different approaches for the solution of the problem can be developed:

I) In the first approach, the photoelastic medium is illuminated with polarized light, and polarization by scattering in the model acts as an analyzer. In this case, the scattered-light intensity is related to the state of polarization of the incident light, so that the state of polarization can be determined from the intensity of the scattered light.

II) In the second approach, the photoelastic medium is illuminated with unpolarized light, and polarization by scattering in the medium acts as a polarizer. In this case, the state of polarization of the scattered light is determined and is related to the relevant photoelastic parameters along the scattering directions.

The equivalence theorem of three-dimensional photoelasticity, developed in Section 8.2, is of great help in finding the solution to the problem using either approach. According to this theorem, any three-dimensional photoelastic medium is optically equivalent to a system of two optical elements, a linear retarder and a pure rotator. By using this theorem, we shall formulate and solve the problem for both the foregoing procedures. We shall use the Mueller calculus.

9.7.2 Polarization by Scattering Used as Analyzer

Let us consider linearly polarized light incident on a three-dimensional photoelastic medium S and let M and N be two points along the incident beam from which the scattering of light is considered (Fig. 9.1). We shall first express the optical-characteristic elements of the section of the photoelastic medium included between points M and N in terms of the corresponding quantities of the sections of the photoelastic medium included between points L, M and L, N. The length of the photoelastic medium included between L and M is optically equivalent, according to the equivalence theorem, to a linear retarder and a pure rotator. According to Mueller calculus, if R_{LM} and T_{LM} are the Mueller matrices of the equivalent linear retarder and rotator, the Mueller matrix M_{LM} of the length LM of the photoelastic medium is given by

$$M_{LM} = R_{LM} T_{LM}, \qquad (9.27)$$

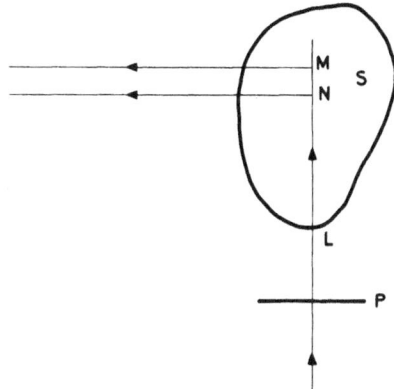

Fig. 9.1. Schematic diagram of linearly polarized light incident on a three-dimensional body S and scattered at points M an N. The scattering property of the medium S acts as the analyzer

where the matrices R_{LM} and T_{LM} are given (as in Table 4.4) by

$$
R_{LM} =
\begin{bmatrix}
1 & 0 & 0 \\
0 & \cos^2 2\theta_{LM} + \sin^2 2\theta_{LM} \cos\delta_{LM} & (1 - \cos\delta_{LM}) \sin 2\theta_{LM} \cos 2\theta_{LM} \\
0 & (1 - \cos\delta_{LM}) \sin 2\theta_{LM} \cos 2\theta_{LM} & \sin^2 2\theta_{LM} + \cos^2 2\theta_{LM} \cos\delta_{LM} \\
0 & \sin 2\theta_{LM} \sin\delta_{LM} & -\cos 2\theta_{LM} \sin\delta_{LM}
\end{bmatrix}
$$

$$
\begin{matrix}
0 \\
-\sin 2\theta_{LM} \sin\delta_{LM} \\
\cos 2\theta_{LM} \sin\delta_{LM} \\
\cos\delta_{LM}
\end{matrix}
\Bigg],
\qquad (9.28)
$$

where θ_{LM} is the angle between the fast axis of the retarder and the Ox axis, δ_{LM} is the retardance of the retarder, and

$$
T_{LM} =
\begin{bmatrix}
1 & 0 & 0 & 0 \\
0 & \cos 2\varphi_{LM} & \sin 2\varphi_{LM} & 0 \\
0 & -\sin 2\varphi_{LM} & \cos 2\varphi_{LM} & 0 \\
0 & 0 & 0 & 1
\end{bmatrix}.
\qquad (9.29)
$$

where φ_{LM} is the optical rotation introduced by the rotator.

Similarly, we have for the Mueller matrix M_{LN} of the length LN of the photoelastic medium

$$
M_{LN} = R_{LN} T_{LN},
\qquad (9.30)
$$

where the matrices R_{LN} and T_{LN} are expressed by (9.28) and (9.29), respectively, with the quantities θ_{LM}, δ_{LM}, and φ_{LM} replaced by the corresponding quantities, θ_{LN}, δ_{LN}, and φ_{LN} for the length LN.

If we take points M and N to be close enough, we can represent the length MN of the photoelastic medium by a linear retarder, whose matrix R_{MN} is expressed by (9.28), with the quantities θ_{LM} and δ_{LM} replaced by the corresponding quantities, θ_{MN} and δ_{MN}.

According to the Mueller calculus we have

$$R_{MN}M_{LM} = M_{LN}, \tag{9.31}$$

or, by taking into account (9.27) and (9.30), we have

$$R_{MN}R_{LM}T_{LM} = R_{LN}T_{LN}. \tag{9.32}$$

By multiplying both members of this relation by T_{LM}^{-1}, we obtain

$$R_{MN}R_{LM} = R_{LN}T_{LN}T_{LM}^{-1}. \tag{9.33}$$

The matrix $(T_{LN}T_{LM}^{-1})$ in (9.33) is a rotation matrix and it is expressed by (9.29) with the angle φ_{LM} replaced by the angle $\varphi \equiv (\varphi_{LN} - \varphi_{LM})$.

By performing the matrix multiplications in both sides of (9.33) and equating the corresponding terms of the matrices, which do not contain the angle φ, we obtain for the inclination θ_{MN} and the retardation δ_{MN} of the element MN the following relations.

$$\tan 2\theta_{MN} = \frac{\sin 2\theta_{LN}\sin\delta_{LN}}{\cos 2\theta_{LN}\sin\delta_{LN} - \sin\delta_{LM}}, \tag{9.34}$$

$$\cos\delta_{MN} = \frac{\cos\delta_{LM}\cos\delta_{LN} + \cos 2\theta_{LN}\sin\delta_{LM}\sin\delta_{LN} - \sin^2\delta_{LM}\sin^2 2\theta_{MN}}{\cos^2\delta_{LM} + \cos^2 2\theta_{MN}\sin^2\delta_{LM}}. \tag{9.35}$$

In (9.34) and (9.35) we assume that the reference axis of the system coincides with the fast axis of the retarder R_{LM} so that $\theta_{LM} = 0$.

These equations express the inclination θ_{MN} and the retardance δ_{MN} of the linear retarder MN in terms of the retardances δ_{LM} and δ_{LN} of the retarders LM and LN and the inclination θ_{LN} of the retarder LN, the inclination θ_{LM} of the retarder LM being equal to zero because the fast axis of this retarder was taken as the reference axis of the system.

Equations (9.34) and (9.35) were derived by using the equivalence theorem in conjunction with the Mueller calculus. They can also be derived by using the equivalence theorem in conjunction with either the Poincaré sphere or the Jones calculus. Equations (9.34) and (9.35) coincide with those given by Gross-Petersen [9.24]. If slightly modified, they yield the equations found by Robert [9.21].

The quantities δ_{LM} and θ_{LM} associated with the equivalent retarder of the length LM can now be determined by measuring the scattered-light intensity from the point M of the photoelastic model. By using a polarizer rotated at a frequency ω, and by taking the observation direction at an angle $(\varphi + 90°)$ with respect to the fast axis of the equivalent retarder of the length considered, we obtain from (9.6) for the intensity I of the scattered light

$$I = kI_0 [1 + \cos 2(\omega t - \theta_{LM} - \varphi_{LM}) \cos 2\theta_{LM}$$
$$+ \sin 2(\omega t - \theta_{LM} - \varphi_{LM}) \sin 2\theta_{LM} \cos \delta_{LM}]. \qquad (9.36)$$

From this expression for I, which consists of a constant and a term that varies sinusoidally with time at a frequency 2ω, the quantities θ_{LM} and δ_{LM} can by determined by use of the methods developed in Chapter 5.

Similarly, the quantities θ_{LN} and δ_{LN} that correspond to the light scattered from the point N are determined. Then from (9.34) and (9.35) the orientation θ_{MN} and the retardance δ_{MN} of the retarder MN may be determined.

By using the outlined procedure and by measuring the scattered-light intensities from all points along the primary beam, we can determine the secondary-stress directions and the stress-optical retardations at all points.

9.7.3 Polarization by Scattering Used as Polarizer

In this case the model is illuminated by unpolarized light and the state of polarization of the scattered light is determined. Let us consider an unpolarized light beam scattered at point M of the photoelastic model, and suppose that the observation direction ML is normal to the primary beam (Fig. 9.2). The scattered light is linearly polarized at the point M and the direction of polarization is normal to the plane of the primary and the scattered light beam. The state of polarization of the linearly polarized light at M, travelling in the photoelastic medium, changes as the scattered light along the direction ML meets the various layers of the medium with different secondary axes and different stress-optical retardations. According to the equivalence theorem,

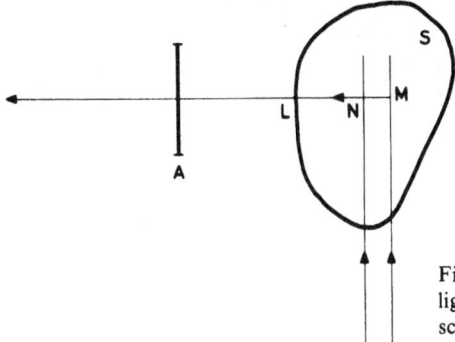

Fig. 9.2. Schematic diagram of an unpolarized light incident on a three-dimensional body S and scattered along the line MN. The scattering property of the medium acts as the polarizer

the part of the photoelastic medium between points M and L is equivalent to a linear retarder and a pure rotator.

Let us now displace the incident unpolarized light beam parallel to itself by the interval MN. The light scattered at point N is linearly polarized and its state of polarization changes continuously along the distance NL. Similarly, the part of the photoelastic model between points N and L is equivalent to a linear retarder and a pure rotator.

If the length MN is small enough so that the secondary stresses do not rotate along it, the length MN is equivalent only to a linear retarder. The orientation θ_{MN} and the retardance δ_{MN} of this linear retarder can be expressed in terms of the orientations θ_{LM}, θ_{LN} and the retardances δ_{LM}, δ_{LN} of the equivalent retarders of the layers LM and LN by (9.34) and (9.35). Therefore, the characteristic elements of the retarder MN can be determined by measuring the state of polarization of the light scattered from points M and N. These measurements can be made by one of the methods developed in Chapter 5. Thus, by displacing the unpolarized light incident on the photoelastic model by small intervals and measuring the state of polarization of the scattered light for each position of the incident light, the corresponding photoelastic quantities along the scattering direction can be determined.

9.8 Application to the Plane Problem

Although the main use of the method of scattered-light photoelasticity is in the investigation of three-dimensional problems, it can be applied successfully also to the solution of two-dimensional problems. In this case, the scattered-light pattern gives immediately the stress normal to the primary beam.

Let us consider a linearly polarized light beam, whose direction of propagation is parallel to the plane of a two-dimensional model. For cases in which conditions of plane stress prevail in the model, this incident light beam is analyzed into the stress component perpendicular to its direction of propagation and into the direction perpendicular to the plane of the model. Because the stress component in the direction perpendicular to the plane of the model is zero, the thus-obtained scattered-light pattern is created by only the stress component, which is perpendicular to the direction of the ray. Therefore, the gradient of the fringes represents the stress component in the direction perpendicular to the incident ray. If the incident light is made to impinge on the two-dimensional model in three different directions, the respective stress components can be determined; these, in turn, permit determination of the principal-stress components and their inclinations. Thus, by recording three scattered-light patterns, the state of stress of a two-dimensional model can be obtained directly.

Furthermore, scattered-light photoelasticity can supplement the usual two-dimensional photoelasticity by providing one more equation for separation of the principal stresses. From the scattered-light patterns for two mutually perpendicular directions, the sum of the principal stresses can be obtained, which, together with their difference, permits separation of the principal stresses. Moreover, by obtaining one scattered-light pattern together with the usual isochromatic and isoclinic patterns, we can completely determine the state of stress in a two-dimensional model.

References

9.1   M. Kerker: *The Scattering of Light and Other Electromagnetic Radiation* (Academic Press, New York 1969)

9.2   R. Weller: J. Appl. Phys. **10**, 266 (1939)

9.3   R. Weller, J. K. Bussey: J. RAeS **44**, 74 (1940)

9.4   R. Weller: J. Appl. Phys. **12**, 610 (1941)

9.5   H. J. Menges: Z. Angew. Math. Mech. **20**, 210 (1940)

9.6   D. C. Drucker, M. M. Frocht: Proc. Soc. Exp. Stress Anal. **5**, 34 (1948)

9.7   D. C. Drucker, R. D. Mindlin: J. Appl. Phys. **11**, 724 (1940)

9.8   E. M. Saleme: Proc. Soc. Exp. Stress Anal. **5**, 49 (1948)

9.9   H. T. Jessop: Brit. J. Appl. Phys. **2**, 249 (1951)

9.10 M. M. Frocht, L. S. Srinath: Proc. 3rd U.S. Natl. Congr. Appl. Mech., ed. by ASME, June 1958, Providence, R.I (1958) pp. 329–337

9.11 L. S. Srinath, M. M. Frocht: Proc. 4th U.S. Natl. Congr. Appl. Mech., ed. by ASME, 1962, Berkeley, Calif., Vol 2 (New York 1962) pp. 775–781

9.12 L. S. Srinath, M. M. Frocht: Proc. Int. Symp. on Photoelasticity, ed. by M. M. Frocht, Illinois Inst. of Technology, Chicago, Ill., 1961 (Pergamon Press, New York 1963) pp. 277–292

9.13 Y. F. Cheng: Exp. Mech. **3**, 275 (1963)

9.14 Y. F. Cheng: Exp. Mech. **7**, 140 (1967)

9.15 Y. F. Cheng: Exp. Mech. **9**, 407 (1969)

9.16 L. S. Srinath: Exp. Mech. **9**, 463 (1969)

9.17 R. W. Aderholdt, J. M. McKinney, W. F. Ranson, W. F. Swinson: Exp. Mech. **10**, 160 (1970)

9.18 W. Shelson, I. W. Smith: Brit. J. Appl. Phys. **7**, 436 (1956)

9.19 J. H. Hemann, R. J. Becherer: Exp. Mech. **12**, 43 (1972)

9.20 A. Robert, E. Guillemet: Brit. J. Appl. Phys. **15**, 567 (1964)

9.21 A. Robert: Exp. Mech. **7**, 224 (1967)

9.22 A. Robert: Int. J. Solids Structures **6**, 423 (1970)

9.23 A. Robert: "*Polarimétrie et Photoélasticimétrie*", Serv. Techn. Const. Armes Navales, Chap. VII (Paris 1972) pp. 145–171

9.24 J. F. Gross-Petersen: Exp. Mech. **14**, 317 (1974)

9.25 J. Cernosek: Exp. Mech. **13**, 273 (1973)

10. Interferometric Photoelasticity

10.1 Introduction

As discussed in Chapter 7, classical two-dimensional photoelasticity is based on Wertheim's stress-optical law, which relates the relative retardation between the two light waves that are linearly polarized in the principal-stress directions of the photoelastic model with the difference of the principal stresses. From analysis of the optical effects obtained by inserting a two-dimensional model in the field of a plane polariscope, which was discussed in Section 7.2, it was concluded that the two quantities of the stress field that appear in Wertheim's law, that is the principal-stress directions and the principal-stress difference, can be determined from the two overlapping families of isoclinics and isochromatics, respectively, obtained in a plane polariscope.

For determination of the individual values of principal stresses, one more equation is needed, besides their difference. For this reason, various auxiliary methods have been developed in classical photoelasticity. All of these methods are described in classical books on photoelasticity [10.1]; they may be classified into four categories: I) methods based on graphical integration of photoelastic data, either along stress trajectories, or along any straight section of the stress field [10.2, 3], II) methods based on various analogies, for example membrane and electrical analogies, which yield the sum of principal stresses [10.4], III) numerical methods of iteration and relaxation [10.1], and IV) the experimental method of oblique incidence of light [10.5]. All of these auxiliary methods are tedious and time consuming in application and involve processing the experimental data in complicated manners, such that experimental errors may be significantly amplified. Therefore, the accuracy of these methods is considerably reduced and at least doubtful.

A more straightforward determination of the values of the stress components can be based on use of the generalized Neumann-Maxwell stress-optic law, which connects the variations of the refractive indices along the principal-stress directions with the values of the principal stresses. Therefore, if the stress-optical retardations along the principal-stress directions can be measured, the stress components may be determined.

The first attempt to measure the values of the stress-optical retardations was made by *Favre* [10.6], who used the Mach-Zehnder interferometer and transparent specimens with optically flat faces. The same principle was afterwards used by *Brahtz* and *Soehrens* [10.7]. All of these investigators found the values of principal stresses at various points of the model. Their approach did not become popular among other investigators because of the difficulties inherent in the use and adjustment of the interferometer, as well as because their method was developed as a point-by-point procedure. Other variations of the same principle were reviewed by *Theocaris* [10.8]. Interferometric methods have also been used to obtain isopachic patterns in thin specimens under generalized plane-stress conditions [10.2, 9]. To overcome difficulties caused by deviations from optical flatness of the faces of the model, *Mesmer* [10.10] considered the interferograms of the loaded and the unloaded model and used the moiré effect to subtract the no-load interferogram of the specimen.

Fabry [10.11] used, instead of a separate interferometer, the specimen itself as an interferometer (the so-called Fabry-Perot interferometer) and obtained the interference pattern formed by successive reflections from the two faces of the specimen and the light rays traversing the specimen. More recently, *Favre* and *Schumann* [10.12] used Fizeau interference fringes and measured absolute variations of the optical paths at various points of the model by photoelectric measurement of the corresponding light intensities. The same principle of Fizeau-interference fringes was subsequently used by *Theocaris* and *Gdoutos* [10.13], who obtained the interference pattern of rays reflected from the front and rear faces of the transparent specimen, by using a laser light source. Interference fringes of high contrast were obtained, because the path difference of the interfering rays is smaller than the coherence length of the laser light.

A more systematic study of the idea of measuring the absolute stress-optical retardations along the principal stresses was reported by *Post* in a series of publications [10.14–20]. He extended the method to a full-field analysis, described its main characteristic features and potentialities, and used a series interferometer to measure absolute stress-optical retardations. *Post* [10.21] also discussed the possibilities of using different interference systems, such as the Fizeau, the Fabry-Perot, and many modifications of them that use mirrors and semi-mirrors, as well as the classical Mach-Zehnder and Michelson interferometers and showed that the optical patterns of all of these systems can be interpreted in a uniform manner. *Nisida* and *Saito* [10.22] used the Mach-Zehnder interferometer for simultaneous determination of the isochromatic and isopachic fringe patterns in two-dimensional stress fields by using an optically flat specimen and ordinary monochromatic light. Also, they showed the feasibility of the method for use with the three-dimensional frozen-stress model. *Theocaris* and *Gdoutos* [10.23], in a recent paper, gave a unified interpretation of the fringe patterns obtained in all interferometric and holographic systems, based on the mechanical interference of the two fundamental optical patterns of the individual principal stress-optical retardations. They showed the complicated character of these patterns and analyzed the factors that influence the interpretation of the interferograms in the most-general case. They concluded that confusion about the interpretation of interferometric and holographic patterns is mainly due to the particular features of each pattern, which may lead researchers to erroneous generalizations. Other interesting discussions of interferometric photoelasticity are referred to the Bibliography.

In the present chapter, the various systems used in interferometric photo-elasticity are discussed and analyzed. The distribution of intensity in the inter-ferometric pattern is found by use of the Jones calculus. The general stress-optical relations for the variation of the stress-optical retardations along the principal-stress directions are established. Finally, at the end of the chapter, a unified interpretation of the interferometric patterns produced by all optical systems is obtained by use of physical concepts, only.

10.2 Stress-Optical Relations

Let us consider a plane transparent specimen with parallel lateral faces, made of a birefringent material, and subjected to a generalized plane-stress field. If a polarized light is normally incident on the specimen, then, according to the Neumann-Maxwell stress-optical law, the incident light is divided into two rays linearly polarized along the principal-stress directions, and the variations of the refractive indices of the material of the specimen along the principal directions are [p. 111, (6.22)]

$$\Delta n_1 = n_1 - n = b_1 \varepsilon_1 + b_2 (\varepsilon_2 + \varepsilon_3)$$
$$\Delta n_2 = n_2 - n = b_1 \varepsilon_2 + b_2 (\varepsilon_1 + \varepsilon_3). \tag{10.1}$$

In these relations, Δn_1 and Δn_2 are the variations of the refractive index along the principal-stress directions 1 and 2, n_1 and n_2 are the refractive indices along the directions 1 and 2 when the specimen is loaded, n is the refractive index of the material of the specimen when it is not loaded, $\varepsilon_1, \varepsilon_2, \varepsilon_3$ are the principal strains, and b_1 and b_2 are the strain-optical constants of the material of the specimen.

We introduce into (10.1) Hooke's stress-strain law, expressed for the gen-eralized plane-stress field by

$$\varepsilon_1 = \frac{1}{E}(\sigma_1 - v\sigma_2)$$

$$\varepsilon_2 = \frac{1}{E}(\sigma_2 - v\sigma_1) \tag{10.2}$$

$$\varepsilon_3 = -\frac{v}{E}(\sigma_1 + \sigma_2),$$

where σ_1 and σ_2 are the principal stresses of the generalized plane-stress field of the specimen and v and E are Poisson's ratio and the modulus of elasticity of the material of the specimen. Equation (10.1) take the form

$$\Delta n_1 = n_1 - n = A\sigma_1 + B\sigma_2 \tag{10.3}$$
$$\Delta n_2 = n_2 - n = B\sigma_1 + A\sigma_2 ,$$

where

$$A = \frac{b_1 - 2\nu b_2}{E} , \quad B = \frac{b_2 - \nu(b_1 + b_2)}{E} . \tag{10.4}$$

Consider the light linearly polarized in the principal-stress direction σ_1, and normally traversing the specimen and let A and B be two reference points on this ray, lying along either side of the specimen (Fig. 10.1). If $AB = L$ and n_0 is the refractive index of the medium that surrounds the specimen, then the optical path s_A between points A and B, when the specimen is unloaded, is given by

$$s_A = Ln_0 + d(n - n_0) , \tag{10.5}$$

where d is the thickness of the specimen.

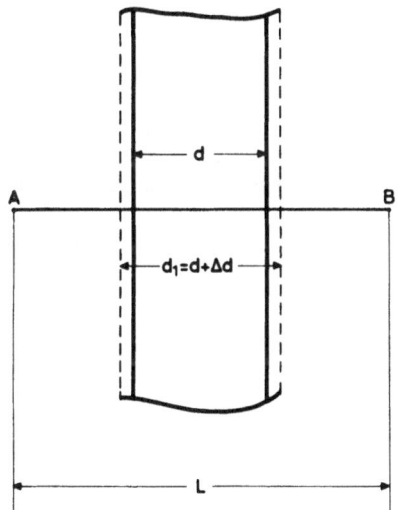

Fig. 10.1. Geometry of an unloaded (continuous lines) and a loaded (dotted lines) specimen traversed by a light ray AB

When the specimen is subjected to a plane-stress field the optical path s_{T_1} between the same reference points A and B is given by

$$s_{T_1} = Ln_0 + d_1(n_1 - n_0) , \tag{10.6}$$

where d_1 is the thickness of the specimen after loading. By putting

$$d_1 = d + \Delta d \tag{10.7}$$

and omitting infinitesimal quantities of second order, we obtain for the variation of the optical path Δs_{t_1} along the σ_1 principal stress

$$\Delta s_{t_1} = s_{T_1} - s_A = d\Delta n_1 + (n - n_0)\Delta d. \tag{10.8}$$

By substituting into (10.8) the value of Δn_1, given by (10.1) or (10.3), and
putting

$$\Delta d = \varepsilon_3 d, \tag{10.9}$$

where ε_3 is given by the third of (10.2), we obtain

$$\Delta s_{t_1} = (a_t\sigma_1 + b_t\sigma_2)d, \tag{10.10}$$

where

$$a_t = \frac{1}{E}[b_1 - 2vb_2 - v(n - n_0)] = A - \frac{v}{E}(n - n_0),$$
$$b_t = \frac{1}{E}[b_2 - v(b_1 + b_2) - v(n - n_0)] = B - \frac{v}{E}(n - n_0). \tag{10.11}$$

Equation (10.10) expresses the variation of the optical path of the linearly
polarized light ray along the direction of the principal stress σ_1 due to the
loading of the specimen.

Similarly, we obtain for the variation of the optical path Δs_{t_2} along the
direction of the σ_2 principal stress

$$\Delta s_{t_2} = (b_t\sigma_1 + a_t\sigma_2)d. \tag{10.12}$$

Equations (10.10) and (10.12) can be put in the form

$$\Delta s_{t_{1,2}} = c_t[(\sigma_1 + \sigma_2) \pm \xi_t(\sigma_1 - \sigma_2)]d, \tag{10.13}$$

where

$$c_t = \frac{a_t + b_t}{2}, \quad \xi_t = \frac{a_t - b_t}{a_t + b_t}. \tag{10.14}$$

For the case of an optically inert material ($b_1 = b_2 = b$), (10.10) and (10.12)
take the form

$$\Delta s_{t_1} = \Delta s_{t_2} = c_t(\sigma_1 + \sigma_2)d, \tag{10.15}$$

with

$$a_t = b_t = c_t = \frac{1}{E}\left[b(1-2v) - v(n-n_0)\right].$$ (10.16)

Consider the case when the light incident on the specimen is reflected from its rear face (Fig. 10.2). Then the optical path length s_A for the ray that traverses the specimen, is reflected from its rear face, and returns to the front face, when the specimen is not loaded, is

$$s_A = 2(L'n_0 + dn).$$ (10.17)

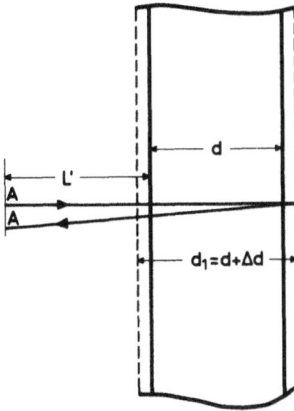

Fig. 10.2. Geometry of an unloaded (continuous lines) and a loaded (dotted lines) specimen and a light ray reflected from its rear face

For the case when the specimen is loaded, the new optical path length s_{T_1} for the light linearly polarized along the direction of the principal stress σ_1 is

$$s_{T_1} = 2\left[\left(L' - \frac{\Delta d}{2}\right)n_0 + d_1 n_1\right].$$ (10.18)

From (10.17) and (10.18) by omitting infinitesimal quantities of second order, we obtain the variation of the optical path length Δs_{r_1} of the light ray along the direction of the σ_1 principal stress,

$$\Delta s_{r_1} = s_{T_1} - s_A = 2\left[(n_1 - n)d + \left(n - \frac{n_0}{2}\right)\Delta d\right].$$ (10.19)

When (10.1) or (10.3) are taken into account and with $\Delta d = \varepsilon_3 d$, (10.19) takes the form

$$\Delta s_{r_1} = 2(a_r\sigma_1 + b_r\sigma_2)d,$$ (10.20)

where

$$a_r = \frac{1}{E}\left[b_1 - 2vb_2 - v\left(n - \frac{n_0}{2}\right)\right] = A - \frac{v}{E}\left(n - \frac{n_0}{2}\right),$$

$$b_r = \frac{1}{E}\left[b_2 - v(b_1 + b_2) - v\left(n - \frac{n_0}{2}\right)\right] = B - \frac{v}{E}\left(n - \frac{n_0}{2}\right). \tag{10.21}$$

Similarly, we obtain for the variation of the optical path length Δs_{r_2} along the direction of the σ_2 principal stress

$$\Delta s_{r_2} = 2(b_r \sigma_1 + a_r \sigma_2)d. \tag{10.22}$$

Equations (10.20) and (10.22) can be put in the form

$$\Delta s_{r_{1,2}} = 2c_r[(\sigma_1 + \sigma_2) \pm \xi_r(\sigma_1 - \sigma_2)]d, \tag{10.23}$$

where

$$c_r = \frac{a_r + b_r}{2}, \qquad \xi_r = \frac{a_r - b_r}{a_r + b_r}. \tag{10.24}$$

For the case of an optically inert material, (10.20) and (10.22) take the form

$$\Delta s_{r_1} = \Delta s_{r_2} = 2c_r(\sigma_1 + \sigma_2)d. \tag{10.25}$$

Either (10.10) and (10.12), or (10.20) and (10.22) suggest that, if the absolute stress-optical retardations $\Delta s_{t_{1,2}}$ or $\Delta s_{r_{1,2}}$ along the principal-stress directions are known, then the principal stresses σ_1 and σ_2 can be determined, provided that the stress-optical constants a_t, b_t or a_r, b_r are known. These constants can be determined from a simple calibration test.

By solving the above relations for σ_1 and σ_2, we obtain

$$\sigma_1 = \frac{a_t \Delta s_{t_1} - b_t \Delta s_{t_2}}{(a_t^2 - b_t^2)d}, \qquad \sigma_2 = \frac{-b_t \Delta s_{t_1} + a_t \Delta s_{t_2}}{(a_t^2 - b_t^2)d}, \tag{10.26}$$

or

$$\sigma_1 = \frac{a_r \Delta s_{r_1} - b_r \Delta s_{r_2}}{2(a_r^2 - b_r^2)d}, \qquad \sigma_2 = \frac{-b_r \Delta s_{r_1} + a_r \Delta s_{r_2}}{2(a_r^2 - b_r^2)d}. \tag{10.27}$$

10.3 Intensity Variation of a Light Ray in a Transparent Material

In some interferometric systems, a light ray that is incident on the specimen passes several times through its thickness. It is therefore necessary to study the distribution of the intensity of the incident light into the various rays that emerge from the front and rear faces of the specimen.

Let us consider a light ray normally incident on a transparent material. One part of this ray is reflected from the front face, while the other passes through the thickness of the specimen. This ray meets the rear face and one part of it is reflected, while the other emerges from the rear face of the specimen. The successive reflections from the two faces of the specimen are repeated, so that an infinity of rays emerges from the front and another from the rear face of the specimen.

The percentage of the intensity of the light is the same at each successive reflection from the front or the rear face of the specimen; let us designate by β the reduction ratio. On the other hand, the reduction coefficient for the refracted light rays differs for the entering or the emerging rays. Let us call the reduction ratio for the entering rays α and the corresponding ratio for the emerging rays α'.

If I is the intensity of the incident light ray, then the intensity of the first reflected ray is given by βI and the intensity of the first refracted ray is equal to αI. This last light ray either emerges from the rear face of the specimen with intensity $\alpha \alpha' I$, or after reflection from this face passes through the thickness of the specimen with intensity $\alpha \beta I$. The distribution of the intensity of light in the first three light rays emerging from the front face and the first two rays emerging from the rear face of the specimen is shown in Fig. 10.3. Assuming that the plate does not absorb light, we get, by applying the law of conservation of energy, that

$$I = I_f + I_r, \tag{10.28}$$

where I_f and I_r are the total intensities of the beams that emerge from the front and rear faces of the specimen, respectively.

Intensities I_f and I_r are given by

$$I_f = \beta I + \alpha \alpha' \beta I (1 + \beta^2 + \beta^4 + \beta^6 + \dots) = I \beta \left(1 + \frac{\alpha \alpha'}{1 - \beta^2}\right), \qquad \beta < 1 \tag{10.29}$$

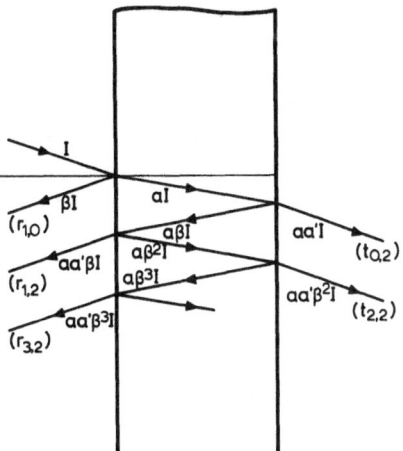

Fig. 10.3. Distribution of the light intensity that emerges from the front and rear faces of a transparent specimen

$$I_r = \alpha \alpha' I (1 + \beta^2 + \ldots) = \frac{I \alpha \alpha'}{1 - \beta^2} . \tag{10.30}$$

Then (10.28) takes the form

$$I = I \beta \left(1 + \frac{\alpha \alpha'}{1 - \beta^2} \right) + \frac{I \alpha \alpha'}{1 - \beta^2} , \tag{10.31}$$

from which it follows that

$$\alpha \alpha' = (1 - \beta)^2 . \tag{10.32}$$

If we represent by $r_{k,l}$ the rays that emerge from the front face of the specimen, where the first index k indicates the number of successive reflections and the second index l indicates the number of the corresponding refractions of this ray $(r_{1,0}, r_{1,2}, r_{3,2}, \ldots)$ and if we represent by $t_{k,l}$ the rays that emerge from the rear face of the specimen, where the indices k, l have the same meaning $(t_{0,2}, t_{2,2}, t_{4,2}, \ldots)$, we obtain for the intensity $I_{r,t,k,l}$ of these rays

$$I_{r,t,k,l} = \beta^k (1 - \beta)^l I , \tag{10.33}$$

where $k = 1, l = 0, 2$ or $k = 3, 5, 7, \ldots, l = 2$ for the rays that emerge from the front face of the specimen and $k = 0, 2, 4, \ldots$ and $l = 2$ for the rays that emerge from the rear face of the specimen.

The value of the coefficient β is given by [10.24]

$$\beta = \left(\frac{n-1}{n+1} \right)^2 , \tag{10.34}$$

where n is the refractive index of the specimen.

For $n = 1.5$, which corresponds approximately to the refractive index of the most common glasses and plastics, (10.34) yields $\beta = 0.04$. Then, by applying (10.33), we obtain for the intensities of the rays that emerge from the front face of the specimen $I_{r_{1,0}} = 0.04 I$, $I_{r_{1,2}} = 0.03686 I$, $I_{r_{3,2}} = 0.00006 I$ and for the rays that emerge from the rear face of the specimen $I_{t_{0,2}} = 0.92160 I$, $I_{t_{2,2}} = 0.00147 I$ [10.24]. It may be concluded from the values of these components of intensities that only $I_{r_{1,0}}$, $I_{r_{1,2}}$, and $I_{t_{0,2}}$ are worthwhile considering for simple cases of applications in interferometry.

10.4 Optical Systems Used in Interferometric Photoelasticity

All optical systems based on interferometry, either for measuring small lengths or for establishing the fundamental metric units, can be used in interferometric photoelasticity. All of these optical systems are based on the concept of di-

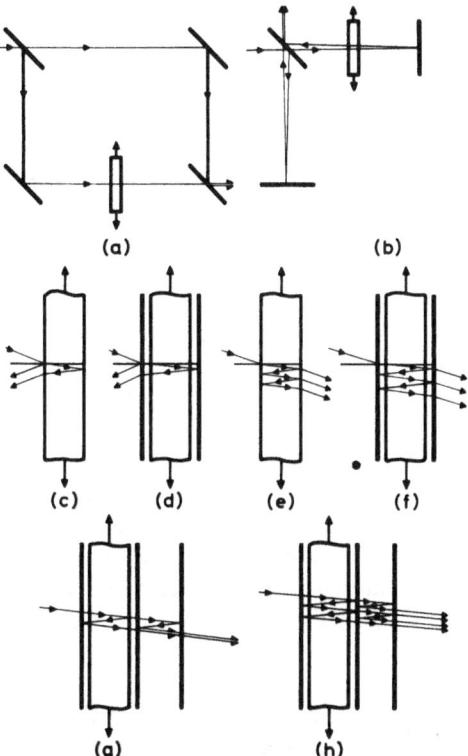

Fig. 10.4a–h. Various types of interferometers (a, b) and interferometric systems (c–h) for obtaining the interferograms of a loaded specimen

viding a light ray into two rays, directing one ray to pass through the specimen under study, while the other ray circumvents the specimen, or by letting the one ray pass through the specimen several times more than the other ray. The ray that passes through the specimen is called the *"active ray"*, because it gathers information from the loaded specimen, whereas the optical path of the other ray, which is called *"the reference ray"*, is independent of the specimen. The optical path of the active ray changes because it passes through the loaded specimen. Therefore the pattern formed by interference of the active and the reference rays contains information about the variations of the optical path of the active ray. If these variations of the optical path can be measured, the values of the principal stresses connected with them can be readily deduced.

All interferometric systems can be classified into two main categories, depending on the principle used to form the interferometric pattern. In the first category belong all systems in which the light beam is divided into two beams, one of which passes through the specimen. This idea is utilized by the various types of interferometers, the most common of which are the Mach-Zehnder and the Michelson interferometers, shown schematically in Fig. 10.4a, b. This figure shows that in the Mach-Zehnder interferometer the optical path of the active ray differs from that of the reference ray by the variation of the optical

path produced by the loaded specimen, whereas in the Michelson interferometer it differs by twice the variation of the optical path produced by the
loaded specimen, because the active ray passes twice through the specimen.
The difference of the optical paths of the active and reference rays is given for
the Mach-Zehnder interferometer by (10.10) and (10.12), whereas for the
Michelson interferometer the difference is twice that given by those equations.
Because the difference of the optical paths is not many times greater than the
wavelength of the light used, a high-quality interference pattern can be obtained by use of both of these interferometers. In order to minimize further
the difference of the optical paths, the whole interferometer is sometimes
immersed in a bath of oil, that has the same refractive index as that of the specimen. In this case, in (10.11), $n = n_0$ and the stress-optical constants a_t and b_t
are equal to A and B, respectively. Thus, the quality of the interference fringes
is greatly enhanced.

In the second category belong interferometric systems in which one ray
is allowed to pass several times more through the specimen than the other
ray, or the other ray does not pass at all through the specimen. Some characteristic interferometric systems of this type are shown in Figs. 10.4c–h.
Figures 10.4c, d represent the Fizeau-type interferometric system; Figs. 10.4e, f
correspond to the Fabry-Perot system. Finally, Figs. 10.4g, h show two-beam
and multiple-beam series interferometers.

In the system of Fig. 10.4c, no external optical element is used and the
optical pattern is formed by interference of the rays reflected from the front
and rear faces of the specimen. In this case, the reference ray does not pass
through the specimen, whereas the active ray passes twice through the thickness of the specimen. In the system of Fig. 10.4d, the mechanism of formation
of fringes is the same as in Fig. 10.4c, with only the difference that the incident
light is not reflected by the faces of the specimen, but from two mirrors placed
one in front and the other behind the specimen. The reflectivities of those two
mirrors are such that the intensities of the two reflected beams are approximately the same, thus yielding good-quality interference fringes. In the Fabry-
Perot interferometric systems of Fig. 10.4e, f, the reference ray passes once
through the thickness of the specimen, whereas the first active ray passes three
times through that thickness, the second active ray passes five times, and so
on. Again, in the system of Fig. 10.4e, no external optical element is used,
whereas in the system of Fig. 10.4f two mirrors are placed, one in front and
the other behind the specimen. In the series interferometers of Figs. 10.4g, h,
the reference ray passes once, whereas the active ray passed three or more
times through the thickness of the specimen.

In the case of Figs. 10.4c–f, the optical-path difference of the interfering
rays, being equal to the thickness or many times the thickness of the specimen,
is very large compared with the wavelength of the light used, so that a light
source of great coherence is necessary to obtain high-quality interference
patterns. In these cases, laser light sources can be used. In contrast with the
systems sketched in Figs. 10.4c–f, in the systems of Figs. 10.4g, h the distance

between the two mirrors behind the specimen is such that the two interfering rays have approximately equal optical-path lengths.

10.5 Analysis of Interferometric Systems

Mach-Zehnder interferometer. The arrangement of the optical elements in the Mach-Zehnder interferometer is shown in Fig. 10.5. A monochromatic source S is used; its light passes through a pinhole P placed at the focus of a lens L_1, so that a parallel light beam is obtained. This light beam impinges on a beam splitter or a half-mirror H_1, so that two perpendicular light beams I and II are obtained, each being a parallel light beam. Beam I impinges on a mirror M_1 and then passes through another beam splitter H_2, whereas beam II, after incidence on mirror M_2, impinges on the beam splitter H_2 and interferes with beam I. The specimen is interposed in the optical path of beam II. Therefore, beam II is the active beam, whereas beam I is the reference beam. The interference pattern formed after passing the two light beams from the beam splitter H_2 is received on a photographic plate Pl.

The optical path lengths of the active and reference rays, before the specimen is placed in the former, are equal; therefore, when a specimen with almost

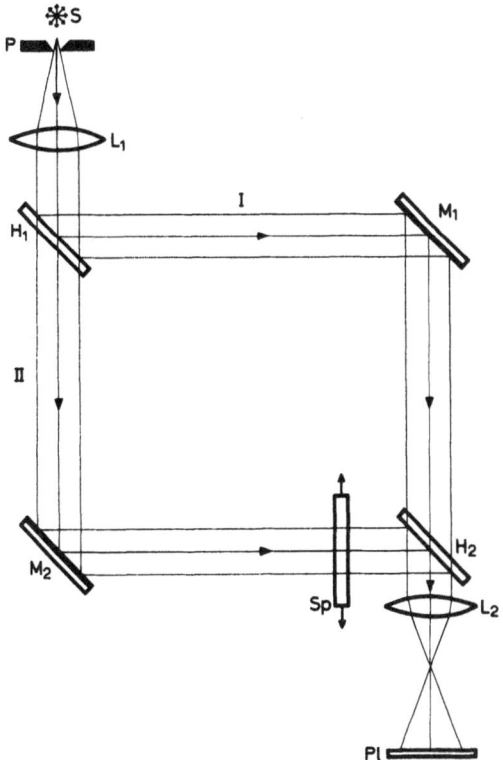

Fig. 10.5. General arrangement of the Mach-Zehnder interferometer

parallel lateral faces is used, no interference pattern is obtained. After the specimen is inserted, the optical path length of the active beam is changed uniformly; therefore, a uniform optical pattern is again formed on the screen behind H_2. However, because none of the commercially used materials for specimens is optically flat with lateral faces exactly parallel, a sparse interferogram always appears on the ground glass of the camera, when the specimen is unloaded.

In order to eliminate that pattern, either a similar specimen is placed in the optical path of the reference ray or, which is the most common case, the whole interferometer is immersed in an oil bath of the same refractive index as that of the specimen. When the specimen is loaded, the variation of the optical path lengths Δs_{t_1}, Δs_{t_2} along its principal-stress directions is given by (10.10) and (10.12). Therefore, the Jones matrix J of the specimen under load, when it is referred to its principal birefringence axes, is given (see Table 4.2) by

$$J = \begin{bmatrix} e^{-i\delta_1} & 0 \\ 0 & e^{-i\delta_2} \end{bmatrix}, \tag{10.35}$$

where

$$\delta_1 = \frac{2\pi}{\lambda} \Delta s_{t_1} = \frac{2\pi}{\lambda} (a_t \sigma_1 + b_t \sigma_2) d$$

$$\delta_2 = \frac{2\pi}{\lambda} \Delta s_{t_2} = \frac{2\pi}{\lambda} (b_t \sigma_1 + a_t \sigma_2) d \tag{10.36}$$

$$\delta = \delta_2 - \delta_1,$$

λ is the wavelength of the monochromatic light used, and δ is the phase difference between the rays that are linearly polarized along the principal birefringence axes of the specimen.

Let us now use a unit-intensity right-handed circularly polarized light. The Jones vector a' of this light is given (see Table 3.2) by

$$a' = \frac{1}{\sqrt{2}} \begin{bmatrix} -i \\ 1 \end{bmatrix}. \tag{10.37}$$

Therefore, the Jones vector a'' of the emerging light beam is

$$a'' = J a' = \frac{1}{\sqrt{2}} \begin{bmatrix} e^{-i\delta_1} & 0 \\ 0 & e^{-i\delta_2} \end{bmatrix} \begin{bmatrix} -i \\ 1 \end{bmatrix} = \frac{1}{\sqrt{2}} \begin{bmatrix} -ie^{-i\delta_1} \\ e^{-i\delta_2} \end{bmatrix}. \tag{10.38}$$

Thus, the Jones vectors a' and a'' of the reference and the active beams are expressed by (10.37) and (10.38), respectively. According to the results of

Section 3.4.5, the Jones vector a of coherent addition of the reference and the active beams is expressed by

$$a = a' + a'' = \frac{1}{\sqrt{2}} \begin{bmatrix} -i(e^{-i\delta_1}+1) \\ (e^{-i\delta_2}+1) \end{bmatrix}. \tag{10.39}$$

The intensity I of the light beam, expressed by the Jones vector a, is

$$I = \tilde{a}\,a = \frac{1}{2}[i(e^{i\delta_1}+1) \quad (e^{i\delta_2}+1)] \begin{bmatrix} -i(e^{-i\delta_1}+1) \\ (e^{-i\delta_2}+1) \end{bmatrix}, \tag{10.40}$$

or

$$I = 2 + \cos\delta_1 + \cos\delta_2 . \tag{10.41}$$

Taking into account (10.36), we get

$$I = 2\left\{1 + \cos\left[\frac{2\pi}{\lambda}\frac{a_t+b_t}{2}(\sigma_1+\sigma_2)d\right]\cos\left[\frac{2\pi}{\lambda}\frac{a_t-b_t}{2}(\sigma_1-\sigma_2)d\right]\right\}. \tag{10.42}$$

Equation (10.42) represents the distribution of intensity in the interferometric pattern obtained by the Mach-Zehnder interferometer. This relation was first derived by *Nisida* and *Saito* [10.22] in another way, by using classical vector analysis to represent light waves. This equation shows that the intensity of the interferometric pattern depends on the values of both the sum and the difference of the principal stresses. Therefore, the interference pattern always contains the superimposed families of isochromatic and isopachic curves, from which the values of the principal stresses at each point of the model can be determined. Some interesting remarks on the formation of both the isochromatic and isopachic fringe patterns can be found in [10.23]. For more details and examples, the interested reader is referred to [10.22].

Michelson interferometer. The arrangement of this interferometer is shown in Fig. 10.6. A monochromatic light beam, after being collimated by passage through lens L_1, impinges on a beam splitter H and is divided into two light beams that follow the paths I and II. Beams I and II impinge on two mirrors M_1 and M_2, placed at equal distances from the beam splitter H; after reflection, they pass through the beam splitter H and interfere. The interference pattern is focussed on a photographic plate Pl by lens L_2. The specimen is inserted in the optical path of either of beams I or II. In Fig. 10.6 the specimen is traversed by beam II, which is therefore the active beam, whereas beam I is the reference beam. Again, in order to get equal optical path lengths for the active and reference rays, a similar specimen is placed in the reference ray or, which is the most common case, the whole interferometer is immersed in oil whose refractive index is the same as that of the specimen. Therefore, if

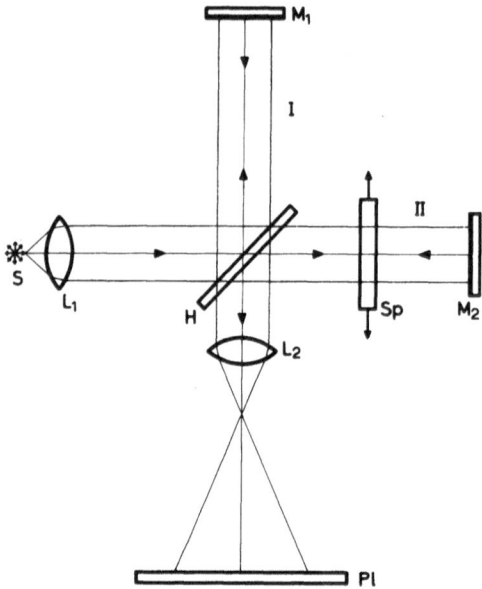

Fig. 10.6. General arrangement of the Michelson interferometer

a specimen with almost parallel faces is used, no interference pattern is obtained when the specimen is unloaded.

The Jones matrix J of the specimen, when it is traversed by the light before reflection from the mirror, is given by (10.35). According to the reversibility principle developed in Section 4.5, the Jones matrix of the specimen traversed by the active beam in the reverse direction after reflection from the mirror M_2 is given by the transpose of the matrix J. Because matrix J is a diagonal matrix, its transpose matrix is equal to this matrix, therefore the Jones matrix of the specimen J_1, traversed twice by the light, is

$$J_1 = J^2 = \begin{bmatrix} e^{-i\delta_1} & 0 \\ 0 & e^{-i\delta_2} \end{bmatrix} \begin{bmatrix} e^{-i\delta_1} & 0 \\ 0 & e^{-i\delta_2} \end{bmatrix} = \begin{bmatrix} e^{-2i\delta_1} & 0 \\ 0 & e^{-2i\delta_2} \end{bmatrix}. \qquad (10.43)$$

Equation (10.43) indicates that the stress-optical retardations along the principal-stress directions of the matrix J_1 are equal to twice the corresponding retardations when the model was traversed only once by the light beam.

Therefore, the intensity I for the Michelson interferometer will again be expressed by (10.42) with only the difference that the angles, whose cosines are obtained in the right-hand side of (10.42), must be doubled.

Fizeau-type interferometer. In this type of interferometer, shown in Fig. 10.4c, the interference pattern is formed from light reflected from the front and rear faces of the specimen. The Jones matrix J of the specimen, traversed twice by the incident light, is

$$J = \begin{bmatrix} e^{-i\delta_1} & 0 \\ 0 & e^{-i\delta_2} \end{bmatrix}, \tag{10.44}$$

with

$$\delta_1 = \frac{2\pi}{\lambda} \Delta s_{r_1} = \frac{2\pi}{\lambda} \cdot 2(a_r \sigma_1 + b_r \sigma_2)d$$

$$\delta_2 = \frac{2\pi}{\lambda} \Delta s_{r_2} = \frac{2\pi}{\lambda} \cdot 2(b_r \sigma_1 + a_r \sigma_1)d, \tag{10.45}$$

where the stress-optical constants a_r and b_r are obtained from (10.21)

Consider a unit-intensity right-hand circularly polarized beam, incident on the specimen with Jones vector given by (10.37). Then, the Jones vector a'' of the light emerging from the specimen, after reflection from the rear face of the specimen, is given by (10.38) with values of δ_1 and δ_2 given by (10.45).

The light reflected from the front face of the specimen undergoes a change of optical path when the specimen is loaded $-\Delta d n_0$ along the two principal birefringence directions, where Δd is the thickness variation of the specimen, due to loading, and n_0 is the refractive index of the medium that surrounds the specimen. Therefore, the Jones matrix of the specimen for the light reflected from the front face of the specimen is given by (10.35), with

$$\delta_1' = \delta_2' = -\frac{2\pi}{\lambda} \Delta d n_0 = \frac{2\pi}{\lambda} \frac{v(\sigma_1 + \sigma_2)d}{E} n_0 . \tag{10.46}$$

Thus, the Jones vector a''' of the light reflected from the front face of the specimen is expressed by (10.38) with the values of δ_1' and δ_2' given by (10.46).

The light rays reflected from the front and rear faces of the specimen, which have Jones vectors a''' and a'', respectively, interfere. Therefore, the Jones vector a of the resulting light is

$$a = a'' + a''' = \frac{1}{\sqrt{2}} \begin{bmatrix} -i(e^{-i\delta_1} + e^{-i\delta_1'}) \\ (e^{-i\delta_2} + e^{-i\delta_2'}) \end{bmatrix}. \tag{10.47}$$

The intensity of the beam with the Jones matrix a (10.47) is

$$I = \tilde{a}a = \frac{1}{2}[i(e^{i\delta_1} + e^{i\delta_1'}) \quad (e^{i\delta_2} + e^{i\delta_2'})] \begin{bmatrix} -i(e^{-i\delta_1} + e^{-i\delta_1'}) \\ (e^{-i\delta_2} + e^{-i\delta_2'}) \end{bmatrix}, \tag{10.48}$$

or

$$I = 2\left\{1 + \cos\left[\frac{2\pi}{\lambda}(a^* + b^*)(\sigma_1 + \sigma_2)d\right]\cos\left[\frac{2\pi}{\lambda}(a^* - b^*)(\sigma_1 - \sigma_2)d\right]\right\}, \tag{10.49}$$

with

$$a^* = \frac{1}{E}(b_1 - 2vb_2 - vn) = A - \frac{v}{E}n$$

(10.50)

$$b^* = \frac{1}{E}[b_1 - v(b_1 + b_2) - vn] = B - \frac{v}{E}n.$$

Equation (10.49) indicates that the interference pattern obtained in the Fizeau-type interferometer is of the same nature as that obtained with the preceding two types of interferometers; the only difference is in the calibration constants, which are expressed by (10.50).

Having analyzed three fundamental cases of interferometric systems, namely the Mach-Zehnder, the Michelson, and the Fizeau-type interferometers, we can derive the intensity distribution in all of the other cases of Fig. 10.4, from the corresponding formulas for these three types of interferometric systems. Thus, the Fizeau type of Fig. 10.4d is equivalent to the Michelson interferometer, because the light traverses the specimen twice and is reflected from optical elements other than the faces of the specimen. The case of Fig. 10.4e is equivalent to that of Fig. 10.4c, because, according to the results of Section 10.3, only the first two light rays that emerge from the rear face of the specimen contribute to the interference pattern obtained, whereas the intensities of the other rays are insignificant. Similarly, the interferometric system of Fig. 10.4f is the same as that of Fig. 10.4d. The arrangements of the other two interferometric systems of Figs. 10.4g and 10.4h are equivalent to those of Figs. 10.4d and 10.4f, respectively.

10.6 Physical Explanation of the Interferometric Patterns

Formation of the patterns. The preceding analysis of the distribution of intensity in the interferometric systems, by use of the Jones calculus showed that all of these patterns are composed of the superposed families of isochromatics and isopachics, representing the loci of the difference and the sum of principal stresses. In the present section, we shall consider the formation of these patterns without deriving them from the intensity distribution formulae, but, by using physical concepts only. The unified character of these patterns in all interferometric systems will be pointed out, as was previously established by consideration of the intensity distribution.

Let us consider an ordinary light ray normally incident on the specimen. This ray, according to the Neumann-Maxwell stress-optic law, is divided into two rays plane-polarized along the directions of the principal stresses σ_1 and σ_2. When the specimen is loaded, the two rays are retarded by different amounts Δs_1 and Δs_2, which, when the light beam either traverses the specimen or is reflected from the rear face of the specimen, are given either by (10.10, .12) or by (10.20, .22), respectively. As indicated by (10.26) or

(10.27), the stresses σ_1 and σ_2 can be determined when the stress-optical re-tardations Δs_1 and Δs_2 are known, provided that the stress-optical constants a_t, b_t or a_r, b_r have been measured from a simple calibration test. A simple way to measure the stress-optical retardations Δs_1 and Δs_2 is to use inter-ference between the active ray that passes through the specimen, which is retarded by Δs_1 and Δs_2 along the principal directions, and the reference ray that does not pass through the specimen. This evaluation, as was indicated previously, is accomplished by use of various types of interferometers whose function is to divide the incident light into reference and active rays. In many types of interferometers, the active ray is directed so as to pass more than once through the specimen (Figs. 10.4e–h). If the active ray passes β times through the specimen, then the optical retardations along the principal direc-tions σ_1 and σ_2 between the active and the reference rays will be $\beta \Delta s_1$ and $\beta \Delta s_2$, respectively.

As pointed out in Section 10.4, optical interference between the active and the reference rays can also be observed without the use of an interferometer, by considering the reflections from the two faces of the specimen. For this, the variation of the optical-path length of the ray, reflected from the front face of the specimen, is

$$\Delta s = - \Delta d n_0, \tag{10.51}$$

where Δd is the thickness variation of the specimen due to loading, whereas the variation of the optical-path length for the ray reflected from the rear face of the specimen is given by (10.20) and (10.22).

Thus, for all interferometric-fringe patterns, the fringe orders N_1 and N_2 along the principal-stress directions, are

$$N_{1,2} = \frac{\beta \Delta s_{1,2}}{\lambda}. \tag{10.52}$$

They can be represented in a uniform manner by

$$N_1 = (A\sigma_1 + B\sigma_2)d \tag{10.53}$$
$$N_2 = (B\sigma_1 + A\sigma_2)d,$$

where

$$A = \frac{\beta a}{\lambda}, \quad B = \frac{\beta b}{\lambda}. \tag{10.54}$$

From (10.10, 12, 53) it can be shown that all interferometric-fringe patterns consist of two overlapping families of fringes, whose orders N_1 and N_2 are expressed by (10.53). Constants A and B in these equations, as can be seen in (10.11) and (10.21), depend on the elastic and stress-optical behavior of the specimen, the particular type of experimental apparatus used, the wavelength of the incident light, and the refractive index of the medium that surrounds the model.

From the foregoing brief discussion, it is concluded that all fringe patterns obtained by the various interferometric techniques consist of two overlapping fringe systems that are the contours of the stress-optical retardations Δs_1 and Δs_2, along the principal stress directions, independently of the specific experimental technique or apparatus used.

Interpretation of the patterns. The two superimposed families of curves $N_1 = N_1(x, y)$ and $N_2 = N_2(x, y)$ do not remain separate but, according to the mechanical interference (moiré) phenomenon [10.8], they combine to form two new families of curves $M_1 = M_1(x, y)$ and $M_2 = M_2(x, y)$, given by

$$M_1 = N_1 - N_2$$
$$M_2 = N_1 + N_2.$$

$$(10.55)$$

Taking into account (10.53), we obtain for M_1 and M_2

$$M_1 = (A - B)(\sigma_1 - \sigma_2)d$$
$$M_2 = (A + B)(\sigma_1 + \sigma_2)d.$$

$$(10.56)$$

Equations (10.56) express the fact that interferometric-fringe patterns consist of two families of curves that express the difference (isochromatics) and the sum (isopachics) of principal stresses. Both of these families form half-tone curves, because they are created by mechanical interference. However, both of these new families are visible only in special cases, under suitable circumstances; in general, only one family of curves is visible under normal experimental conditions.

The complicated nature of moiré patterns formed by two superimposed dissimilar grid systems has been shown by *Zandman* et al. [10.25]. They derived geometrically the equations that govern the moiré patterns formed by two linear periodic, mutually inclined systems, for the most-general case when the opaque strips of each system are not equal to the corresponding transparent strips (transmittance different than 50 percent). They also established criteria for production of either of these moiré-fringe systems. They showed, for example, that for the case of the "positive plus positive" original grids, that is, when $(b/c) = (b'/c')$, where b and c are the opaque and transparent widths, respectively, of the first system and b' and c' the corresponding quantities for the second system, a single moiré pattern is formed when the following inequality is satisfied:

$$c \leq b \leq 2c.$$

They indicated, however, that apart from the fact that two moiré patterns can be derived by theoretical considerations, under certain circumstances, in actual cases, under normal experimental conditions, only one moiré pattern is formed. Furthermore, they derived the interesting fact that, for the case of mixed systems, composed of a fine grid system placed between a coarse grid system, moiré fringes interspersed with discontinuities can be produced.

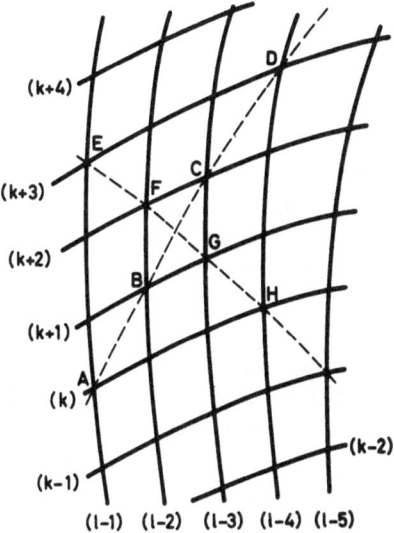

Fig. 10.7. Schematic representation of indicial formation of subtractive (along diagonals *EFGH*) and additive (along diagonals *ABCD*) moiré patterns

For the usual case of gratings that are used in experimental mechanics, which are composed of lines of equal opaque and transparent widths, *Theocaris* [10.8] showed that, although theoretically two moiré patterns, the subtractive and the additive, are formed, however, the effective or visible moiré pattern is that in which the interfringe spacing is the longest, that is, the pattern in which the moiré fringes coincide with the shortest diagonals of the individual quadrangles formed by the intersection of gratings (Fig. 10.7). Furthermore, he indicated that because the shape and dimensions of the individual quadrangles change with the relative displacement of the two gratings, the effective moiré pattern may alternate from the subtractive to the additive type and vice versa. These two different types of moiré patterns are separated by the so-called *commutation moiré boundary* or simply the *moiré boundary* [10.8]. This boundary is defined by

$$\Psi(x,y) = \frac{\partial N_1}{\partial x} \frac{\partial N_2}{\partial x} + \frac{\partial N_1}{\partial y} \frac{\partial N_2}{\partial y} = 0, \tag{10.57}$$

where $N_1 = N_1(x,y)$ and $N_2 = N_2(x,y)$ are the parametric families of the combined systems.

By taking into account (10.53), we obtain

$$\Psi(x,y) = AB\left[\left(\frac{\partial\sigma_1}{\partial x}\right)^2 + \left(\frac{\partial\sigma_1}{\partial y}\right)^2 + \left(\frac{\partial\sigma_2}{\partial x}\right)^2 + \left(\frac{\partial\sigma_2}{\partial y}\right)^2\right] d$$
$$+ (A^2 + B^2)\left(\frac{\partial\sigma_1}{\partial x}\frac{\partial\sigma_2}{\partial x} + \frac{\partial\sigma_1}{\partial y}\frac{\partial\sigma_2}{\partial y}\right) d .$$

(10.58)

When $\Psi(x,y) > 0$, the subtractive moiré pattern defined by $N = N_1 - N_2$ is effective, whereas, when $\Psi(x,y) < 0$, the additive moiré pattern, defined by $N = N_1 + N_2$, is effective.

The foregoing discussion shows that the interpretation of interferometric-fringe patterns is rather complicated. These patterns consist of one or two families of half-tone curves that represent isochromatics and/or isopachics. The appearance of one family is common; two families appear only under special experimental conditions. However, for the common case in which only one family of curves is formed, their nature, as isochromatics or isopachics, cannot be decided with certainty, nor in what region of the specimen either of these families appears.

However, although generally speaking the two families N_1 and N_2 of retardations Δs_1 and Δs_2 along the principal-stress directions combine to form isochromatic and isopachic fringe patterns, when these two families are too sparse or when their pitches have a large mismatch, their mechanical interference is rather poor, so that the families of N_1 and N_2 can be more visible than the resulting mechanical interference. These results complicate even more the interpretation of the interferometric patterns.

General rules. From the foregoing discussion of interpretation of the interferometric-fringe patterns, the following general rules can be deduced:

1) The unified character of all interferometric, and, consequently, holographic, fringe patterns, which represent isochromatics and isopachics as half-tone fringes, was shown, using physical concepts. Thus, the interpretation of interferometric patterns formulated by *Nisida* and *Saito* [10.22], who used a Mach-Zehnder interferometer, that half-tones are identified as isochromatics and dark-bright bands as isopachics is not always correct. Similarly, the discussion by *Post* [10.21] of the interpretation of interferometric patterns can be considered only a crude approximation. He mentioned that, for different types of interferometers, different kinds of curves, including isochromatics, isopachics, or contours of constant optical path lengths along the two principal-stress directions, or combinations of them, can be obtained. However, when *Post* says that contours of constant optical-path lengths can be obtained, he means, in reality, a point by point procedure and not a full-field method. Indeed, he used light plane polarized along the principal stress directions. Therefore, his patterns represent contours of constant optical-path lengths only along the isoclinic whose parameter coincides with the angle of inclination of the plane-polarized light used. The present analysis showed that contour lines of constant

optical path lengths can be obtained in the final pattern only when these two families are quite sparse, so that their mechanical interference is quite poor.

2) The most common case, under normal experimental conditions, is that only one family of curves, representing either isochromatics or isopachics, is obtained. It is possible, however, for the family of curves that appears in some region of the pattern to represent isochromatics and in another region isopachics.

3) Even for a given apparatus and model material, it cannot be predicted in advance, whether and in what region of the model the family of fringes shown by the pattern, represents isochromatics or isopachics. The type of the family of fringes that appears depends on a) the stress-optical constants, incorporating the mechanical and the optical constants of the model material [see (10.19) and (10.21)], b) the type of apparatus used, c) the wavelength of the light used, d) the medium that surrounds the model, e) the photographic procedure used to record the interferometric patterns (by varying the exposure time and the developing procedure, the widths of the dark bands in fringe patterns can be varied, so that the characteristics of the moiré patterns created can be varied), and f) the state of stress induced in the specimen by loading.

Although the influence of the first five of these factors on the interferometric patterns can be controlled and predicted in advance, and general laws can be established, no such prediction can be made for the last factor, because the stress distribution in a given model is generally unknown. This last factor makes impossible the formulation of general laws on the nature of interferometric patterns.

4) The great confusion and conflict that has arisen in the literature about the interpretation of interferometric patterns results from the fact that the six factors listed in 3), as influencing these patterns, take different values in the various experimental techniques. This has resulted in different interpretations of the corresponding patterns. The various authors restricted themselves to the interpretation of their particular patterns, obtained under special conditions, and did not attempt a general interpretation of these patterns.

5) For the more favorable case, when both isochromatics and isopachics are obtained in interferometric fringe patterns, a fact which necessitates special precautions being taken during application of the method, difficulties may arise from the uncertainty of deciding what family of curves represents isochromatics or isopachics in each region of the specimen. As was shown, no invariant identification of isochromatics or isopachics with half-tones or dark-bright bands, respectively, or vice-versa, can be made.

6) It is, therefore, concluded that interferometric-fringe patterns must be supplemented by isochromatics obtained by classical photoelasticity or by isopachics obtained by interferometry, using optically inert materials. This is necessary even in the case when both isochromatics and isopachics are obtained, for identification of what family of the obtained fringes represents either isochromatics or isopachics.

7) Interpretation of interferometric-fringe patterns that is based on the mechanical interference of the two original patterns that correspond to the stress-optical retardations along the principal-stress directions, is dependent on the assumption that the bright and dark bands of both these patterns are of equal width. In this case, the moiré fringes and the moiré boundary are expressed by (10.55) and (10.57), respectively. However, when the exposure times of the unstressed and stressed models in interferometry are different, the widths of the dark and bright bands of the original patterns obtained can be quite different, and the interpretation should be modified.

8) The foregoing discussion showed that simultaneous appearance of both isochromatics and isopachics in interferometric patterns necessitates special photographic techniques. The usual case is that over the whole specimen only isochromatics or isopachics, or sometimes isochromatics in one region and isopachics in another region, predominate. By applying interferometric methods to typical plane-stress problems, it can be shown that the transition from isochromatics to isopachics, or the movement of the moiré boundary, for the case when isochromatics predominate in one region and isopachics in another, can be caused for one model material by changing the refractive index n_0 of the medium that surrounds the model. This suggests that isochromatics and isopachics can be recorded separately from the same model by changing the refractive index of the oil in which the specimen is immersed. Such oil baths are generally used in interferometry, in order to have the refractive index of the medium surrounding the model approximate that of the model, so as to eliminate the undesired effects of refraction of the rays as they enter the model. Therefore, a new interferometric method may be developed, by which isochromatics and isopachics can be obtained separately, by changing the refractive index of the medium that surrounds the model.

For more details of the above-described unified interpretation of the interferometric, as well as the holographic, fringe patterns and for some illustrative examples, the interested reader is referred to [10.23].

References

10.1   M. M. Frocht: *Photoelasticity*, Vols. I. II (Wiley and Sons, New York 1941, 1948)
10.2   M. M. Frocht: Proc. Soc. Exp. Stress Anal. **6**, 39 (1948)
10.3   E. G. Coker, L. N. G. Filon: *A Treatise on Photoelasticity*, 2nd ed. (Cambridge University Press, London 1957)
10.4   M. Hetényi: *Handbook of Experimental Stress Analysis*, 1st ed. (Wiley and Sons, New York 1950)
10.5   D. C. Drucker: J. Appl. Mech., Trans. ASME **10**, A156 (1943)
10.6   H. Favre: Rev. Opt. **8**, 193, 241, 289 (1929)
10.7   J. H. A. Brahtz, J. E. Soehrens: J. Appl. Phys. **10**, 242 (1939)
10.8   P. S. Theocaris: *Moiré Fringes in Strain Analysis*, 1st ed. (Pergamon Press, London 1969)
10.9   P. S. Theocaris: J. Mech. Phys. Sol. **11**, 181 (1963)
10.10 G. Mesmer: Proc. Soc. Exp. Stress Anal. **13**, 21 (1956)
10.11 C. Fabry: C. R. Acad. Sci. **190**, 457 (1930)

10.12 H. Favre, W. Schumann: Proc. Int. Symp. on Photoelasticity, ed. by M. M. Frocht, Illinois Inst. of Technology, Chicago, Ill., 1961 (Pergamon Press, New York 1963) pp. 3–25
10.13 P. S. Theocaris, E. E. Gdoutos: J. Phys. D: Appl. Phys. **7**, 472 (1974)
10.14 A. A. Wells, D. Post: Proc. Soc. Exp. Stress Anal. **16**, 69 (1958)
10.15 D. Post: J. Opt. Soc. Am. **48**, 309 (1958)
10.16 D. Post: Proc. Soc. Exp. Stress Anal. **13**, 119 (1956)
10.17 D. Post: Proc. Soc. Exp. Stress Anal. **12**, 191 (1954)
10.18 D. Post: Proc. Soc. Exp. Stress Anal. **12**, 143 (1955)
10.19 D. Post: Proc. Soc. Exp. Stress Anal. **12**, 99 (1954)
10.20 D. Post: J. Opt. Soc. Am. **44**, 243 (1954)
10.21 D. Post: Exp. Mech. **7**, 233 (1967)
10.22 M. Nisida, H. Saito: Exp. Mech. **4**, 366 (1964)
10.23 P. S. Theocaris, E. E. Gdoutos: J. Strain Anal. **13**, 95 (1978)
10.24 P. S. Theocaris: Appl. Opt. **10**, 2240 (1971)
10.25 F. Zandman, G. S. Holister, V. Brcic: J. Strain Anal. **1**, 1 (1965)

11. Holographic Photoelasticity

11.1 Introduction

Holography or *wavefront reconstruction* is an optical method for complete recording of a given wavefront, that is, for the recording of both the amplitude and the phase of the light wave. The term "holography" comes from the Greek word "holos" which means "the whole", and "graphein" which means recording, and this explains that holography is concerned with the complete recording of the characteristics of a light wave. The term "wavefront reconstruction" is also used, because, by the holographic method the wavefront of an illuminated object, which reflects, diffuses, refracts or diffracts the light, is reconstructed by means of an optical process.

Before holography, the only way of recording and retaining as a permanent record the picture of an object was the process of photography. In the photographic process, the wavefront emitted by an object is transformed by the optical lens and impinges on a photosensitive plate, which responds to the intensity of the incident light. Thus, by the process of photography only the amplitude-not the phase-of the light wave can be recorded. However, in the holographic process the emitted light wave from an object is recorded in such a way that, by subsequent illumination of the obtained record, the original object wave can be reconstructed. Visual observation of this reconstructed wave results in perception of the object in space, with appropriate perspectives from changing points of view.

To record the phase of a light wave in the holographic method, the optical interference phenomenon is used. Interference patterns that are stable in time are formed by two coherent light beams. They incorporate the phase difference of the interfering beams. Thus, to record simultaneously both the amplitude and the phase of a light wave in holography, the interference pattern of the wavefront emitted by the object, called the *object wave*, and another known wavefront, called the *reference wave*, are recorded. For this reason, the object is illuminated by a light source and the wavefront of the light scattered or diffracted by the object, the object wave, is received on a photographic plate. On the same photographic plate also impinges the wavefront emitted by the same light source, the reference wave. The object wave and the reference wave are mutually coherent because they both come from the same source. The interference pattern produced by the combination of the object and reference waves is recorded on the photographic plate. After development, it is called the *hologram*. It consists of a complex distribution of transparent and opaque areas that correspond to the recorded interference fringes. When it is viewed with ordinary light, the photographic deposit does not seem to have any relation to the object from which it was generated.

The formation of the hologram constitutes the first step of the holographic process. To reconstruct the object wave, the hologram is illuminated by a light beam similar to the reference wave used for the formation of the hologram, and the transmitted wave is observed. Because the hologram is in essence a transmittance grating, formed by interference of the object and reference waves, when it is illuminated by either of these waves, the first-order diffraction pattern from the grating represents the other wavefront. Thus, by illuminating the hologram by the reference wave, a duplicate of the object wave is obtained as the diffraction pattern produced by the hologram. In this way, the original object wave is reconstructed from the hologram. The formation of a duplicate of the original wavefront, which, when it is conveniently observed, gives an exact replica of the appearance of the original object, constitutes the second step in the holographic process.

Thus, holography is a two-step process, by which an image of an object is formed without using lenses. In the first step, the required information is stored in the hologram; in the second step, the hologram is suitably illuminated, so that the obtained wavefront is that originally emitted by the object. Owing to the complete reconstruction of the wavefront of the object, holography forms a three-dimensional image of the object.

The general idea of the two-step lensless imaging process of wavefront reconstruction was invented by *Gabor* [11.1–3] in 1948. He first conceived the idea that, if on the wavefront emitted by an object, another coherent reference wavefront is superimposed, then both the amplitude and the phase of the object wavefront can be recorded, despite the fact that the photographic film responds only to intensity. Furthermore, he demonstrated that an image of the object can be obtained by illuminating the recorded interference pattern of the object and reference wavefronts by the reference wave. However, in 1948, a sufficiently intense coherent-light source was not available, so holography could not be practically realized. Therefore, with rather limited results, holography almost abandoned by its first investigators.

The renaissance of holography, some fifteen years later, was mainly due to the invention of lasers, which played an essential role in the evolution of modern holography.

In the first holographic arrangements used by *Gabor* the light source, the object, and the recording medium (photographic plate) were located along the same axis. The object wavefront was that diffracted by the object, which had to be transparent. Thus, on the photographic plate two waves were superimposed, the strong, uniform plane reference wave and the weak object wave. The resulting interference pattern, after photographic development, was then illuminated by the reference wave, to obtain the image of the object. The Gabor hologram, called the *in-line hologram*, suffered from two serious limitations that restricted its applicability. First, the object had to be highly

transparent. Otherwise, the coexistent reference wave was so strong compared with the object wave that the object wave was entirely obliterated.
Second, with the *in-line* arrangement, two twin images were formed, which
were difficult to separate. However, in the Gabor hologram a common light
source could be used.

Following the experiments of *Gabor*, many other attempts were made to
eliminate or weaken the effect of the twin wave. This problem was solved by
Leith and *Upatnieks* [11.4, 5] in 1962. They introduced a new approach to
optical holography, based on communication-theory techniques. In 1964,
they obtained convincing evidence that holography may have practical
applications provided that a coherent light beam of sufficient intensity, emitted
from a laser, can be used. They developed a fairly simple means for elimination
of the twin wave, consisting of illuminating the photographic plate with an
off-axis reference beam. In the reconstruction a corresponding off-axis wave
is used, which results in the separation of the real and the virtual images.
However, for this *off-axis hologram* a coherent light beam was indispensable;
such light was provided by the invention of the gas laser. Two years later
Leith and *Upatnieks* [11.6] introduced diffuse-illumination holography, thus
eliminating the necessity of using transparent specimens.

After *Leith* and *Upatnieks*, holography progressed in an accelerated
manner. A large number of papers and articles have been devoted to improving
holographic techniques and to widening the range of their applications.
At the present time, holography is so extended that the full description of its
theory and its applications would require a separate book, completely devoted
to the subject.

In the present chapter, we shall introduce the Jones calculus as it applies to
holography and show its possible applications to only those areas that are
related to the determination of quantities that are involved in the mechanics
of deformable bodies, i. e., displacements, stresses, and strains. Furthermore,
in this chapter we shall restrict ourselves first to showing by the Jones calculus
the general principles of holography and then we shall refer to holographic
interferometry and its applications in photoelasticity. The reader interested
in further details of the theory, the experimental arrangements used, and the
various applications of holography, as well as in a complete bibliography
of this subject is referred to specialized books indicated in [11.7–10].

11.2 Basic Holographic Equations

As has already been stated, holography is a two-step process for the complete
recording of a wavefront emitted by an object. In the first step, called *the
recording process*, the object wave interferes with a known, usually plane,
reference wave and the resulting interference pattern is stored in a recording
medium, usually a photographic plate with a high-resolution emulsion. Thus,
the amplitude and phase of the object wave are converted into intensity

variations of the interference pattern, which, after development, is called the *hologram*. In the second step, called the *reconstructing process*, the hologram is illuminated by the reference wave and the resulting diffraction pattern has the complete characteristics, that is the amplitude and the phase, of the object wave. We shall now give the basic equations that describe the recording and the reconstructing processes of holography. All of the optical transformations involved in both processes will be described by use of the Jones calculus.

Recording Process. Let a_1 and a_2 be the Jones vectors that describe the beams emitted when objects 1 and 2 are illuminated by a coherent light beam, namely light from a gas laser (Fig. 11.1a). The light wave emitted by the object numbered 1 may represent the reference wave, whereas that emitted by the object numbered 2 represents the object wave. Let the Jones vectors a_1 and a_2 be

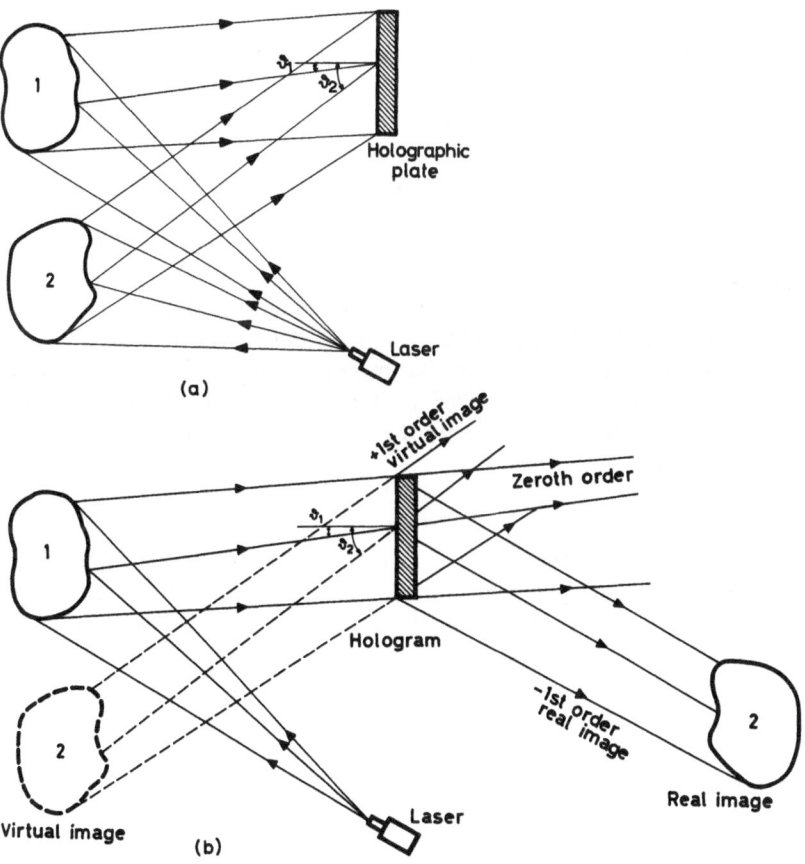

Fig. 11.1 a and b. Formation of hologram of an object during the recording process (a) and corresponding reconstructing process of the wavefront of object 2 by illuminating the hologram with the wavefront of object 1 (b)

$$\boldsymbol{a}_1 = \begin{bmatrix} A_{x_1}\, e^{i\delta_{x_1}} \\ A_{y_1}\, e^{i\delta_{y_1}} \end{bmatrix} \qquad \boldsymbol{a}_2 = \begin{bmatrix} A_{x_2}\, e^{i\delta_{x_2}} \\ A_{y_2}\, e^{i\delta_{y_2}} \end{bmatrix}. \tag{11.1}$$

If both these waves $\boldsymbol{a}_1$ and $\boldsymbol{a}_2$ are incident on the recording medium, namely the holographic plate of Fig. 11.1a, at angles ϑ_1 and ϑ_2, respectively, then the Jones vectors $\boldsymbol{a}_1'$ and $\boldsymbol{a}_2'$ of the incident reference and object beams will be

$$\boldsymbol{a}_1' = \boldsymbol{a}_1\, e^{-ikx\sin\vartheta_1} \qquad \boldsymbol{a}_2' = \boldsymbol{a}_2\, e^{-ikx\sin\vartheta_2}. \tag{11.2}$$

The two waves $\boldsymbol{a}_1'$ and $\boldsymbol{a}_2'$ form an interference pattern, whose Jones vector $\boldsymbol{a}'$, according to the additive property of Jones vectors, will be

$$\boldsymbol{a}' = \boldsymbol{a}_1' + \boldsymbol{a}_2'.$$

The light intensity I corresponding to the Jones matrix $\boldsymbol{a}'$ is (Sect. 3.4.5)

$$I = \tilde{\boldsymbol{a}}'\boldsymbol{a}' = \tilde{\boldsymbol{a}}_1'\,\boldsymbol{a}_1' + \tilde{\boldsymbol{a}}_2'\,\boldsymbol{a}_2' + \tilde{\boldsymbol{a}}_1'\,\boldsymbol{a}_2' + \tilde{\boldsymbol{a}}_2'\,\boldsymbol{a}_1' \tag{11.3}$$

where $\tilde{\boldsymbol{a}}'$ is the hermitian conjugate matrix of matrix $\boldsymbol{a}'$.

Reconstructing Process. In the reconstructing process, the holographic plate exposed to the light intensity I, after development, is illuminated by the reference wave emitted by the object numbered 1 (Fig. 11.1 b). For the quantitative description of these operations, we must define the response of the emulsion of the holographic plate to the impinging light intensity. A photosensitive plate is usually characterized by the so-called $H - D$ curve, which gives the variation of the optical density D of the developed plate,

$$D = \log(1/J), \tag{11.4}$$

where J is the transmittance intensity, that is, the ratio of the intensity of light transmitted through the emulsion to that incident on it, versus the logarithm of the exposure E, defined as the product of the exposure time t and the impinging intensity $I\,(E = t \cdot I)$. Figure 11.2 gives a typical form of the $H - D$ curve [11.10].

If γ is the slope of the straight part of the $H - D$ curve, then we have

$$\log(1/J) = \gamma(\log E - \log K)$$

or

$$J = K^\gamma E^{-\gamma} = K^\gamma t^{-\gamma} I^{-\gamma}, \tag{11.5}$$

where $\log K$ is the intercept of the linear portion of the $H - D$ curve with the

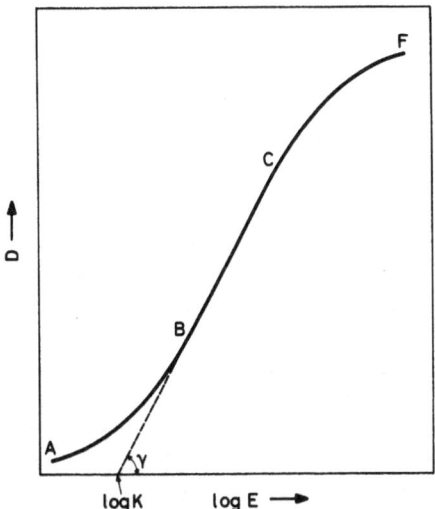

Fig. 11.2. Typical form of the $H-D$ curve

$\log E$ — axis. For the cases studied here, we are generally interested on the amplitude transmittance τ, rather than J, given by

$$\tau = J^{1/2} .$$

If the holographic plate is developed as a negative, then we have the amplitude transmittance τ_n, expressed by

$$\tau_n = K^{\gamma_n/2} t_n^{-\gamma_n/2} I^{-\gamma_n/2} = k I^{-\gamma_n/2} . \tag{11.6}$$

When the plate is developed as positive, we have the amplitude transmittance

$$\tau_p = J_p^{1/2} \propto (J_n)^{-\gamma_p/2} \propto (I^{-\gamma_n})^{-\gamma_p/2} \propto k I^{\gamma/2} , \tag{11.7}$$

where $\gamma = \gamma_n \gamma_p$, and the equality sign is replaced by the proportionality symbol $\propto$.

By using (11.3) for the light intensity I, we obtain the amplitude transmittance τ of the hologram,

$$\tau = k(\tilde{a}_1' a_1' + \tilde{a}_2' a_2' + \tilde{a}_1' a_2' + \tilde{a}_2' a_1')^{\gamma/2} . \tag{11.8}$$

If the hologram is illuminated with the reference wave a_1' and we assume for simplicity that $\gamma = 2$ without loss of generality, we obtain from (11.8) for the Jones vector of the light transmitted through the hologram,

$$a_T = k a_1' (\tilde{a}_1' a_1' + \tilde{a}_2' a_2' + \tilde{a}_1' a_2' + \tilde{a}_2' a_1') , \tag{11.9}$$

or, by taking into account (11.2), we obtain for the intensity of the transmitted light,

$$I = k \left[(A_{x_1}^2 + A_{y_1}^2) + (A_{x_2}^2 + A_{y_2}^2) \right] a_1 e^{-ikx\sin\vartheta_1} + k a_1 \tilde{a}_1 a_2 e^{-ikx\sin\vartheta_2}$$
$$+ k a_1 \tilde{a}_2 a_1 e^{-ikx(2\sin\vartheta_1 - \sin\vartheta_2)} \, .$$

(11.10)

When $\gamma \neq 2$, terms occur that are of order higher than the first. When entered in the expression of τ, they do not change the conclusions derived for $\gamma = 2$. Therefore, this case will not be considered in the present analysis.

Equation (11.10) consists of three terms. The first represents the product of the reference wave a_1' and the sum of the intensities of the reference and the object waves, which can approximately be assumed constant all over the field. Therefore, this term represents a reconstructed image of the reference wave. It corresponds to the zero-order diffraction pattern from the hologram.

The second term, which corresponds to the first-order diffraction pattern, represents a light beam that propagates in the same direction as the original object beam; its Jones vector is proportional to $a_1 \tilde{a}_1 a_2$. By taking into account (11.1) and making the matrix multiplications, this vector takes the form

$$a_1 \tilde{a}_1 a_2 = A_{x_1}^2 \begin{bmatrix} A_{x_2} e^{i\delta_{x_2}} \\ A_{y_2} e^{i\delta_{y_2}} \end{bmatrix} + (A_{y_1}^2 - A_{x_1}^2) \begin{bmatrix} 0 \\ A_{y_2} e^{i\delta_{y_2}} \end{bmatrix}$$
$$+ A_{x_1} A_{y_1} \begin{bmatrix} A_{y_2} e^{i(\delta_{y_2} - \delta_{y_1} + \delta_{x_1})} \\ A_{x_2} e^{i(\delta_{x_2} - \delta_{x_1} + \delta_{y_1})} \end{bmatrix}$$

The first term of this expression is proportional to the original object wave a_2 which is therefore reconstructed by the hologram. The other two terms of the above expression represent two light waves that are focused in the same direction as that of the object wave a_2. The influence of these two waves on the reconstruction of the object wave is to add some vagueness in the image of the object, which is more pronounced in the regions where the above two vectors of the corresponding waves take significant values. Elimination of the second term of the above expression can be achieved by using a circularly polarized light as reference wave.

Finally, the third term of (11.10) represents a light beam that for the case of small angles ϑ_1, ϑ_2 ($\sin\vartheta_{1,2} \simeq \vartheta_{1,2}$) propagates in a direction that makes an angle $(\vartheta_1 - \vartheta_2)$ with the reference beam. By arguments analogous to those of the second term it can be proved that the third term contains the conjugate vector $\tilde{a}_2$ of vector a_2. Therefore, by illuminating the hologram with the reference wave, three new waves arise: the *reference wave*, the *real object wave*, and the *conjugate object wave*. The second and third of these waves give the virtual and real images of the object. The reconstructing process in which the hologram is illuminated by the reference wave and the virtual and the real images of the object are formed, is shown in Fig. 11.1b. This figure shows that, in the reconstructing process, the virtual image of the object is formed in the original location of the object.

Figure 11.1 shows that, in the case when the reference and the object waves are not on the same axis, the reference wave and the two object waves that arise by illuminating the hologram with the reference wave do not overlap. They are completely separated. This arrangement of the off-axis hologram was invented by *Leith* and *Upatnieks* and represents a major contribution to the method of holography, after its discovery by *Gabor*. The arrangement of the off-axis holography necessitates, however, highly coherent light sources to compensate the large optical-path differences in the two interfering object and reference waves. This was achieved only by the invention of the laser. Thus, the renaissance of holography in the earlier sixties followed immediately the invention of lasers in the same period.

In the usual holographic arrangements, in order that the reference wave be sufficiently uniform, it is provided by reflection from a plane mirror. In this case, the reference wave is a plane wave. When it is incident at a small angle ϑ_x relative to the Ox axis, it has the form

$$a_1 = \begin{bmatrix} A_x e^{i\delta_x} \\ A_y e^{i\delta_y} \end{bmatrix} e^{-i\delta_{x_1}} \tag{11.11}$$

where

$$\delta_{x_1} = \frac{2\pi}{\lambda} \vartheta_x .$$

Schematic diagrams of the optical arrangements for the recording and the reconstructing processes of holography, when a plane mirror is used to provide the reference wave, are shown in Fig. 11.3 and 11.4, respectively. Figure 11.5 shows these arrangements when the object is transparent and a beam splitter is used to provide the reference and the object-illumination waves.

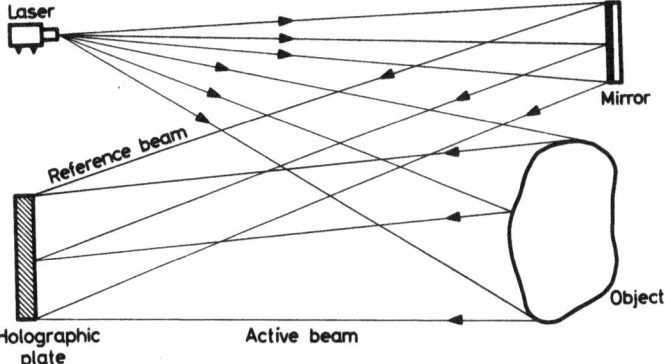

Fig. 11.3. Optical arrangement for recording an object by use of a plane wave reflected from a mirror as reference wave

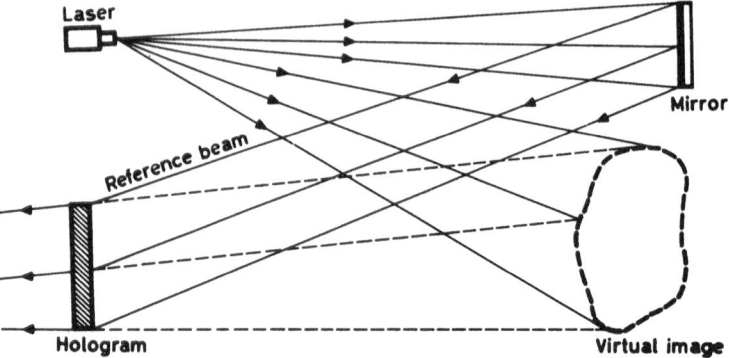

Fig. 11.4. Optical arrangement for the reconstructing process of the object of Fig. 11.3 by illuminating the hologram with the reference wave

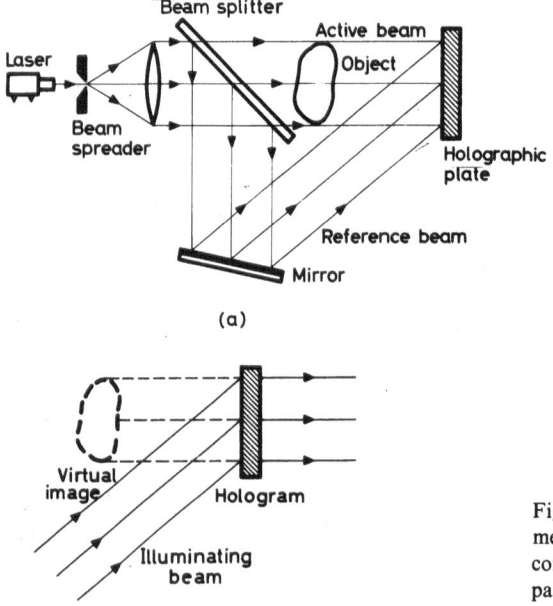

Fig. 11.5a and b. Optical arrangements for the recording (a) and reconstructing (b) processes of a transparent object by use of a beam splitter

11.3 Physical Explanation of Holography

In the preceding section, the formation of the image of an object by the two-step method of holography was analyzed by use of the Jones calculus for the description of the optical phenomena that occur in both the recording and the reconstructing processes. In the present section, a physical explanation of the holographic process will be provided, based on the relationship between the phenomena of interference and diffraction of light.

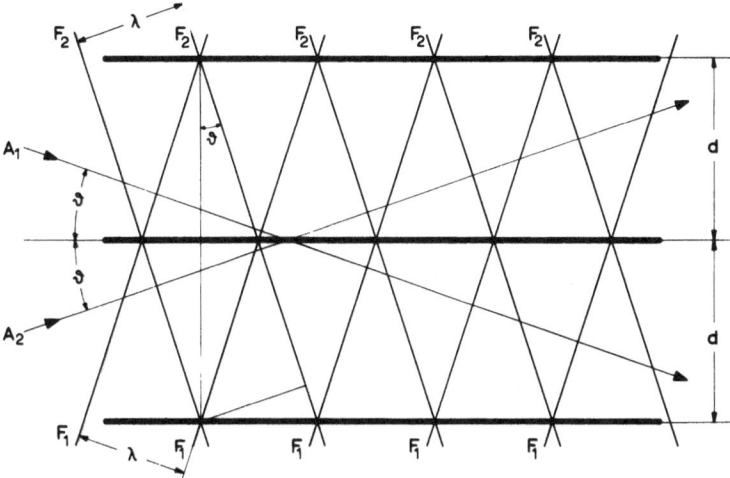

Fig. 11.6. Geometry of the interference pattern of two monochromatic plane waves with wavefronts F_1 and F_2

Let us first consider the interference of two plane, monochromatic, coherent waves travelling along the directions A_1 and A_2, which are at an angle 2ϑ to each other, with wavefronts F_1 and F_2, respectively (Fig. 11.6). These two waves form an interference pattern, consisting of straight equidistant lines at distances d from each other. From the geometry of Fig. 11.6 it follows that

$$d = \frac{\lambda}{2\sin\vartheta},\tag{11.12}$$

where λ is the wavelength of the interfering waves.

If that interference pattern is recorded by a photosensitive medium, developed, fixed, and printed as positive, a diffraction grating with pitch d is formed. Let this grating be illuminated by the one of the interfering beams, say the beam with wavefront F_1 (Fig. 11.7). Then, according to the laws of diffraction of light, the diffracted pattern formed by the grating will consist of three waves, the illuminating wave, which is the zero-order diffraction pattern, and the real and virtual images F_2, F_2' of the other interfering beam, which are the two first-order diffraction patterns formed by the grating. Thus, as a result of this two-step process, the first of which consists of the formation of the interference pattern of the two beams and its recording and developing, and the second corresponds to the illumination of the thus-formed grating by one of the initial beams, the other initial beam is reconstructed.

The same procedure is followed in the holographic method to reconstruct the image of an object. One of the two interfering waves is the object wave; the other is the reference wave. Both of these waves interfere; their interference

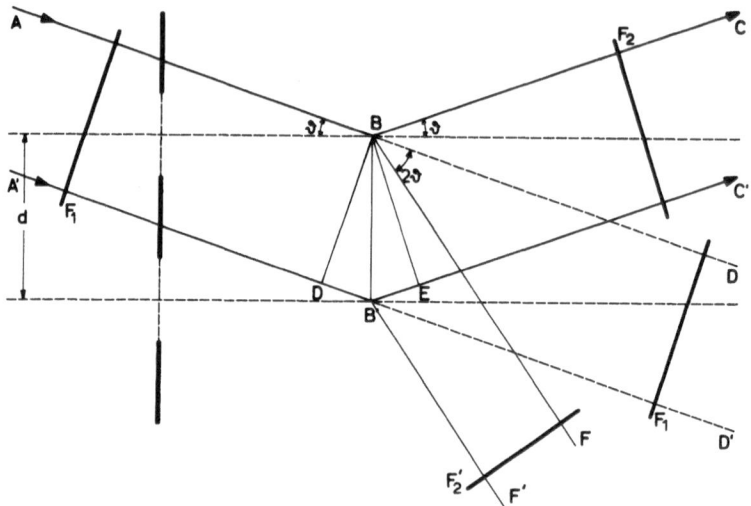

Fig. 11.7. Geometry of the diffraction pattern of a grating illuminated by a plane wave

pattern, recorded on the holographic plate, after development, forms the hologram, which is in essence a diffraction grating with varying pitch and orientation. If, now, the hologram is illuminated by one of the interfering waves, namely, the reference wave, the resulting diffraction pattern gives as the zero-order diffraction wave the reference wave, while the two first-order diffraction waves form the real and virtual images of the object. Thus, the object wave is reconstructed by the two-step process of holography.

Another simple explanation of the recording and the reconstructing processes of the holographic method, based on the mechanical-interference (moiré) phenomenon, was provided in [11.11–13]. Let us consider two super-imposed gratings whose indicial equations are

$$S(x,y)=s,$$
$$R(x,y)=r,$$

(11.13)

where s and r are index parameters that take all integer values from minus to plus infinity. Each separate curve of the family of each grating is specified by the particular value assigned to the parameter s or r. The resultant mechanical interference, or moiré, pattern of the superposition of these two gratings is two new indexed families of curves, with indicial equations of the form

$$M(x,y)=m, \quad \text{and} \quad \bar{M}(x,y)=\bar{m},$$

(11.14)

where the parameters m and $\bar{m}$ satisfy

$$s-r=m,$$
$$s+r=\bar{m}.$$

(11.15)

The family of the moiré pattern for which the equation is $s-r=m$ or $s+r=\bar{m}$ is called the subtractive or the additive moiré pattern, respectively. The first of (11.15) can be written in three forms,

$$s-r=m,$$

$$s-m=r, \quad \text{or} \qquad (11.16)$$

$$r+m=s.$$

Equations (11.16) indicate that the subtractive moiré pattern of the families s and r gives the family m; the subtractive pattern of the families s and m gives the family r; and the additive pattern of the families r and m gives the family s. All of these three equations, including two subtractive and one additive moiré patterns, are satisfied by any two overlapping families of curves and the induced new family of their mechanical interference.

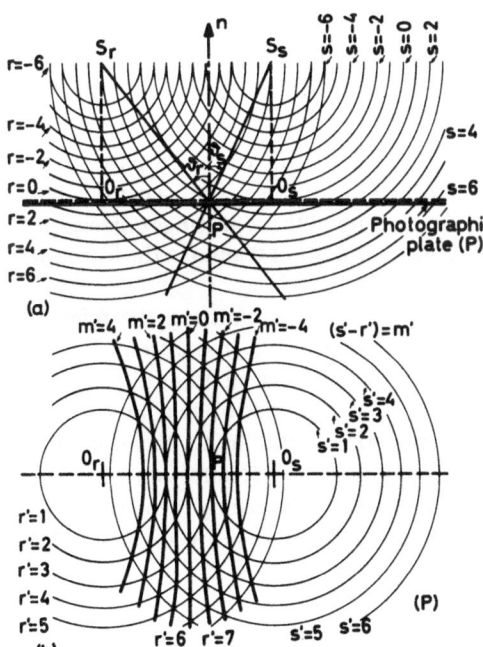

Fig. 11.8 a and b. Geometrical interpretation of the recording process of the spherical waves emitted from sources S_r and S_s by using the principles of the mechanical-interference (moiré) phenomenon

Let us now consider two point light sources S_r and S_s that emit spherical waves and let Fig. 11.8 represent the intersections of the wavefronts of these two waves with a plane that passes through the points S_r and S_s. These intersections are concentric circles with centers S_r and S_s; they can be represented by equations of the forms

$$a_s = A_s \cos\left(\omega t - \frac{2\pi}{\lambda} s\right),$$

$$a_r = A_r \cos\left(\omega t - \frac{2\pi}{\lambda} r\right),$$

(11.17)

where s and r are the distances of the points S_s and S_r from a given point P, on each separate family. Interference of the two waves produces dark and bright fringes on a photosensitive plate, whose trace on the plane of the concentric circles is shown in Fig. 11.8; points of the plate that have the same illumination, are represented by

$$s - r = m,$$

where m is an integer.

To either point light source S_s or S_r corresponds a circular grating with indices s' and r', respectively; as a result of mechanical interference of these two gratings two new gratings with equations of the form

$$s' - r' = m',$$

$$s' + r' = \bar{m}',$$

(11.18)

which correspond to the subtractive and to the additive moiré patterns are formed. The intersections of these gratings with the photosensitive plate give two gratings on it; recording and developing of these gratings corresponds to the recording process of the holographic method that results in the formation of the hologram.

Either of the sources S_s or S_r can be regarded as emitting the reference wave, say S_r, while the other one, S_s, emits the object wave. In the reconstructing process, the two gratings formed on the photosensitive plate are illuminated with the reference wave emitted by source S_r. As a result of mechanical interference of the gratings indexed m' or $\bar{m}'$ (corresponding to the hologram) with the grating indexed r' (corresponding to the reference wave) and in accordance with the indicial equations (11.16) concerning any two overlapping gratings, two new gratings with indices s' and $\bar{s}'$ are created. These greatings are expressed by

$$s' = m' + r'$$

$$\bar{s}' = \bar{m}' + r'.$$

(11.19)

The grating that is indexed s' gives the real object wave, whereas that indexed $\bar{s}'$ corresponds to the virtual object wave.

Thus, the formation of both the real and the virtual object waves obtained by the reconstructing process of holography can be explained by using the mechanical interference (moiré) phenomenon.

11.4 Holographic Interferometry

11.4.1 Introduction

Holographic interferometry provides one of the most elegant and powerful applications of holography. The method completes and extends the range of classical interferometry. Ever since the wave theory of light was generally accepted, interferometry has played an important role in the accurate measurement of small lengths, testing optical elements, and generally detecting optical-path-length variations. However, classical interferometry is restricted to examining highly polished surfaces of relatively simple shapes and requires the use of accurately flat optical elements. All of these limitations of classical interferometry are removed by holographic interferometry, which is one of the most revolutionary innovations of classical interferometry. Thus, holographic interferometry permits measurements to be made on any three-dimensional object that can be interferometrically detected at two separate times and does not require the use of high-quality optical components.

The method is based on the fundamental property of a hologram to store and reconstruct a given wave. In it, a given wave, generated at some earlier time and stored in the hologram, is later reconstructed and can interfere with another wave. Also, on the same hologram various waves can be stored that, after their reconstruction, can interfere. These two basic ideas constitute the key points of the two main techniques of holographic interferometry, that is the *real-time* and *double-exposure* procedures.

The development of holographic interferometry began a short time after the renaissance of holography by *Leith* and *Upatnieks* in 1962. The inventor of holography, *Gabor* [11.14], first conceived the idea that the interferometric effect can be produced by superposition of the wavefronts either of the object and its holographic image or of two different holographic images. *Horman* [11.15] applied the basic principles of the wavefront reconstruction to interferometry and gave an explanation of the method, by using only physical concepts. *Powell* and *Stetson* [11.16] used holographic interferometry for vibration measurements, while *Haines* and *Hildebrand* [11.17–19] applied the method for detecting the topography of three-dimensional objects.

Holographic interferometry was applied to the determination of small displacements of bodies caused by a system of applied loads. This is of particular importance in the mechanics of deformable bodies and is related to the calculation of the induced strain or stress field in the body. In real-time holographic interferometry, a hologram is formed with the specimen used as the object in the basic holographic arrangement. The hologram is replaced in exactly its previous position and the specimen is subjected to a system of loads. If the hologram is illuminated by both the laser light reflected or transmitted by the specimen and the reference wave derived from the same laser light, the object wave of the unloaded specimen is reconstructed and interferes with the object wave of the loaded specimen. The fringes that result from interference of the two object waves form contour lines of the displacements of the specimen in a given direction.

In double-exposure holographic interferometry, the holographic plate is subjected to two exposures, one with the specimen unloaded and the other with the specimen loaded, so that the hologram stores the object waves of both the unloaded and the loaded specimen. Then, the hologram is placed in its original position and is illuminated only by the reference wave. The two waves from the object, the one with the specimen loaded and the other with the specimen unloaded, are reconstructed by the hologram; they interfere, as in the case of real-time holographic interferometry.

Among the first researchers who applied the method of holographic interferometry for displacement measurements in two- and three-dimensional strain fields were *Sampson* [11.20], *Sollid* [11.21], *Gottenberg* [11.22] et al. (see also [11.23–26]). They gave also the relevant equations that relate the interferometric fringes with the displacements. Other interesting applications of holographic interferometry in the analysis of the displacement field in many problems of the mechanics of deformable bodies can be found in [11.27–30].

The idea of *multiple-exposure* holographic interferometry was very early extended to the limiting case of a continuum of exposures which, result, for example, when the object moves continuously. By this technique, the range of holographic interferometry was extended to the study of vibrations. The method was first suggested by *Powell* and *Stetson* [11.16] and was further elaborated and applied to the solution of vibration problems by *Monahan* and *Bromley* [11.31], *Neumann* et al. [11.32], and *Aprahamian* et al. [11.33–35] (see also [11.36, 37]).

Holographic interferometry has also been applied to photoelasticity and used for the solution of many problems of mechanics. Such applications of holographic interferometry resulting in a new method of experimental stress analysis, *holographic photoelasticity*, will be the subject of Section 11.5.

11.4.2 Real-Time Holographic Interferometry

By use of the basic procedure of wavefront reconstruction by the two-step holographic method described in Sections 11.2 and 11.3, the interference of the object and reference waves is recorded in the hologram, which when afterwards is illuminated by the reference wave, reconstructs the original object wave. If we leave the object in the position where we had placed it in order to create the hologram, under the same loading conditions, and illuminate both the hologram and the object with the same reference light beam used to create the hologram, then the observer behind the hologram will see only the image of the object and not an interferogram. Failure to create an interferogram is because both the reconstructed virtual image of the object and the object itself are formed at exactly the same place.

This phenomenon is analogous to the phenomenon of classical interferometry, in which two optically flat mirrors split the light beam in such a way as to create identical optical paths. In this case again no interferogram is viewed at the observation point of the interferometer. However, it is an exclusive property of holographic interferometry to confound two waves scattered from arbitrary surfaces, and not create an interferogram, while in classical interferometry these surfaces have to be accurately flat optical surfaces.

In the case in which the initially created hologram is of the positive type (γ = positive), the superposition of the virtual image with the object itself results in an increase of contrast, while for a negative hologram (γ = negative) the contrast is diminished.

Let us now assume that in the reconstructing process the shape of the object is changed, owing, for example, to the application of a system of loads; therefore, the hologram is illuminated by the reference wave and by the new form of the object wave. By the previously established principle of the relationship between the phenomena of interference and diffraction of light, the illumination of the hologram with the reference wave reconstructs the original object wave, whereas the illumination of the hologram with the new form of the object wave gives as the zero-order diffraction pattern this new form of the object wave. The two forms of the object wave that correspond to the original and to the deformed shapes of the object, when the displacement of the object is less than the coherence length of the light used, interfere; the resulting interferogram consists of fringes that are loci of equal displacement of the object.

The situation in real-time holographic interferometry can be easily compared with that in a Mach-Zehnder interferometer (Fig. 11.9 a). Let us consider

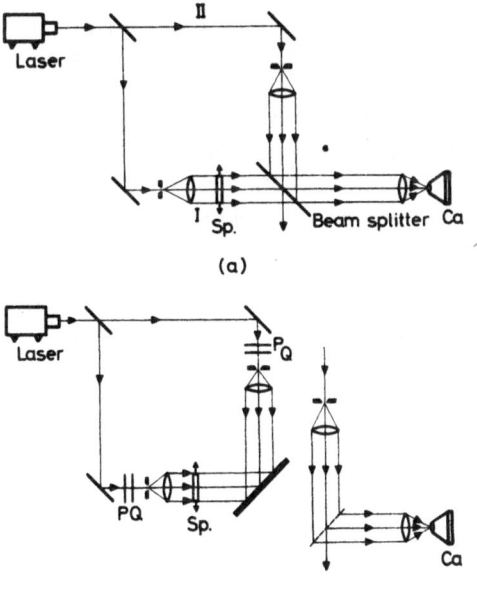

Fig. 11.9a and b. Mach-Zehnder interferometer and corresponding recording and reconstructing arrangements for holographic interferometry

that the two optical paths I and II in the interferometer are exactly equal and that the two mirrors have identical optical surfaces. If an object (Sp) is put in one optical path, then the interference pattern represents the surface irregularities of the object. Assume now that the object is deformed, for example, as a result of application of a system of loads. The new interference pattern corresponds to the irregularities of the form of the surface after deformation. The two interference patterns can be compared by using the subtractive property of the mechanical-interference (moiré) phenomenon. Thus, if the two interferograms are superimposed, the fringes in the resulting moiré pattern are loci of equal displacement between the initial and the new form of the object.

Let the holographic plate be put in place of the beam splitter in the interferometer of Fig. 11.9a and exposed to the reference and the object beams when the specimen is unloaded (Fig. 11.9b). When, in the reconstructing process, the hologram is illuminated by both the reference and the object beam, and the object is loaded, the object waves for the loaded and the unloaded specimen are simultaneously obtained and interfere to give fringes that are loci of equal displacement of the specimen.

The foregoing comparison of the Mach-Zehnder interferometer and real-time holographic interferometry shows that, although with the Mach-Zehnder interferometer the two interferograms obtained when the object is loaded and unloaded, must be superimposed to obtain the loci of equal displacements of the object, real-time holographic interferometry subtracts the loaded and unloaded interferograms automatically. This is the main advantage of holographic interferometry, compared to classical interferometry.

Automatic subtraction of the two object waves that correspond to the loaded and the unloaded specimen is of particular importance in practical applications, because many difficulties arise in mechanical superposition of interferograms [11.38].

Besides the foregoing explanation of the optical transformations that occur in holographic interferometry, which is based on the relationship between the phenomena of interference and diffraction, these transformations can also be explained by use of the other two manners of explaining the holographic phenomenon, that is, by use of the basic equations that govern these transformations, and the concepts of the mechanical-interference (moiré) phenomenon. In order to see how the basic holographic equations are applied in the case of holographic interferometry, consider the simple case when both the reference and object waves are one-dimensional, that is, when their corresponding Jones vectors are of the form

$$
\boldsymbol{a}_1 = \begin{bmatrix} A_{x_1} e^{i\frac{2\pi}{\lambda}\vartheta_x x} \\ 0 \end{bmatrix} \qquad \boldsymbol{a}_2 = \begin{bmatrix} A_{x_2} e^{i\delta_x} \\ 0 \end{bmatrix}, \tag{11.20}
$$

with the reference wave incident at a small angle ϑ_x with respect to the Ox axis.

The transmitted light intensity I from the hologram will be

$$I = k(\tilde{a}_1 + \tilde{a}_2)(a_1 + a_2) = k\left\{A_{x_1}^2 + A_{x_2}^2 + A_{x_1}A_{x_2}\left[e^{i\left(\delta_x - \frac{2\pi}{\lambda}\vartheta_x x\right)} + e^{-i\left(\delta_x - \frac{2\pi}{\lambda}\vartheta_x x\right)}\right]\right\},$$

(11.21)

where k is a constant.

Let us consider that the hologram is illuminated by both the reference wave and the new form of the object wave, when the object has been deformed. If a_2' is the Jones vector of the new form of the object wave,

$$a_2' = \begin{bmatrix} A_{x_2}' e^{i\delta_x'} \\ 0 \end{bmatrix},$$

(11.22)

we obtain for the wave transmitted by the hologram,

$$a_T = \begin{bmatrix} A_{x_T} e^{i\delta_{x_T}} \\ 0 \end{bmatrix}$$

(11.23)

and, therefore

$$A_{x_T} e^{i\delta_{x_T}} = I\left(A_{x_1} e^{i\frac{2\pi}{\lambda}\vartheta_x x} + A_{x_2}' e^{i\delta_x'}\right)$$

$$= k\left[A_{x_2}'A_{x_2}^2 e^{i\delta_x'} + A_{x_2}'A_{x_1}^2 e^{i\delta_x'} + A_{x_2}'A_{x_1}A_{x_2} e^{i\left(\delta_x + \delta_x' - \frac{2\pi}{\lambda}\vartheta_x x\right)}\right.$$

$$+ A_{x_2}'A_{x_1}A_{x_2} e^{-i\left(\delta_x - \delta_x' - \frac{2\pi}{\lambda}\vartheta_x x\right)} + A_{x_1}A_{x_2}^2 e^{i\frac{2\pi}{\lambda}\vartheta_x x}$$

$$\left. + A_{x_1}^3 e^{i\frac{2\pi}{\lambda}\vartheta_x x} + A_{x_2}A_{x_1}^2 e^{i\delta_x} + A_{x_1}^2 A_{x_2} e^{-i\left(\delta_x - \frac{4\pi}{\lambda}\vartheta_x x\right)}\right].$$

(11.24)

Equation (11.24) shows that the light transmitted by the hologram consists of many waves that travel in several directions. Of all these waves, those transmitted in approximately the direction of the original object wave are represented by the first two and the seventh terms on the right-hand side of (11.24). Thus, we obtain for the transmitted component $A_{x_T}' e^{i\delta_{x_T}'}$.

$$A_{x_T}' e^{i\delta_{x_T}'} = k\left[A_{x_2}'(A_{x_1}^2 + A_{x_2}^2)e^{i\delta_x'} + A_{x_2}A_{x_1}^2 e^{i\delta_x}\right].$$

(11.25)

The intensity that corresponds to the wave given by (11.25) is

$$I' = k^2\left[(A_{x_1}^2 + A_{x_2}^2)^2 A_{x_2}'^2 + A_{x_1}^4 A_{x_2}^2 + 2A_{x_1}^2 A_{x_2}(A_{x_1}^2 + A_{x_2}^2)A_{x_2}'\cos(\delta_x - \delta_x')\right].$$

(11.26)

Therefore, the optical pattern obtained has equal intensities for points for which

$$\delta_x - \delta'_x = m\frac{\pi}{2}\,;\qquad\qquad\qquad (11.27)$$

with $m=(2l+1)$ for the dark fringes, and $m=2l$, for the bright fringes, where l is an integer.

Equation (11.27) represents the previously found result, that fringes in the optical pattern obtained by the method of holographic interferometry are loci of equal optical-path differences between the original and the new forms of the object wave.

The main characteristic of the method of real-time holographic interferometry is that a series of pictures of an object can be obtained and compared interferometrically with an initial form of the object. For example, the initial form of the object can be that of its unloaded state, whereas its successive forms are obtained by application of an increasing system of loads. The hologram of the unloaded object is first recorded and then it is illuminated by the reference beam and the object beams that correspond to the successive forms of the object during deformation. For each form of the object, a corresponding interferogram is obtained instantaneously; hence the name: *real-time holographic interferometry*.

Experimental difficulties in the method arise from the fact that the hologram is made at one time and is illuminated by the new form of the object beam at another time. All optical elements, including the hologram and the object itself, must be at exactly the same positions at both times. Another source of error is that the emulsion of the hologram may shrink during development and drying and may therefore reconstruct a distorted initial object wave. The method is characterized by a particularly large flexibility in obtaining a series of interferograms of the object, corresponding to its various forms during deformation.

11.4.3 Double-Exposure Holographic Interferometry

In the method of double-exposure holographic interferometry, the holographic plate is exposed to the reference wave and the object wave that correspond to the original state of the specimen, and also to the reference wave and the object wave that correspond to the new state of the specimen. In both cases, the reference wave is the same. In the reconstruction process, the hologram is exposed only to the reference wave, which results in the reconstruction of the object waves that correspond to the original and the new forms of the specimen. If the displacements of the points of the specimen are less than the coherence length of the light used, the two reconstructed object waves interfere. The fringes in the resulting interferogram are loci of equal displacement of points on the surface of the specimen.

The formation of the interference pattern in the method of double-exposure holographic interferometry can also be explained by use of the relevant equations that describe the recording and the reconstructing processes of holography, in the same way as was done for real-time holographic interferometry. In both cases, the formation of patterns can also be interpreted in terms of the mechanical-interference (moiré) phenomenon used in Section 11.3 to explain the basic holographic process.

Compared with real-time holographic interferometry, the double-exposure method has the disadvantage that only two forms of the specimen can be recorded in the hologram and compared in the reconstructing process, whereas, in the real-time method, the hologram constitutes the basis for comparison of one state of a specimen with a whole series of other states of the same specimen, which may result, for example, from application of a progressively increasing load. However, in the double-exposure method, all difficulties that result from the fact that, in real-time holographic interferometry, the holographic plate is exposed at one time to the original object wave and is illuminated with the new object wave at another time are eliminated. Also, the object is not used in the reconstructing process. Moreover, the shrinkage of the emulsion of the holographic plate is the same for both waves; therefore the reconstructed waves are not distorted, relative to each other. In addition, the ratio of the light intensities that illuminate the undeveloped holographic plate and then the developed hologram is not significant, whereas it is particularly important in the real-time method, because the contrast of the interference fringes depends on the ratio of the intensities of these two beams. Therefore, in the double-exposure method, the interference between the original and the new forms of the object wave is made under ideal conditions.

In double-exposure holographic interferometry the original and the new forms of the object wave are recorded on the same photographic plate and therefore reconstructed simultaneously when illuminated afterwards by an identical reference beam. Therefore, the waves obtained are both $+1$ order virtual images of the same object at different loading steps. In the real-time method the reconstructed wave corresponding to the initial shape of the object is in general its $+1$ order virtual image, while the wave corresponding to the new form of the object is the same object wave. Therefore, both the original and the new forms of the object waves are the same, while in the real-time method and for negative hologram they subtract from each other and reduce the contrast of the image of the object, in the double-exposure method they always add and give a bright image of the object.

11.4.4 Interpretation of the Fringe Patterns

In both real-time and double-exposure holographic interferometry, the object waves that correspond to the initial S_i and the new S_n forms of the specimen are reconstructed and are interferometrically compared. Let us consider a light bundle incident on the surface of the specimen, in the direction of the unit vector n_i, which is scattered, reflected, or diffracted through the new form of the surface in the direction of the unit vector n_0 (Fig. 11.10). Let A and A' be the initial and the new locations of a point on the surfaces S_i and S_n, respectively. The light rays that originate from the points A and A' interfere; the resulting

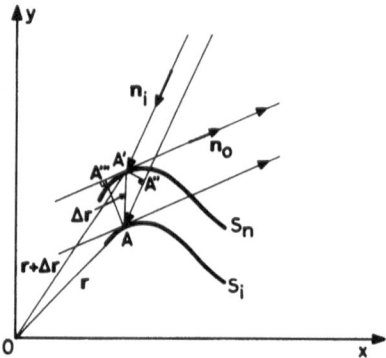

interferometric pattern depicts the difference of the optical path s for the light ray incident at point A and emanating from point A'. Figure 11.10 shows that

$$s = AA'' - A'A''' = \Delta r \cdot n_i - \Delta r \cdot n_0 = \Delta r \cdot (n_i - n_0). \tag{11.28}$$

The phase difference Δ, corresponding to the difference s of the optical path, is

$$\Delta = \frac{2\pi}{\lambda} s = \frac{2\pi}{\lambda} \Delta r \cdot (n_i - n_0), \tag{11.29}$$

where λ is the wavelength of the light.

Therefore, the interference patterns of both the real-time and the double-exposure holographic interferometric methods depict contour lines of the quantity expressed by (11.29). For complete determination of the displacement of a surface, three different interferograms, corresponding to three different directions, must be obtained, so that the components of the displacement vector of the surface in these three directions can be determined.

11.5 Holographic Photoelasticity

11.5.1 Introduction

The method of holographic photoelasticity is the application of holographic interferometry, developed in the previous section, to photoelasticity. As was shown, in the method of holographic interferometry the object waves that correspond to two different forms of the object are reconstructed and then interfere. For the case of the mechanics of deformable bodies, the two different forms of the object result from the application of a system of loads to the object. Hence, the relative displacements of the object are generally small compared to its dimensions. The interferometric pattern depicts the contour lines of the relative displacement of the object.

The situation is completely analogous in the case when the object is made of a photoelastic material and the light is allowed to pass through the object. In this case, when the object is stressed, according to the photoelastic phenomenon, discussed in Chapter 6, two principal polarization directions appear at every point of the object, across which the light is propagated with different velocities. Therefore, by applying the method of holographic interferometry to photoelasticity, when the original and the new forms of the object correspond to the specimen unloaded and loaded, respectively, fringes in the interferogram correspond to the loci of equal values of the two optical-path differences in the principal-stress directions at each point of the object. Thus, by the method of holographic photoelasticity, contour lines of optical-path differences in the principal-stress directions are obtained. These two superimposed families of curves, as discussed in Section 10.6, interfere mechanically and form two new families of curves, namely the isochromatics and isopachics, that correspond to the difference and the sum of the principal stresses.

Thus, application of holographic interferometry to photoelasticity gives interference patterns that depict the superimposed families of isochromatics and isopachics, as in the application of classical interferometry developed in Chapter 10. However, although the final optical patterns in both classical and holographic interferometry methods are the same, there is a great difference in the manner by which these patterns are obtained. In this respect, we must distinguish the specimens used into two categories: those with plane parallel surfaces and those with irregular initial surfaces, the irregularities being of the order of some wavelengths of the light used. The latter type of specimens constitutes the usual case encountered in photomechanics. As developed in Chapter 10, when the initial surfaces of the specimen are quite parallel, the interferogram obtained by the method of classical interferometry gives the families of isochromatics and isopachics. Therefore, in this case, these two patterns are obtained directly in both classical and holographic interferometry. However, in the more usual case of initially irregular specimens, the pattern obtained in classical interferometry with the specimen loaded does not represent the isochromatics and isopachics but is distorted by the initial irregularities of the specimen. In order to get the families of isochromatics and isopachics, the interferograms that correspond to the unloaded and loaded specimens must be superimposed, and the initial irregularities of the specimen must be subtracted, as may be done automatically by the mechanical-interference (moiré) phenomenon. But, because there are two families of curves in the interferogram, the superposition of the interference patterns with the specimen loaded and unloaded is difficult and the pattern obtained gives the isochromatics and isopachics only under especially favorable circumstances. Holographic photoelasticity circumvents these difficulties and gives immediately the families of isochromatics and isopachics.

Besides difficulties that result from nonparallel surfaces of models in classical interferometry, those that arise from non-ideal quality of the optical elements used must also be considered. Therefore, holographic photoelasticity constitutes the only practical method for obtaining, simultaneously, valuable isochromatic and isopachic patterns by using polarized light.

The idea of using holography to record photoelastic patterns was first introduced by *Rogers* [11.39] in 1966. However, the method of holographic photoelasticity to record simultaneously both the isochromatic- and isopachic-fringe patterns in two-dimensional stress fields was first presented by *Fourney* [11.40] in 1967. He demonstrated the basic holographic arrangements for obtaining isochromatics, isopachics, and isoclinics and gave some applications of the method. The mathematical concepts of the method for recording isochromatics and isopachics were then formulated by *Hovanesian* et al. [11.41] and were further developed by *Fourney* and *Mate* [11.42]. A detailed analysis of the formation of the optical patterns in holographic photoelasticity was presented by *Hosp* and *Wutzke* [11.43, 44]. The idea of recording the state of polarization of the light scattered by the object was first suggested by *Lohmann* [11.45] in 1965. He pointed out that this can be achieved by using two orthogonally polarized reference light beams. *Bryngdahl* [11.46] gave an example of the application of the method. The method was first applied for recording isoclinics in two-dimensional problems by *Fourney* et al. [11.47]. Further applications of the method were reported in [11.42–44]. An interesting review of the method of holographic photoelasticity was presented by *Fourney* [11.48].

The analysis of the optical patterns obtained by holographic photoelasticity in [11.40–44] showed that these patterns consist of superimposed families of isochromatics and isopachics. The coexistence of these two families causes many difficulties in interpreting the patterns of holographic photoelasticity. Thus, *Hovanesian* et al. [11.41] remarked that they do not agree with *Fourney*'s [11.40] interpretation of the holographic pattern obtained from a photoelastic material. They reported that they had experimentally noted that discontinuity of isopachics occurs at intersections with isochromatics and have further predicted the phenomenon mathematically, whereas *Fourney* [11.40] concluded that isopachics are continuous and do not undergo a half-order change upon intersection with isochromatics. *O'Regan* and *Dudderar* [11.49] remarked that the isochromatic and isopachic fringe patterns "are not completely independent, but interact in such a way as to make an interpretation difficult at critical regions of the model". They further observed that, in the case in which the spatial frequencies of isochromatics and isopachics are nearly equal, interpretation of the fringe patterns can be confusing. *Holloway* and *Johnson* [11.50], in discussing the interpretation of fringes in double-exposure holographic photoelasticity, wrote that "an interpretation of the photoelastic pattern as a simple combination of isochromatic and isopachic-fringe patterns is shown to be possible only under certain conditions". *Sanford* and *Durelli* [11.51], showed, by using computer-generated holographic interference patterns, that the interpretation of these patterns, presented in previous studies, as consisting of the independent superposition of the isopachic and isochromatic family "is not always valid and can result in serious errors in some cases". They have further shown that, "the position and even the existence of the fringes are affected by the interaction of isopachics and iso-chromatics", and that "this effect is most pronounced when the two families of fringes are nearly parallel and of approximately the same spatial frequency". Moreover, they proved that the independent-superposition interpretation is most accurate when the two families of fringes are orthogonal, whatever the ratio of spatial frequencies might be.

From this brief review of the salient points of the most interesting works in holographic photoelasticity it appears that there is confusion about the interpretation of the patterns obtained in holographic photoelasticity. A unified interpretation of these patterns was presented by *Theocaris* and *Gdoutos* [11.52], based on the mechanical-interference (moiré) of the two fun-damental optical patterns of the stress-optical retardations along the principal stress directions. They also showed the complicated character of the patterns obtained and concluded that the existing confusion is mainly due to the partic-ular features of each particular arrangement of holographic photoelasticity discussed in the literature, which may have led the investigators to erroneous generalizations. Furthermore, they attempted to identify and analyze the factors, that influence the interpretation of the interferograms for the most general case. In order to show the main features of the interpretation they presented two examples of typical photoelastic plane problems, i. e., the circular disk in diametral compression and the infinite plate with a circular hole, in tension, from which they concluded that the interferograms obtained may consist of the isochromatics, isopachics, or both of these families. They also indicated the factors that influence the formation of the corresponding optical pattern. The main results of this interpretation, which is common to both classical interferometric and holographic photoelasticity, were presented in Section 10.6.

The difficulties in interpreting stress-optical patterns in holographic photoelasticity, which consist of superimposed families of isochromatics and isopachics, led many researchers to the idea of separating these two patterns. Indeed, the flexibility provided by holographic photoelasticity, in contrast with the method of interferometric photoelasticity developed in Chapter 10, makes possible many techniques for separation of isochromatic- and isopachic-fringe patterns. Thus, *Chau* [11.53] utilized a rotator in the object beam and obtained only the isopachics. *O'Regan* and *Dudderar* [11.49] used an arrangement similar to that of *Chau* to obtain independently the isochromatics and isopachics. *Sciammarella* and *Quintanilla* [11.54] suggested a method for determining the patterns of the stress-optical retardations along the principal-stress directions. Other interesting studies for the purpose of separa-tion of the isochromatic- and isopachic-fringe patterns are listed in [11.24, 55–59].

In this section, an application of the Jones calculus to holographic photo-elasticity will be presented. First, we shall study the formation of the combined isochromatic and isopachic pattern, the formation of isoclinics, and, finally, the various techniques developed for the separation of isochromatics and isopachics.

11.5.2 Combined Isochromatic-Isopachic Patterns

I) *Basic Equations.* We shall now derive the equations for the light-intensity distribution in the interference-fringe patterns formed by the methods of real-time and double-exposure holographic photoelasticity. Let us first consider the case of real-time holographic photoelasticity. According to the analysis of Section 11.2, if a_1' and a_2' are the Jones vectors of the reference and object beams, respectively, that impinge on the holographic plate, then the light intensity I transmitted by the hologram will be

$$I = k(\tilde{a}_1' + \tilde{a}_2')(a_1' + a_2'),$$

where k is a constant.

In the reconstructing process, the hologram is illuminated by both the reference and the new form of the object beams, whose Jones vectors are expressed by a_1' and a_3', respectively. Then, the beam that emerges from the hologram has a Jones vector a_T given by

$$a_T = I(a_1' + a_3') = k(\tilde{a}_1' + \tilde{a}_2')(a_1' + a_2')(a_1' + a_3'). \tag{11.30}$$

The relevant terms of this expression, which contribute to the formation of the interferogram, are characterized by the Jones vector

$$a_T' = k(\tilde{a}_1' a_1' + \tilde{a}_2' a_2')a_3' + k\tilde{a}_1' a_2' a_1'. \tag{11.31}$$

In this expression, the quantities $\tilde{a}_1' a_1'$ and $\tilde{a}_2' a_2'$ represent the intensities of the reference beam and the original form of the object beam, respectively. Therefore, they can be taken to a high degree of approximation to be constant all over the field. Therefore, (11.31) can be written in the form

$$a_T' = k\tilde{a}_1' a_2' a_1' + k_2' a_3',$$

where k_2' is a new constant.

The intensity I' that corresponds to the Jones vector a_T', expressed by the above relation is

$$I' = \tilde{a}_T' a_T'. \tag{11.32}$$

Let us now consider that a circularly polarized light is used in both the reference and the object beams. The initial form of the object beam, with Jones vector a_2', passes through the unloaded specimen, whereas the new form of the object beam, with Jones vector a_3', passes through the loaded specimen. According to Table 4.2, the Jones matrix J of the loaded specimen, when the absolute, instead of the relative, retardation along its principal axes is introduced, is

$$J = \begin{bmatrix} e^{i\delta_1} \cos^2 \vartheta + e^{i\delta_2} \sin^2 \vartheta & (e^{i\delta_1} - e^{i\delta_2}) \cos \vartheta \sin \vartheta \\ (e^{i\delta_1} - e^{i\delta_2}) \cos \vartheta \sin \vartheta & e^{i\delta_1} \sin^2 \vartheta + e^{i\delta_2} \cos^2 \vartheta \end{bmatrix}, \tag{11.33}$$

where δ_1 and δ_2 are the absolute stress-optical retardations along the principal-stress directions, [see (10.10–12)]

$$\delta_1 = \frac{2\pi}{\lambda} \Delta s_{t_1} = \frac{2\pi}{\lambda} (a_t \sigma_1 + b_t \sigma_2) d,$$

$$\delta_2 = \frac{2\pi}{\lambda} \Delta s_{t_2} = \frac{2\pi}{\lambda} (b_t \sigma_1 + a_t \sigma_2) d, \tag{11.34}$$

with

$$\delta = \delta_1 - \delta_2,$$

where δ is the relative stress-optical retardation in the expression for J in Table 4.2.

The unloaded model is isotropic; therefore its Jones matrix is the unit matrix. According to Table 3.2, the normalized Jones vector a'_1 of the right-circularly polarized light is

$$a'_1 = \frac{1}{\sqrt{2}} \begin{bmatrix} -i \\ 1 \end{bmatrix}. \tag{11.35}$$

Because the original form of the object beam with Jones vector a'_2 passes through the unloaded specimen which has a unit Jones matrix, it is circularly polarized; it is, therefore,

$$a'_2 = \frac{1}{\sqrt{2}} \begin{bmatrix} -i \\ 1 \end{bmatrix}. \tag{11.36}$$

The new form of the object beam, with Jones vector a'_2, passes through the loaded specimen, whose Jones matrix J is given by (11.33). Therefore,

$$a'_3 = J a'_1. \tag{11.37}$$

Introducing the values of a'_1, a'_2, and a'_3 from (11.35–37) into (11.32) and taking into account (11.33), we obtain for the variable part of the light intensity that contributes to the formation of the interferogram,

$$I'_1 = C \cos\left[\frac{2\pi}{\lambda} \frac{a_t + b_t}{2}(\sigma_1 + \sigma_2)d\right] \cos\left[\frac{2\pi}{\lambda} \frac{a_t - b_t}{2}(\sigma_1 - \sigma_2)d\right], \qquad (11.38)$$

where C is a constant.

Equation (11.38) gives the same results as (10.42), obtained from the analysis of the field of a Mach-Zehnder interferometer in Section 10.5. Equation (11.38) was also obtained in another way in [11.39].

In double-exposure holographic photoelasticity, the hologram is exposed to both the original and the new forms of the object wave. Therefore, the intensity transmitted by the hologram will be

$$I = k_1(\tilde{a}'_1 + \tilde{a}'_2)(a'_1 + a'_2) + k'_1(\tilde{a}'_1 + \tilde{a}'_3)(a'_1 + a'_3). \qquad (11.39)$$

In the reconstructing process, the hologram is illuminated by only the reference beam. Therefore, the Jones vector a_T of the beam that emerges from the hologram, will be

$$a_T = I a'_1 = [k_1(\tilde{a}'_1 + \tilde{a}'_2)(a'_1 + a'_2) + k'_1(\tilde{a}'_1 + \tilde{a}'_3)(a'_1 + a'_3)] a'_1. \qquad (11.40)$$

The terms of this expression that contribute to the formation of the interferogram are characterized by the Jones vector

$$a'_T = k_1 \tilde{a}'_1 a'_2 a'_1 + k'_1 \tilde{a}'_1 a'_3 a'_1 \qquad (11.41)$$

The corresponding intensity I' is

$$I' = \tilde{a}'_T \tilde{a}'_T. \qquad (11.42)$$

By taking into account (11.36) and (11.37) in combination with (11.33), we obtain for I' (assuming $k_1 = k'_1$)

$$I' = C\left\{1 + 2\cos\left[\frac{2\pi}{\lambda} \frac{a_t + b_t}{2}(\sigma_1 + \sigma_2)d\right] \cos\left[\frac{2\pi}{\lambda} \frac{a_t - b_t}{2}(\sigma_1 - \sigma_2)d\right]\right.$$
$$\left. + \cos^2\left[\frac{2\pi}{\lambda} \frac{a_t - b_t}{2}(\sigma_1 - \sigma_2)d\right]\right\}, \qquad (11.43)$$

where C is a constant. Equation (11.43) was derived in [11.42] by using vector analysis.

Equations (11.38) and (11.43), which express the intensity distribution in the interference-fringe patterns obtained by real-time and double-exposure holographic photoelasticity, suggest that the optical patterns obtained consist of two families of curves superimposed on each other, namely the isochromatics and isopachics. From these two patterns, the values of the principal stresses can be immediately determined.

II) *Physical Interpretation.* Besides the foregoing mathematical analysis of the optical patterns of real-time and double-exposure holographic photoelasticity, these patterns can also be interpreted by use of physical concepts, only. The same procedure developed in Section 11.3 for the explanation of the holographic phenomenon based on the relationship between interference and diffraction of light, will be used. Let us first consider the case of the real-time holographic photoelasticity.

The holographic plate, in this case, is exposed to the reference beam and to the object beam that corresponds to the unloaded specimen. In the reconstruction, the hologram is illuminated by both the reference beam and the new form of the object beam, which corresponds to the loaded specimen. Thus, in the reconstructing process, the illumination of the hologram with the reference beam gives as the first-order diffraction pattern the object beam that corresponds to the unloaded specimen, whereas illumination of the hologram with the object beam that corresponds to the loaded specimen gives as the zero-order diffraction pattern the beam itself. Therefore, two waves, one that corresponds to the specimen loaded, and the other that corresponds to the specimen unloaded, which is reconstructed from the hologram, interfere. Furthermore, the wave that corresponds to the loaded specimen consists of two linearly polarized waves along the principal-stress directions, which express the absolute stress-optical retardations along these directions. These waves interfere with the wave that corresponds to the unloaded specimen; therefore, the irregularities of the unloaded specimen are removed from the final pattern. The interference pattern consists of two superimposed families of curves that are the loci of equal stress-optical retardations along the principal-stress directions. These two families of curves, as was explained in Section 10.6, interfere mechanically and give, as the final optical pattern of real-time holographic photoelasticity, superimposed families of isochromatics and isopachics.

Similarly, in double-exposure holographic photoelasticity, the holographic plate is exposed once to the reference beam and the object wave that corresponds to the unloaded specimen, and at another turn to the reference beam and to the new form of the object wave that corresponds to the loaded specimen. In reconstruction, the hologram is illuminated by the reference beam only. Thus, the object waves that correspond to the unloaded and loaded specimens are reconstructed and the optical pattern consists of the superimposed families of isochromatics and isopachics, as in the case of real-time holographic photoelasticity.

The optical patterns in both real-time and double-exposure holographic photoelasticity can also be explained by the interpretation used for the explanation of the basic holographic phenomenon in Section 11.3 based on the mechanical interference (moiré) phenomenon.

11.5.3 Determination of Isoclinics

The ability of holography to record and reconstruct the state of polarization of the light scattered by the object was first suggested by *Lohmann* [11.45] and was further elaborated by *Bryngdahl* [11.46]. The method was then applied to the case of a photoelastic model by *Fourney* et al. [11.42, 47] and resulted in the determination of the family of isoclinics from the hologram. The capability of holography to reconstruct not only the amplitude and the phase of polarized light, but also its complete polarization form, further justified its name as *total reconstruction*, the term total referring to all characteristics, that is, amplitude, phase, and state of polarization.

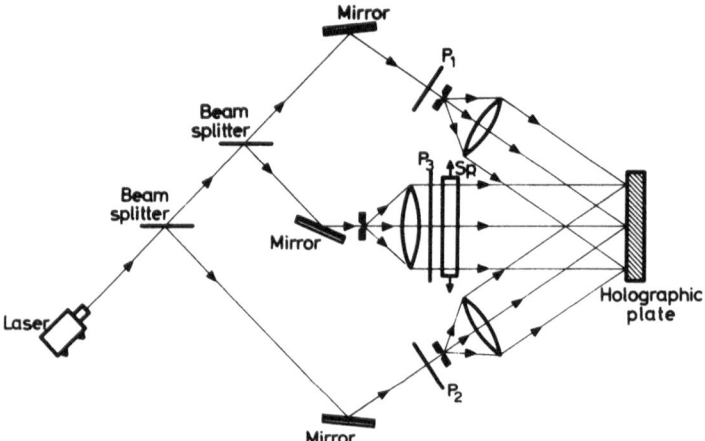

Fig. 11.11. Optical arrangement for recording the state of polarization of the light scattered by the object (S_p) by use of two reference light beams whose planes of polarization are mutually perpendicular

Recording of the state of polarization of the light scattered by the object is based on the use of two reference light beams whose planes of polarization are mutually perpendicular (Fig. 11.11). The holographic plate is illuminated by the object beam and the two reference beams. In the reconstructing process, the hologram is illuminated by two mutually orthogonally polarized reference beams and reconstructs the polarization state of the light emitted by the object.

The recording and reconstructing of the state of polarization of the light emitted by the photoelastic model by the process described are explained as follows. According to the photoelastic phenomenon, when a light beam is incident on the photoelastic model, at each point it is split into two components linearly polarized along the principal-stress directions at the point considered. In the recording process, the components that are linearly polarized along the directions of the mutually perpendicularly principal stresses interfere with the similarly polarized reference beam and form two superimposed interferograms on the holographic plate. In the reconstructing process, when the hologram is illuminated by the two reference light beams, the components of the two linearly polarized light beams along the principal-stress directions, emitted by the object, are reconstructed as the first-order diffraction patterns of the hologram, and combine vectorially. As a result of that combination, the state of polarization of the light emitted by the photo-elastic model is reconstructed. To detect and display the polarization form, in the form of fringes, the pattern reconstructed by the hologram is viewed through a linear analyzer.

In order to get a quantitative picture of the thus-formed optical pattern, let us consider that the model is illuminated by a linearly polarized beam and that the linear analyzer is oriented with its pass axis at an angle ϑ with the Ox axis. Then, if δ_1 and δ_2 are the absolute stress-optical retardations of the photoelastic model along its principal-stress directions, given by (11.34), the Jones vector a of the light emitted by the object is

$$a = \begin{bmatrix} A_1 e^{i\delta_1} \\ A_2 e^{i\delta_2} \end{bmatrix}. \tag{11.44}$$

The Jones matrix J of the linear analyzer is given by (see Table 4.1)

$$J = \begin{bmatrix} \cos^2 \vartheta & \sin \vartheta \cos \vartheta \\ \sin \vartheta \cos \vartheta & \sin^2 \vartheta \end{bmatrix}. \tag{11.45}$$

Then, the Jones vector a' of the light that emerges from the analyzer is

$$a' = J a = \begin{bmatrix} A_1 e^{i\delta_1} \cos^2 \vartheta + A_2 e^{i\delta_2} \sin \vartheta \cos \vartheta \\ A_1 e^{i\delta_1} \sin \vartheta \cos \vartheta + A_2 e^{i\delta_2} \sin^2 \vartheta \end{bmatrix}. \tag{11.46}$$

The light intensity I that corresponds to the vector a' is

$$I = \tilde{a}' a', \tag{11.47}$$

where $\tilde{a}'$ is the hermitian conjugate matrix of a'. From (11.46) and (11.47),

$$I = A_1^2 \cos^2 \vartheta + A_2^2 \sin^2 \vartheta + A_1 A_2 \sin 2\vartheta \cos(\delta_1 - \delta_2) \tag{11.48}$$

In order to avoid the introduction of a bias in the direction of polarization of the light emitted by the object, it is preferable to use a circularly, instead of a linearly polarized beam, to illuminate the specimen. In this case, a 90° phase difference must be added to one of the phases δ_1 or δ_2 in (11.44). Following the same procedure and putting $A_1 = A_2 = A$, we obtain for the intensity

$$I = A^2 [1 + \sin 2\vartheta \sin(\delta_1 - \delta_2)]. \tag{11.49}$$

From either (11.48) or (11.49) it follows that the angle ϑ between the analyzer and the principal-stress directions at the point of the photoelastic model considered enters into these relations, which give the distribution of the intensity in the interference-fringe pattern. In the same relations the relative stress-optical retardation along the principal-stress directions also appears. Thus, the pattern consists of superimposed families of isochromatics and isoclinics.

In this process for holographic recording of isoclinics, the locations of the hologram and the reference beam must be more exactly reproduced in the reconstructing process than in ordinary holography. This is because a small deviation from the exact reproduction of the recording conditions influences greatly the final results. To ensure that the hologram is not moved from its original position, it is usually clamped to a rigid frame and developed in place.

11.5.4 Separation of Isochromatics and Isopachics

As was shown in Section 11.5.2, the direct application of holographic interferometry to photoelasticity results in the creation of a pattern in which the families of isochromatics and isopachics are superimposed. However, as was deduced from [11.41, 42, 51–54], many difficulties arise in the interpretation of the optical patterns of holographic photoelasticity. For this reason, many attempts have been made to separate the families of isochromatics and isopachics.

The first fruitful and efficient idea for separating the families of isochromatics and isopachics was introduced by *Chau* [11.53], in 1968; it was further elaborated in [11.49, 55, 57]. In this method, a 90° rotator is introduced in the object beam, which results in a 90° rotation of the plane of polarization of the two linearly polarized light beams that emerge from the specimen. The rotated polarized light beam is then passed through the specimen again, so that, during the second pass, the component linearly polarized along the direction of one of the two principal stresses encounters the other principal-stress direction and vice versa. In this way, the stress-optical retardation along each of the two principal-stress directions is the same and is equal to the sum of the optical retardations along the principal-stress directions of the photoelastic model. These two linearly polarized components combine to form a light vector whose plane of polarization is perpendicular to the plane of polarization of the original incident light.

Thus, in the recording process, the holographic plate is illuminated by the reference beam, which is linearly polarized in the same direction as the original object beam, and by the new object beam that is linearly polarized at right angles to the reference beam. By illuminating the hologram with the reference beam in the reconstructing process, the object beam is obtained. In order to detect and transform the object beam into a fringe pattern, a linear polarizer must be placed after the hologram. Thus, when the hologram is viewed through an analyzer with its pass axis in the same direction as the polarization of the

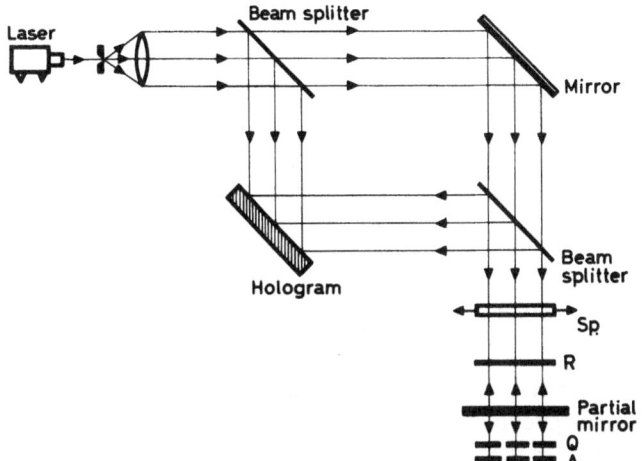

Fig. 11.12. Optical arrangement for separating the isopachic and isochromatic fringe patterns in holographic photoelasticity

incident light, the family of isopachics is formed. A schematic diagram of the experimental apparatus used for obtaining only the isopachic fringe pattern is shown in Fig. 11.12.

For a simultaneous recording of the family of isochromatics, circularly polarized light is used in both the reference and the object beams and the experimental arrangement of Fig. 11.12 is complemented by a quarter-wave plate and a polarizer (shown in Fig. 11.12 with dotted lines), which are inserted in the object beam which traverses the specimen only once. All these elements are arranged in the same manner as in the circular polariscope.

When the object beam that traverses the specimen is viewed through the analyzer, the family of isochromatics is obtained, whereas the hologram detects the family of isopachics. In this case, when circularly polarized light is used, the hologram does not need to be viewed through an analyzer, as is necessary when linearly polarized light is used.

The formation of the isopachic pattern alone by use of a rotator, which was explained by use of physical arguments only, can also be interpreted by calculating the intensity distribution in the interference pattern. Let us consider the case of the circularly polarized reference and object beams.

In order to get 90° rotation of the plane of polarization of the beam that passes through the specimen for the second time, a 45° rotator must be used in the optical arrangement shown in Fig. 11.12. The Jones matrix R of a 90° rotator is, according to Table 4.2,

$$R(90°) = \begin{bmatrix} 0 & 1 \\ -1 & 0 \end{bmatrix}. \tag{11.50}$$

The Jones matrix a'_3 of the beam that corresponds to the loaded specimen is

$$a'_3 = J R J a'_1 , \qquad (11.51)$$

where J is the Jones matrix of the loaded specimen and a'_1 is the Jones vector of the incident circularly polarized light, given by (11.33) and (11.35), respectively.

The Jones matrix of the unloaded specimen is the unit matrix. Therefore the Jones vector a'_2 of the light that emerges from the unloaded specimen is given by (11.36). Introducing the values of a'_2 and a'_3 from (11.36) and (11.51) into (11.32), we obtain for the intensity

$$I'_1 = C \left\{ 1 + \cos \left[\frac{4\pi}{\lambda} \frac{a_t + b_t}{2} (\sigma_1 + \sigma_2) d \right] \right\} . \qquad (11.52)$$

Equation (11.52) confirms the previously found result that the optical pattern consists of the family of isopachics only and that the sensitivity is doubled, because the object beam passes twice through the model.

Another simple idea for determining the family of isopachics alone is to record, by using holographic interferometry, the thickness variation of the specimen due to loading, which, according to Poisson's effect, is directly proportional, in the elastic plane-stress case, to the sum of the principal stresses. For this purpose, one part of the object beam is reflected from the front face of the specimen and forms the hologram with the reference beam, while the other part of the object beam traverses the specimen, beyond which a quarter-wave plate and a linear analyzer are inserted. The arrangement of the optical elements for the light rays that traverse the specimen is similar to that of a circular polariscope, whereas the arrangement for the light rays that are reflected from the front surface of the specimen is similar to that used in holographic interferometry. A schematic diagram of the optical arrangement for simultaneous, but separate, determination of the families of isochromatics and isopachics is shown in Fig. 11.13. This method was first used by *Holloway* et al. [11.24] for the study of static and dynamic problems.

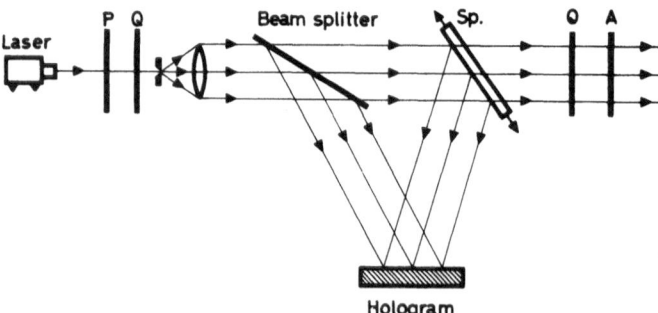

Fig. 11.13. Separate determination of the families of isopachics and isochromatics by recording the thickness variation of the specimen by holographic interferometry and using the typical arrangement of the circular polariscope

Ajovalasit [11.58] and *Ajovalasit* and *Bardi* [11.59] introduced another approach for separating the principal stresses in plane-stress elastic problems, by determining the families of the absolute stress-optical retardations Δs_1 and Δs_2 along the principal-stress directions. If these quantities are determined, the principal stresses σ_1 and σ_2 can be calculated from (10.26), provided that the stress-optical constants a_t and b_t are known from a calibration experiment. In the method described in [11.58], which is based on real-time holographic photoelasticity, the hologram of the model is recorded by use of circularly polarized light for the reference and the object beams. In the reconstructing process, the hologram is illuminated by the reference and object beams and is viewed through a rotating linear analyzer. At the points of the model where the principal-stress directions are parallel to the pass axis of the analyzer, the corresponding family of the stress-optical retardation Δs_1 or Δs_2 is observed. Similarly, in the method of [11.59] (Fig. 11.14), which is also based on real-time holographic photoelasticity, the hologram is recorded with linearly polarized light of the same orientation for both the reference and object beams. In the reconstructing process, the hologram is illuminated by the linearly polarized reference and object beams that correspond to various polarization directions. For each such direction, the families of Δs_1 and Δs_2 are observed at the points of the model whose principal axes are along the directions of the linearly polarized reference and object beams.

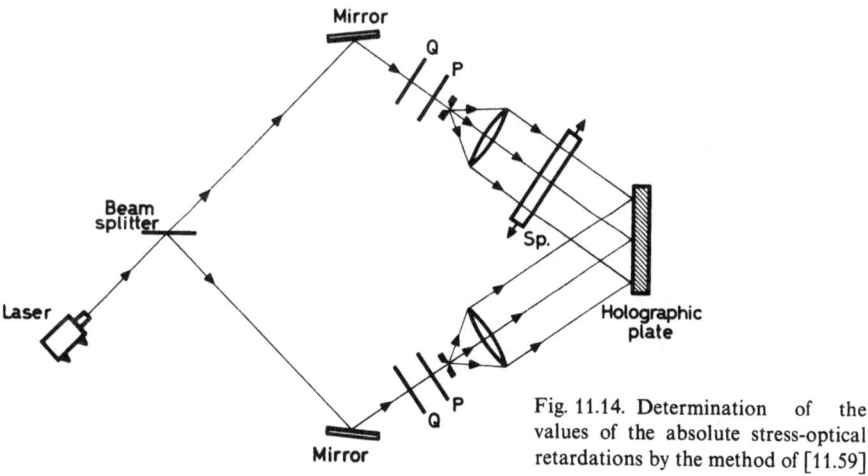

Fig. 11.14. Determination of the values of the absolute stress-optical retardations by the method of [11.59]

References

11.1   D. Gabor: Nature **161**, 777 (1948)
11.2   D. Gabor: Proc. R. Soc. London A **197**, 454 (1949)
11.3   D. Gabor: Proc. R. Soc. London B **64**, 449 (1951)
11.4   E. N. Leith, J. Upatnieks: J. Opt. Soc. Am. **52**, 1123 (1962)
11.5   E. N. Leith, J. Upatnieks: J. Opt. Soc. Am. **53**, 1377 (1963)
11.6   E. N. Leith, J. Upatnieks: J. Opt. Soc. Am. **54**, 1295 (1964)
11.7   G. W. Stroke: *An Introduction to Coherent Optics and Holography* 2nd ed. (Academic Press, New York 1969)

11.8 J. B. Develis, G. O. Reynolds: *Theory and Applications of Holography* (Addison-Wesley, Reading, Mass. 1967)

11.9 H. M. Smith: *Principles of Holography* (Wiley-Interscience, New York 1969)

11.10 R. J. Collier, C. B. Burckhardt, L. H. Lin: *Optical Holography* (Academic Press, New York 1971)

11.11 J. Pastor, G. E. Evans, J. S. Harris: Opt. Acta **17**, 81 (1970)

11.12 P. S. Theocaris: Tech. Ann. (in Greek) **10**, 861 (1971)

11.13 V. K. Der, D. C. Holloway, W. L. Fourney: Appl. Opt. **12**, 2552 (1973)

11.14 D. Gabor, G. W. Stroke, R. Restrick, A. Funkhouser, D. Brumm: Phys. Lett. **18**, 116 (1965)

11.15 M. H. Horman: Appl. Opt. **4**, 333 (1965)

11.16 R. L. Powell, K. A. Stetson: J. Opt. Soc. Am. **55**, 1593 (1965)

11.17 K. A. Haines, B. P. Hildebrand: Phys. Lett. **19**, 10 (1965)

11.18 K. A. Haines, B. P. Hildebrand: Appl. Opt. **5**, 595 (1966)

11.19 B. P. Hildebrand, K. A. Haines: J. Opt. Soc. Am. **57**, 155 (1967)

11.20 R. C. Sampson: Exp. Mech. **10**, 313 (1970)

11.21 J. E. Sollid: Appl. Opt. **8**, 1587 (1969)

11.22 W. G. Gottenberg: Exp. Mech. **8**, 405 (1968)

11.23 R. Mark, R. O'Regan: Exp. Mech. **12**, 332 (1972)

11.24 D. C. Holloway, W. F. Ranson, C. E. Taylor: Exp. Mech. **12**, 461 (1972)

11.25 S. K. Dhir, J. P. Sikora: Exp. Mech. **12**, 323 (1972)

11.26 T. R. Hsu: Exp. Mech. **14**, 408 (1974)

11.27 C. A. Sciammarella, T. Y. Chang: Exp. Mech. **14**, 217 (1974)

11.28 A. D. Wilson: Appl. Opt. **10**, 908 (1971)

11.29 J. D. Hovanesian: *Proc. 4th Int. Conf. Exp. Stress Analysis,* Cambridge, 1970, ed. by M. L. Meyer (The Institution of Mechanical Engineers, London 1971)

11.30 L. H. Taylor, G. B. Brandt: Exp. Mech. **12**, 543 (1972)

11.31 M. A. Monahan, K. Bromley: J. Acoust. Soc. Am. **44**, 1225 (1968)

11.32 D. B. Neumann, C. F. Jacobson, G. M. Brown: Appl. Opt. **9**, 1357 (1970)

11.33 R. Aprahamian, D. A. Evensen: J. Appl. Mech. **37**, 287 (1970)

11.34 R. Aprahamian, D. A. Evensen: J. Appl. Mech. **37**, 1083 (1970)

11.35 R. Aprahamian, D. A. Evensen, J. S. Mixson: Exp. Mech. **10**, 421 (1970)

11.36 C. R. Hazell, S. D. Liem: Exp. Mech. **13**, 339 (1973)

11.37 J. P. Sikora, F. T. Mendenhall: Exp. Mech. **14**, 230 (1974)

11.38 P. S. Theocaris: *Moiré Fringes in Strain Analysis* (Pergamon Press, London 1969)

11.39 G. L. Rogers: J. Opt. Soc. Am. **56**, 831 (1966)

11.40 M. E. Fourney: SESA Spring Meet. (Ottawa, Ontario, May 16–19, 1967); Exp. Mech. **8**, 33 (1968)

11.41 J. D. Hovanesian, V. Brcic, R. L. Powell: Exp. Mech. **8**, 362 (1968)

11.42 M. E. Fourney, K. V. Mate: Exp. Mech. **10**, 177 (1970)

11.43 E. Hosp, G. Wutzke: Materialprüfung **11**, 409 (1969)

11.44 E. Hosp, G. Wutzke: Materialprüfung **12**, 13 (1970)

11.45 A. W. Lohmann: Appl. Opt. **4**, 1667 (1965)

11.46 O. Bryngdahl: J. Opt. Soc. Am. **57**, 545 (1967)

11.47 M. E. Fourney, A. P. Waggoner, K. V. Mate: J. Opt. Soc. Am. **58**, 701 (1968)

11.48 M. E. Fourney: *Applications of Holography in Mechanics* (ASME Publ., New York 1971) pp. 17–38

11.49 R. O'Regan, T. D. Dudderar: Exp. Mech. **11**, 241 (1971)

11.50 D. C. Holloway, R. H. Johnson: Exp. Mech. **11**, 57 (1971)

11.51 R. J. Sanford, A. J. Durelli: Exp. Mech. **11**, 161 (1971)

11.52 P. S. Theocaris, E. E. Gdoutos: J. Strain Analysis **13**, 95 (1978)

11.53 H. H. M. Chau: Rev. Sci. Instr. **39**, 1789 (1968)

11.54 C. A. Sciammarella, G. Quintanilla: Exp. Mech. **12**, 57 (1972)

11.55 A. Assa, A. A. Betser: Exp. Mech. **14**, 502 (1974)

11.56 J. D. Hovanesian: Exp. Mech. **14**, 233 (1974)

11.57 K. Gasvik: Exp. Mech. **16**, 146 (1976)

11.58 A. Ajovalasit: J. Strain Analysis **10**, 148 (1975)

11.59 A. Ajovalasit, A. Bardi: Exp. Mech. **16**, 273 (1976)

12. The Method of Birefringent Coatings

12.1 Introduction

The method of birefringent coatings allows the evaluation of the strain distribution on the surface of an opaque body, by using the principles of two-dimensional photoelasticity developed in Chapter 7. This method may be considered as an application and extension to opaque materials of two-dimensional photoelasticity, which is applicable to only transparent models. The method of birefringent coatings consists of cementing a piece of transparent material (the birefringent coating) on the surface of the body under investigation and measuring the amount and the principal directions of the temporary birefringence developed in the coating when the main object is subjected to a system of loads.

If the bond between the coating and the opaque body is perfect, the surface strains of the body are transferred to the coating and may be determined from the accompanying birefringence of the coating, by use of the conventional methods of two-dimensional photoelasticity. Therefore, the methods of birefringent coatings and two-dimensional photoelasticity differ only in the manner by which the state of stress is developed in the birefringent coating. In the method of birefringent coatings, the state of stress is developed in the main body and is transferred to the coating, whereas in two-dimensional photoelasticity, the state of stress is developed in the photoelastic model itself. Another difference between the two methods lies in the fact that, by the method of classical two-dimensional photoelasticity, measurements are made only on a model of the real structure, whereas, by the method of birefringent coatings, the determination of the surface strains of a structural member can also be made in situ.

The idea of using birefringent coatings to measure the surface strain of an opaque body was first conceived by *Mesnager* [12.1], in 1930. He also established the basic principles of the method. *Oppel* [12.2], some years later, applied the method to the case of a photoelastic strain gauge for local determination of the state of stress in a body. However, due to the low optical sensitivity of the material of the coating used by *Oppel*, many difficulties were encountered in the interpretation of the optical patterns obtained. Almost two and a half decades after its invention, the method of birefringent coatings received considerable attention by many investigators who further elaborated and developed the method as a potential tool for stress analysis. Many publications aiming to establish and apply the method to engineering problems were published in the late fifties [12.3–9].

The method of birefringent coatings can be compared with the common methods of measuring surface strains, such as those that use various types of extensometers (mechanical, optical, electrical). However, the great advantage of the coating method over the conventional methods of measuring strain is that the strain distribution over an extended area can be determined, whereas, by the usual extensometers, the strain is determined only at a given point.

In the method of birefringent coatings great difficulty arises in the interpretation of the stress-optical patterns obtained. When the strain on the surface of the body under test is uniform or changes only slowly over distances at least of the same order of magnitude as the coating thickness, the birefringence of the coating is directly proportional to the difference of the principal stresses at the corresponding point on the main body. However, when the strain distribution on the surface of the body changes rapidly, the deformation of the coating will be variable through its thickness. Therefore, the birefringence obtained will depend on the magnitude, as well as on the gradient of the strain. Therefore, the thickness of the coating is extremely critical in the interpretation of the fringe patterns. The choice of the thickness of the coating is determined by two contradictory factors, the number of observed stress-optical fringes, which influences the sensitivity of the method, and the parasitic effects that occur in the transition of the surface strains from the surface of the main body to the coating. As the coating becomes thinner, the number of the observed fringes decreases, whereas as the thickness of the coating increases the parasitic effects, which make difficult the interpretation of the induced fringe patterns, increase.

The effect of the thickness of the coating on the accuracy of the results was studied by many investigators. The following factors influence the interpretation of the fringe pattern: a) the reinforcing effect of the coating; b) the transition of strains from the surface of the body to the surface of the coating; c) the strain gradients over the surface in question; and d) the difference between the Poisson ratios of the materials of the main body and the coating. All of these factors reduce the accuracy of interpretation of the fringe patterns. Their significance increases as the thickness of the coating increases. The influence of all of these factors on the accuracy of the results obtained has been thoroughly studied by various authors [12.10–17]. The basic results of all these investigations, which are critical for the accuracy of the method of birefringent coatings, will be presented in a subsequent section of this chapter.

After the establishment of its basic features and characteristics, the method of birefringent coatings was applied to a great number of engineering problems and became a powerful tool in the hands of both the researcher and the engineer, for investigating strain distributions developed in metallic structures. The method presents the great advantages, when compared with two-dimensional photoelasticity, that no model need be made of the structural member under investigation and that it can be used for strain determination of structures in situ. Also, the method can be applied for the determination of plastic-strain distributions in metallic surfaces, provided that the magnitude of the strains does not exceed the linear range of the material of the birefringent coating. Measurement of plastic states of stress is difficult by two-dimensional photoelasticity, where a model material that simulates the metal under investigation must be used. Because it is possible to select the polymer used as coating so as to have the appropriate yield or fracture limits, convenient for measurement of the large plastic strains anticipated in the metallic structure, it is clear that the method of birefringent coatings is a versatile one and convenient for measuring plastic strains as they have developed in structures. The possibility of changing the mechanical properties of polymers is achieved by adding to the prepolymer an amount of plasticizer. Thus, a preliminary study of the properties of polymers convenient for coatings is necessary for accurate evaluation of the variation of the mechanical and optical properties of these materials with addition of plasticizer. Epoxy polymers, which are most convenient materials for coatings, were studied extensively in this manner by *Theocaris* and his co-workers [12.18–20].

On the other hand, many attempts have been made to extend the principles of photoelasticity to investigate plastic-stress or strain distributions and thus to establish a new method, well known as photoplasticity. However, this method has not made much progress, up to now, because of inherent difficulties and inconsistencies. For a critical review of the limitations of the method of photoplasticity, the interested reader is referred to [12.21, 22]. Among the great number of applications of the method of birefringent coatings, can be mentioned investigation of strain distribution in curved surfaces [12.23], study of plastic stress-concentration problems with the analysis of the evolution of the elastic-plastic boundary as a function of the applied load [12.24–27], analysis of residual stresses [12.28], fracture-mechanics problems [12.29, 30], and orthotropic materials [12.31]. Many investigations that deal with the application of the method of birefringent coatings to the solution of engineering problems are included in the Bibliography.

The method of birefringent coatings was further improved by Russian investigators who applied the concepts of photoelastic interferometry developed in Chapter 10 to birefringent coatings and thus achieved complete separation of the strain field [12.32]. The principles of holographic interferometry, developed in Chapter 11, have also been applied to the birefringent-coating method [12.33].

An interesting application of the method of birefringent coatings is in photoelastic strain gauges, which were first introduced by *Oppel* [12.2, 34]. A photoelastic strain gauge is a thin, small piece of a birefringent material that is cemented to the surface of the body and is used to measure the state of strain at the location where it is cemented. The principle of measuring the strain components is the same as that used in the birefringent-coating method, that is, the gauge is illuminated by a polarized light beam and the emerging light is viewed through an analyzer. Two types of strain gauges, linear and circular, for measuring uniaxial and biaxial states of strain, respectively, are usually used. Owing to their many advantages, photoelastic strain gauges became popular among experimenters and are frequently used in practical applications.

In the present chapter, the characteristic features of the method of birefringent coatings will be presented in the frame of this monograph. Particular emphasis will be given to interpretation of the induced fringe patterns, as well as to the accuracy of the method. For more technical information about the method and for the vast number of its applications the reader is referred to other related works [12.35–39].

12.2 Interpretation of the Birefringent-Coating Fringe Patterns

In the method of birefringent coatings, a thin layer of a photoelastic material is bonded on the surface of the body under consideration and is then examined by an ordinary plane or circular polariscope. When the main body studied is made of an opaque material, a reflection-type polariscope is used. The arrangement of the optical elements in the reflecting polariscope is the same as that in the usual transmitting polariscope. That is, in plane polariscopes, the pass axes of the polarizer and analyzer are mutually perpendicular or parallel, to obtain dark- or bright-field polariscopes, respectively. In circular polariscopes, the principal axes of the quarter-wave plates are arranged at $\pm 45°$ with respect to the axes of the polarizer or the analyzer, respectively. The only difference between transmitting and reflecting polariscopes is that, in the former, all optical elements are arranged on the same axis and on both sides of the specimen, whereas, in reflecting polariscope, the optical elements all lie on the same side of the specimen. Figure 12.1 shows two common types of reflecting polariscope. Although in Fig. 12.1a different filters are used as polarizer, analyzer, and quarter-wave plates, in Fig. 12.1b by use of a beam splitter, the same polarizing filter plays the role of both polarizer and analyzer and the same quarter-wave plate serves both purposes of the two shown in Fig. 12.1a. According to either of these arrangements, the light rays pass twice through the thickness of the coating and are reflected by the surface of the main body. To obtain high reflectance, the surface of the main body must be polished or a reflective type of glue must be used to bond the coating to the metallic surface.

When the main body is stressed and the birefringent coating is perfectly bonded to the surface of the body, continuity conditions of the displacement field are satisfied on the interface and the in-plane displacements of the surface of the main body are transmitted to the coating. Owing to the displacements transmitted to the coating, a state of stress is developed in it. Therefore, differences of the optical-path lengths along the principal-stress or strain directions of the coating are induced. The values and directions of the principal stresses or strains of the coating can be determined by using the same methods of analysis as when the coating is considered alone. That is, the principles of two-dimensional photoelasticity, developed in Chapter 7, for determination of the

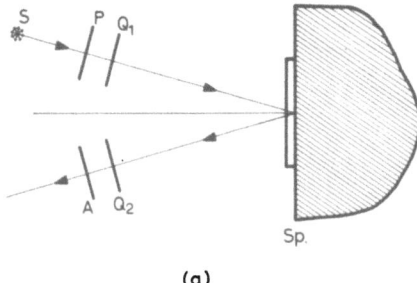

Fig. 12.1a and b. Two typical types of reflecting polariscope used in the method of birefringent coatings

(a)

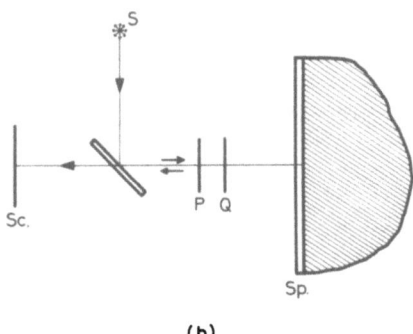

(b)

directions and difference of the principal stresses, as well as the principles of interferometric and holographic photoelasticity, developed in Chapters 10 and 11 for determination of the individual values of the principal stresses can be applied to the coating. Therefore, all of the methods and techniques of two-dimensional, interferometric, and holographic photoelasticity can be applied to the method of birefringent coatings. Figure 12.2a represents the formation of the Fizeau-type interference pattern, when the reflectances of the surface of the main body and of the coating are of the same order of magnitude, whereas in Fig. 12.2b an inclined semi-mirror is used for fringe multiplication.

Let us now formulate mathematically the condition of continuity of strains at the interface of the metallic body and the birefringent coating, and examine how the surface stresses of the body can be determined by measurement of the stress-optical retardations Δs_1 and Δs_2 of the coating along its principal-stress directions. The continuity condition of the strains is

$$\varepsilon_1^c = \varepsilon_1^s ,$$
$$\varepsilon_2^c = \varepsilon_2^s ,$$

(12.1)

where ε_1 and ε_2 are the principal strains and the indices c and s refer to the coating and the specimen, respectively.

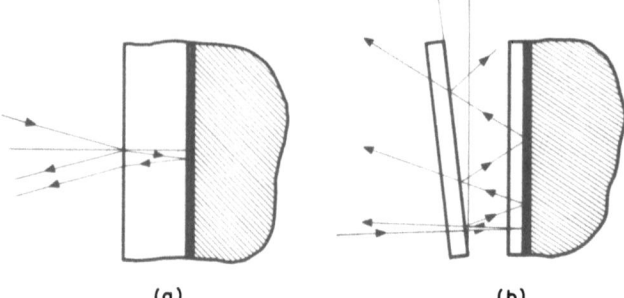

Fig. 12.2a and b. Optical arrangements for the formation of the Fizeau-interference pattern (a) and for fringe multiplication (b) in the method of birefringent coatings

(a)   (b)

Because plane-stress conditions are assumed in both the coating and the surface of the specimen, the principal strains $\varepsilon_1^c, \varepsilon_2^c$ and $\varepsilon_1^s, \varepsilon_2^s$ of the coating and the specimen can be expressed in terms of the corresponding principal stresses σ_1^c, σ_2^c and σ_1^s, σ_2^s by use of Hooke's stress-strain relationships,

$$\varepsilon_1^c = \frac{1}{E^c}(\sigma_1^c - v^c \sigma_2^c) \qquad \varepsilon_2^c = \frac{1}{E^c}(\sigma_2^c - v^c \sigma_1^c),$$

$$\varepsilon_1^s = \frac{1}{E^s}(\sigma_1^s - v^s \sigma_2^s) \qquad \varepsilon_2^s = \frac{1}{E^s}(\sigma_2^s - v^s \sigma_1^s). \tag{12.2}$$

In these equations E and v denote the modulus of elasticity and Poisson's ratio, respectively, while indices c and s are referred to the material of the coating and the specimen, respectively.

Substituting (12.2) into (12.1) and solving for σ_1^s and σ_2^s, we obtain

$$\sigma_1^s = \frac{E^s}{E^c[1-(v^s)^2]} \left[(1-v^c v^s)\sigma_1^c + (v^s - v^c)\sigma_2^c\right],$$

$$\sigma_2^s = \frac{E^s}{E^c[1-(v^s)^2]} \left[(v^s - v^c)\sigma_1^c + (1-v^c v^s)\sigma_2^c\right]. \tag{12.3}$$

The values σ_1^c and σ_2^c of the principal stresses of the coating are expressed in terms of the stress-optical retardations Δs_1 and Δs_2 along the principal stress directions of the coating as follows [see (10.26)]:

$$\sigma_1^c = \frac{a_t \Delta s_1 - b_t \Delta s_2}{(a_t^2 - b_t^2)d},$$

$$\sigma_2^c = \frac{a_t \Delta s_2 - b_t \Delta s_1}{(a_t^2 - b_t^2)d}, \tag{12.4}$$

where a_t, b_t are the stress-optical constants and d is the thickness of the coating.

Introducing the values of σ_1^c and σ_2^c from (12.4) into (12.3), we obtain

$$\sigma_1^s = \frac{E^s}{E^c[1-(v^s)^2](a_t^2-b_t^2)d}[(1-v^c v^s)(a_t \Delta s_1 - b_t \Delta s_2)+(v^s-v^c)(a_t \Delta s_2 - b_t \Delta s_1)],$$

$$(12.5)$$

$$\sigma_2^s = \frac{E^s}{E^c[1-(v^s)^2](a_t^2-b_t^2)d}[(v^s-v^c)(a_t \Delta s_1 - b_t \Delta s_2)+(1-v^c v^s)(a_t \Delta s_2 - b_t \Delta s_1)].$$

From (12.5) and by determining the values of the stress-optical retardations Δs_1 and Δs_2 along the principal-stress directions, by use of the methods of interferometric and holographic photoelasticity, the values of the principal stresses σ_1^s and σ_2^s on the surface of the opaque body can be determined.

Let us now consider that the birefringence of the coating is measured by use of the principles of two-dimensional photoelasticity developed in Chapter 7, which result in determination of the difference of the principal stresses of the coating $(\sigma_1^c - \sigma_2^c)$. Then we obtain, from (12.3),

$$(\sigma_1^s - \sigma_2^s) = \frac{E^s}{E^c} \frac{1+v^c}{1+v^s}(\sigma_1^c - \sigma_2^c).$$

$$(12.6)$$

Equation (12.6) shows that the difference of the principal stresses $(\sigma_1^s - \sigma_2^s)$ on the surface of the opaque body is directly proportional to the difference of the principal stresses $(\sigma_1^c - \sigma_2^c)$ in the birefringent coating. Because the quantity $(\sigma_1^c - \sigma_2^c)$ can be determined from the order of the photoelastic fringes in a typical isochromatic fringe pattern, the difference $(\sigma_1^s - \sigma_2^s)$ of the principal stresses on the surface of the opaque body can be determined by use of (12.6).

Therefore, (12.5) and (12.6) show that the values and the difference of the principal stresses in the opaque body are linearly related to the stress-optical retardations along the principal axes and the birefringence of the birefringent coating. Thus, the surface strains in the opaque body can be determined by use of all methods suitable for determination of two-dimensional stress field in the coating. These methods, namely, two-dimensional photoelasticity and interferometric and holographic photoelasticity were developed in Chapters 7, 10, and 11 of this book by use of the modern methods of describing polarization-optics problems, based on the Poincaré sphere and the Mueller and Jones calculi. Therefore, these methods can also be applied for the determination of the surface strains of opaque bodies studied by the method of birefringent coatings.

For determination of the values of the stress components σ_1^s and σ_2^s, when the induced stress-optical birefringence of the coating is measured, the well-known methods of stress separation of two-dimensional photoelasticity can be employed. Among them, of particular importance is the method of oblique incidence of light, according to which the light that passes through the coating is inclined relative to the normal to the interface of the surface of the specimen and the coating. If $(\pi/2 \pm \varphi)$ is the angle of incidence of the light beam, with respect to the direction of the principal stress σ_2^c, then the induced stress-optical retardation δ is

$$\delta = (\sigma_1^c - \sigma_2^c \cos^2 \varphi) \frac{2d}{\cos \varphi}, \tag{12.7}$$

where d is the thickness of the coating.

From (12.7) and by use of the measured value of δ, the quantity $(\sigma_1^c - \sigma_2^c \cos^2 \varphi)$ is determined, which, in combination with the quantity $(\sigma_1^c - \sigma_2^c)$, measured under normal incidence yields the individual values of the stress components σ_1 and σ_2. This method was applied to coatings by *Theocaris* [12.24] to determine the distribution of stress in an elastic-perfectly plastic notched specimen in tension. The light was incident at 45° from the normal to the metallic surface; the strains were completely separated all over the field. By use of the strain field determined at successive loading steps, and application of the Prandtl-Reuss stress-strain relations, the stress distribution and the evolution of the elastic-plastic boundary of the metal were determined. The results obtained for the strain components by the method of oblique incidence were compared with those derived by determining the sum of the principal strains by use of the electrical analog method and the method of relaxation. Good agreement was obtained between all of these results, which established the accuracy of the method of birefringent coatings in combination with the method of oblique incidence of light for determination of the strain components in an elastic-plastic body [12.24].

12.3 Accuracy of the Strains Measured by the Method of Birefringent Coatings

In the method of birefringent coatings, a thin piece of photoelastic material is bonded on the surface of the body under investigation and the strains on the surface of the body are determined by measuring the birefringence induced in the coating when the body is stressed. Because the strains on the surface of the body are determined from the birefringence of the coating, the accuracy of their determination will depend on the accuracy of measurement of the birefringence of the coating. Therefore, in this respect, the accuracy of the method of birefringent coatings will be the same as the accuracy of two-dimensional photoelasticity.

However, the surface strains of the body are transferred to the coating by shear forces developed at the interface of the structure and the coating. This transfer of load from structure to coating results in a relaxation of loads carried by the structure in the coated area. The amount of this relaxation depends on the thickness of the coating, on the mechanical properties of both materials, on the strain gradients through the thickness of the coating, as well as on the mode of strain distribution on the structural part of the ensemble. Thus, these factors influence the accuracy of the method.

First of all, we consider the phenomenon of relaxation of load in the structure due to the transfer of a part of the load to the coating. This phenomenon is called the *reinforcing effect* of birefringent coating. It is self-evident that the amount of reinforcing influences the accuracy of measurements with coatings. Therefore, it is necessary to derive correction factors that take into consideration the reinforcing effect, so that the strains measured by the coating may be correctly related to the corresponding strains of the uncoated part of structure.

Another effect that influences the accuracy of the results is the variation of strain through the thickness of the coating. This variation influences the observed birefringence and therefore the accuracy of the measured strains. The influence of the thickness of the coating, which is called the *thickness effect*, on the evaluation of strains in the uncoated part of the structure is taken care by another correction factor. With the thickness effect is also intimately related the influence of the *strain gradients* in the structure, which distort the fringe pattern of the coating to different amounts that depend on the thickness of the coating.

Finally, *the difference of the values of Poisson's ratios* for the coating material and the body considered, and the influence of *the free edges of the coating*, further affect the accuracy of the method.

The influence of all these effects on the accuracy of the method of birefringent coatings increases as the thickness of the coating increases. On the other hand, the thickness must be great enough to produce a sufficient number of photoelastic fringes for an accurate evaluation of the fringe pattern. These two conflicting considerations must be taken into account in the determination of the thickness of the coating. We shall now briefly examine the influence of each of the effects on the accuracy of the method of birefringent coatings.

12.3.1 Reinforcing Effect of Birefringent Coatings

This effect has been studied by *Zandman* et al. [12.10], who also gave numerical results for some simple cases of loading. The results of this investigation were presented in the form of correction factors, and curves were given for different ratios of the thicknesses of the coating and the metal and for different combinations of materials of the coating and of the metal under investigation.

In order to get an idea how the reinforcing effect can be taken into account in the evaluation of the optical patterns for the determination of the surface stresses of the body studied, let us examine the typical case of a body subjected to plane-stress conditions. Let us consider an infinitesimal coated element of the body and let h^s and h^c be the thicknesses of the body and of the coating, respectively. We compare the total stress developed on both the body and the coating with the stress created at the uncoated body only. If σ_x^s, σ_y^s and σ_x^c, σ_y^c are the normal stresses developed in the body and the coating, respectively, and σ_x^0, σ_y^0 are the normal stresses developed in the uncoated body, we obtain, by equating the total force in both the coated and the uncoated body, the following pair of equations along the Ox and Oy axes

$$h^s dy \sigma_x^s + h^c dy \sigma_x^c = h^s dy \sigma_x^0 ,$$

$$h^s dx \sigma_y^s + h^c dx \sigma_y^c = h^s dx \sigma_y^0 ,$$

or

$$\sigma_x^s + \frac{h^c}{h^s} \sigma_x^c = \sigma_x^0$$

$$\sigma_y^s + \frac{h^c}{h^s} \sigma_y^c = \sigma_y^0 .$$
(12.8)

Using in (12.8) the strain components ε_x^s and ε_y^s, ε_x^c and ε_y^c, and ε_x^0 and ε_y^0, expressed in terms of the corresponding stresses for the case of plane stress,

$$\sigma_x = \frac{E}{1-v^2}(\varepsilon_x + v\varepsilon_y)$$

$$\sigma_y = \frac{E}{1-v^2}(\varepsilon_y + v\varepsilon_x) ,$$
(12.9)

and taking into account the boundary conditions on the interface of the surface of the body and the coating, expressed by (12.1), we obtain

$$\varepsilon_x^0 - \varepsilon_y^0 = \left(1 + \frac{h^c}{h^s} \frac{E^c}{E^s} \frac{1+v^s}{1+v^c}\right)(\varepsilon_x^c - \varepsilon_y^c) .$$
(12.10)

Defining

$$C = \frac{h^c}{h^s} \frac{E^c}{E^s} \frac{1+v^s}{1+v^c} ,$$
(12.11)

we obtain

$$\varepsilon_x^0 - \varepsilon_y^0 = (1+C)(\varepsilon_x^c - \varepsilon_y^c) .$$
(12.12)

Equation (12.12) expresses the fact that the difference $(\varepsilon_x^c - \varepsilon_y^c)$ of the surface strains in the coated body is obtained by dividing the difference $(\varepsilon_x^0 - \varepsilon_y^0)$ of the true stresses by the factor $(1+C)$. Therefore factor C is the correction factor that takes into account the reinforcing effect of the coating when plane stress is developed on the surface of the structural part. Equation (12.11) shows that the correction factor C increases as the thickness h^c of the birefringent coating increases.

Similar expressions for the correction factors C were obtained for other typical states of stress, including flexure of a plate, combined plane stress and flexural loading, torsion of circular shafts, and pressurized cylindrical tubes. For an analysis of these cases the interested reader is referred to [12.38], which contains graphs that show the influence of characteristic elements of the coating and the body on the correction factor C.

The theory and graphs in [12.38] show that different correction factors for the thickness effect are required for different parts of the strain field. These variations depend on the various strain gradients in the metallic part. However, for a very thin coating, the correction factor due to the thickness effect tends to zero and the observed birefringence approaches its real value.

The foregoing analysis is valid only when the following assumptions are fulfilled: I) The metal and the coating are isotropic, elastic materials.

II) There is identical deformation of the metal and the coating at the interface.

III) Plane stress conditions prevail in the structure and the coating.
In such cases, the factors required to correct for the reinforcing effect for metallic plates and polymeric coatings that have elastic modulus E of the order of 4×10^5 psi are rather small and do not exceed a few percent. Finally, in all cases where plane-stress conditions prevail for the coating and the metal, measurements on the coating yield the strains integrated through the thickness, which approach very closely the true surface conditions.

12.3.2 Edge Effect of Birefringent Coatings

The displacements of the surface of the structural part are transferred to the birefringent coating by means of shear forces developed at the interface between the metal and the coating. These shear forces are large along a strip of variable width along the boundaries of the metallic specimen. The width of this strip is of the order of a few times the thickness of the coating; the concentration of the shear force there is due to the progressive transfer of forces from the metallic specimen, which is directly loaded, to the birefringent coating, which is strained indirectly by the structural part. The length of this transition strip, where the stress trajectories in the coating are not approximately parallel to the interface, depends mainly on the thickness of the coating and on the gradient of the strains in the metallic structure that are normal to its boundary.

At the boundary of the metallic plate in which elongation strains are normal to the boundary, the birefringent coating bulges in by amounts that are larger for greater thicknesses of the coating and for greater elongations of the metal. On the other hand, at regions of the boundary of the metallic structure where there are compressive strains, the coating bulges out. Therefore, at boundaries at which there are both types of strains, there is a point where the boundary of the coating is perpendicular to the interface. For boundaries without such transitions of strains normal to the boundary from elongations to compressions, such vertical boundaries do not exist. However, for the general plane-stress problem, the distribution of strains normal to the boundaries is always variable. Therefore the width of this transition strip is also variable. This indicates that, in general, the boundary of the metallic plate covered by the coating and the coating itself during loading are not identical.

It can be proved that complete identity of boundaries between the metallic plate and the coating is in general not feasible. The fringe orders shown in the coating are directly proportional to the surface strain of the structural part only if the strain on the metallic part is either uniform or varies linearly with distance. With all other types of deformation, strains evaluated directly from birefringent coatings contain some error, which may be large.

This error is further aggravated in studies of plasticity problems, in which birefringent coatings have their greatest advantages. In problems where plastic enclaves appear and propagate in the metallic structure, these enclaves, defined by the elastic-plastic boundary, create zones where there are transitions in the properties of the material. Although in the elastic regions the situation in the coating is similar to the elastic strain distribution in the metallic plate and both states of strain are governed by the same elastic laws, in the plastic enclaves not only the mechanical properties of the metallic materials change drastically, but also the stress and strain distributions are governed by much more complicated laws. Therefore, it is reasonable to expect internal disturbances of the strain distributions in the coating, rotation of the strain tensor, and, in general, a situation similar to this along the boundaries, due to the edge effect.

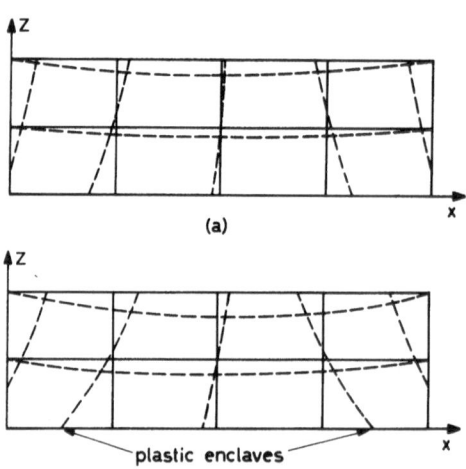

Fig. 12.3a and b. Schematic view of the distortion of the coating (a) and the creation of a plastic enclave in some region of the structural part (b)

Figure 12.3a presents a schematic view of the distortion of the coating due to a variable interface elastic displacement. The network consisting of perpendiculars and parallels to the interface traced before loading becomes oblique and distorted when an elastic load is applied to the structural part. If plasticity is engendered in the metal, internal disturbances, indicated by further distortions of the original perpendiculars, are created, which are larger at zones that neighbor the elastic-plastic boundary. Figure 12.3b indicates a plastic enclave in some region of the structural part and the local distortion of the network of the coating.

To these primary effects, another effect is added that aggravates the situation shown in Figs. 12.3 a,b, for which it was assumed that the interface remains plane after loading. This is certainly not the case. The thickness of the metal plate changes during loading, owing to its lateral deformation, because of the Poisson effect. This thickness variation is most pronounced in plastic enclaves (where Poisson's ratio increases to 0.50) and in areas of high stress concentration. Thus, the substratum of the coating is distorted. To this distortion of the interface, another distortion due to the change of thickness of the coating must be added, which is different than the distortion of the metal surface especially in regions where plastic enclaves develop in the structural plate. This aggravates the deviation from flatness of the upper (free) surface of the coating and introduces another source of errors.

Finally, the Poisson-ratio effect, which has some bearing on the edge effect, should be mentioned. Indeed, the metal specimens, which are usually made of aluminum alloys or of steel, have a Poisson ratio that varies between 0.30 and 0.33, whereas the polymers used as coatings have generally greater values of Poisson's ratio, higher than 0.36 (for instance, cold-setting epoxy polymers, which constitute excellent coatings, when unplasticized have a Poisson ratio $v = 0.365$ at ambient temperature; in their plasticized state, they have Poisson ratios that approach 0.5, depending on the amount of plasticizer). This difference of Poisson ratios causes different surface deformations of the metal and the coating. Because the coatings have, in general, greater values of Poisson's ratio than the metals, all surface deformations of the coating are greater than the deformations of the metal plate; therefore the edge effect is aggravated. This situation is attenuated in plastic regions, where the values of Poisson's ratio for metals reach the limiting value of 0.5, whereas the coatings retain the lower values of the elastic range.

This long discussion of the factors that influence the accuracy of strain measurements with coatings was deemed necessary to elucidate this complicated and asymmetric problem of three-dimensional deformation of a plate (the birefringent coating) submitted to a field of displacements along only one of its lateral faces. However, if the thickness of the plate is reduced to a thin layer, all of these contributions to errors are drastically diminished and the approximation of equating the average, through the thickness of the coating, strains, measured photoelastically, with the actual strains at the interface of the metal-coating is satisfactory.

In cases in which we want to measure average deformations in a restricted area of the strain field, where no steep strain gradients exist, small disks of coatings have been used successfully as photoelastic gauges. There are many applications of such gauges. In these cases, the average strain at the gauge position is sought and the disk yields satisfactory results for locally uniform strain fields. Although there is a striplike region along the boundaries of the disk where the deformations measured by the coating are not equal to the actual deformations of the structural part, at some distance from these boundaries the equivalence of these two groups of values is improved, especially for strain fields in which there are no steep strain gradients.

12.3.3 Strain Gradient and Curvature Effects

When the body under test is strained either elastically of plastically, the displacements of its surface are transmitted to the birefringent coating; the range of the method extends up to the limit of the range in which the induced deformations of the coating are elastic. For strains that exceed this range, the birefringence in the coating must be analyzed by use of the principles of the theory of photoplasticity. Such analysis, however, has not made much progress up to now. Consequently, the deformations induced in the coating should never exceed its elastic limit; the analysis in the coating is always restricted to the elastic case. When large strains of the metal structure have to be measured, care should be taken to use a coating material that will remain elastic at the largest deformations that will be encountered, even with plastic deformations of the metal structure. This can be done by using an appropriate amount of plasticizer in the epoxy polymer used as the coating material. For a thorough study of the properties (mechanical and optical) of plasticized epoxy polymers see [12.18, 40–42]. Therefore, the problem of determination of the stress field in the coating is a problem of the theory of elasticity, where an elastic strip is subjected to a given displacement field along one of its faces. Generally speaking, the state of stress in the birefringent coating is three dimensional and each layer of the coating is subjected to a different stress field. Thus, the usual assumption in the interpretation of the fringe pattern of the coating, that the state of stress does not alter through the thickness of the coating, is not generally fulfilled. However, when the state of strain on the surface of the body is almost uniform and the thickness of the coating is comparatively small, the stress in the coating can be taken as uniform, with high accuracy.

The problem of the triaxial state of stress in the coating, which complicates the interpretation of the fringe patterns obtained, is greatly aggravated when the state of stress in the body studied is triaxial and produces appreciable stress variations in directions normal to the surface of the body. Such stress gradients, which occur near geometrical discontinuities, continue into the coating and are responsible for the triaxial state of stress in the coating. The problem is further aggravated when high strain gradients occur in the surface of the body and when the surface studied is curved. In all of these cases, the average strain distribution through the thickness of the coating is not equal to the strain at the coating-body interface.

The influence of the strain gradients and the surface curvature on the state of stress in the coating has been studied by *Duffy* [12.12], who expressed his results in the form of two factors that correct for the strain gradients and the surface curvature for the case in which the surface displacement varies along only one direction. He concluded that strain gradients and surface curvatures greatly influence the birefringence of the coating and that appreciable errors may be introduced if these factors are not taken into account.

However, the correction factors given by *Duffy* exceed the normally accepted values in usual cases. This is mainly due to the fact that the uniaxial distributions of strains considered by *Duffy*, when he evaluates his correction factors, are rather unrealistically represented by trigonometric functions.

Such types of strain distributions never happen in actual cases of metal plates loaded in the elastic as well as in the plastic range. This situation was discussed in a review paper by *Theocaris* and *Dafermos* [12.16], who concluded that correction of experimental results obtained with thick coatings, by use of *Duffy*'s correction method, may lead to exaggerated corrections and significant errors. *Lee* et al. [12.13] studied the case in which the surface displacements present a radially symmetrical distribution along the coating-body interface and suggested a method for determining the surface displacements from the measured birefringence of the coating when the surface curvature is known or can be determined. *Theocaris* and *Dafermos* [12.43] studied the case of a coating of finite dimensions, based on the analysis of a rectangular strip loaded uniformly along its edge, and concluded that because the thickness of the strip was less than one-tenth of its shorter dimension, the state of stress along the greater part of the strip is uniform. The same authors [12.16], discussing *Duffy*'s results, introduced a new criterion for the accuracy of the results derived from observed birefringence of coatings, depending on the distance of the center of gravity of the area under the steeper portion of the displacement curve $u = u_0(x)$ from the u axis. The influence of the triaxial state of stress of the specimen upon the accuracy of measurement of its surface strains was considered by *Post* and *Zandman* [12.11], who studied the case of a thick plate with a hole. They concluded that, in this case, when shear forces are developed in the coating-body interface, an error of the order of 5 percent is introduced in the surface strain determination.

For the accurate evaluation of the influence of the various factors on the accuracy of the determination of stresses in the structure interesting experiments were executed by *Theocaris* [12.44]. This series of tests consisted of creating a structure-coating sandwich, where the place of the metal plate was occupied by a pure-epoxy polymer plate and the coating was represented by a highly plasticized sheet. The ratio of elastic moduli between the two plates was of the order of 50 times, which closely resembled many real situations between metals and coatings. A reflective surface was created at the interface between coating and plate before cementing the two plates; therefore it was possible to study the stress distribution either directly in the plate by looking on the one side of the sandwich, or through the plasticized coating by looking on the other side. The thickness of each of the two plates was 5 mm and each had a circular hole. The sandwich of the two plates was subjected to tension and the strains in each plate were measured either directly or through the coating. By comparing the results of both groups of measurements it was shown for such thicknesses of the coatings that the accuracy in measuring strains in metal plates was high, the error not exceeding 5 to 7 percent at a thin strip along the boundaries.

From the above studies, it can be concluded that although many attempts have been made to take into account the influence of the thickness of the coating on the accuracy of the determination of surface strain by the method of birefringent coatings, the problem has not yet been completely solved. A practical method for circumventing the problem is to determine surface strains by using various coating thicknesses and to extrapolate the results obtained for zero coating thickness. This procedure was introduced by *Post* and *Zandman* [12.11].

12.3.4 Effect of Different Coating-body Poisson Ratios

Equality of the Poisson ratios for both the coating and the metal plate is necessary for development of identical deformations at the coating-body interface. The equality of the Poisson ratios is particularly required at the edges of the coating, to assure exact transmission of the surface deformations to the coating. The influence of different values of Poisson's ratios for the coating and the body has been studied by *Post* and *Zandman* [12.11]. From this study, they concluded that mismatch of the Poisson ratios for the coating and the body exerts only a small influence on the accuracy of surface-strain measurement by the method of birefringent coatings. However, the situation is aggravated when plastic deformation occurs in the metal plate, when the Poisson ratio tends to 0.50, whereas other regions in the plate are deformed elastically, where the Poisson ratio retains its elastic value.

12.4 Photoelastic Strain Gauges

A photoelastic strain gauge is a small piece of birefringent material that is bonded to the surface of the opaque body studied, and whose function is to measure characteristic quantities of the strain field at the point where it is bonded. Therefore, the principle of operation of the photoelastic strain gauge is the same as that of the method of birefringent coatings. Two forms of strain gauges, uniaxial and biaxial gauges, are mainly used.

The uniaxial strain gauge has the form of a longitudinal strip; it responds only to longitudinal strains parallel to the axis of the strip. This is because, the width of the strip being relatively small, its response to strains perpendicular to the axis of the strip is limited. When the ratio of the thickness to width of the strip does not exceed 3, the response of the strip to lateral strains is less than 7 percent. For a general two-dimensional state of strain at a point of the surface of a body, three strain gauges oriented in three different directions suffice for complete determination of the state of strain at that point. Alternatively, the results of a single gauge can be combined with the isoclinic and isochromatic fringe patterns obtained by the method of birefringent coatings for separation of the principal stresses at the point where the strain gauge is bonded.

The two-strip coating method introduced by *Mönch* [12.45] must be mentioned. In this method, a thin photoelastic layer that plays the role of a substrate contains parallel and equally spaced strips of the same coating. The surfaces between the basic substrate and the strip are polished, to reflect the light. The photoelastic effect produced in the substrate gives the difference of the principal strains at the surface of the body; the outer strips respond only to the uniaxial strains parallel to their axes. By using an orthogonally crossed pair of strips and measuring the order of the isochromatics in the substrate and the two strips, the difference and the longitudinal components of the surface strains along the axes of the strips are determined. By this method, which consists of a combination of the birefringent-coating and uniaxial-strain-gauge methods, the strain field at the surface of the body is completely determined.

The biaxial strain gauge has the form of a small circular ring; it is used to determine the principal strain directions and the relative magnitude of the principal strains at the point where it is attached. The principal strain directions are determined from the axes of symmetry of the fringe pattern, whereas the relative magnitude of the principal strains is determined from the distortion of the fringe pattern.

To increase the sensitivity of the strains measured by the coating, initial fringes are usually introduced by a previous frozen-stress cycle in both types of strain gauges. Both of these types are combined with a linear- or circular-polarizer sheet, to facilitate determination of the obtained fringe patterns.

Another technique that uses uniaxial or biaxial strain gauges, which normally are illuminated by a light beam parallel to the surface of the body, was introduced by *Duffy* and *Lee* [12.6] and *Zandman* [12.46]. The basic principle of the method is shown in Fig. 12.4. This method has the advantage over

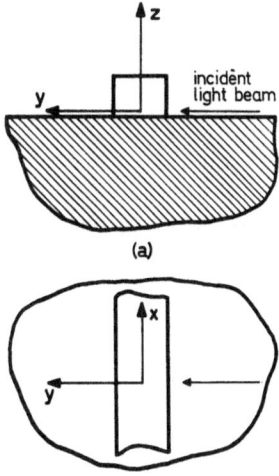

(a)

(b)

Fig. 12.4a and b. Principle of strain-gauge technique, using a strip illuminated by a light beam parallel to the surface of a metallic body

conventional photoelastic strain gauges of giving a fringe pattern that is independent of the thickness of the strip; thus, all problems associated with the thickness of the coatings are avoided.

References

12.1   A. Mesnager: C. R. Acad. Sci. **190**, 1249 (1930)
12.2   G. U. Oppel: Z. VDI **81**, 803 (1937)
12.3   J. D'Agostino, D. C. Drucker, C. K. Liu, C. Mylonas: Proc. Soc. Exp. Stress Anal. **12**, 123 (1955)
12.4   J. D'Agostino, D. C. Drucker, C. K. Liu, C. Mylonas: Proc. Soc. Exp. Stress Anal. **12**, 115 (1955)
12.5   F. Zandman, M. R. Wood: Prod. Eng. (London) **27**, 167 (1956)
12.6   J. Duffy, T. C. Lee: Exp. Mech. **1**, 109 (1961)
12.7   K. Kawata: J. Sci. Res. Inst. Tokyo **52**, 17 (1958)
12.8   F. Zandman: Anal. des Contrs (G.A.M.A.C.) **2**, 3 (1956)
12.9   R. Fleury, F. Zandman: C. R. Acad. Sci. **238**, 1559 (1954)
12.10 F. Zandman, S. S. Redner, E. I. Riegner: Exp. Mech. **2**, 55 (1962)
12.11 D. Post, F. Zandman: Exp. Mech. **1**, 21 (1961)
12.12 J. Duffy: Exp. Mech. **1**, 74 (1961)
12.13 T. C. Lee, C. Mylonas, J. Duffy: Exp. Mech. **1**, 134 (1961)
12.14 J. Duffy, C. Mylonas: *Proc. Int. Symp. on Photoelasticity*, 1961, ed. by M. M. Frocht (Pergamon Press, London 1963) pp. 27–42
12.15 E. E. Day, A. S. Kobayashi, C. N. Larson: Exp. Mech. **2**, 115 (1962)
12.16 P. S. Theocaris, K. Dafermos: Exp. Mech. **4**, 271 (1964)
12.17 C. H. Yew, B. R. Blackburn: Exp. Mech. **8**, 91 (1968)
12.18 P. S. Theocaris, C. Mylonas: J. Appl. Mech. **28**, 601 (1961)
12.19 P. S. Theocaris: Rheol. Acta **2**, 92 (1962)
12.20 P. S. Theocaris: J. Mech. Phys. Sol. **12**, 125 (1964)
12.21 P. S. Theocaris: In *Proc. IUTAM Symp. on the Photoelastic Effect and Its Applications*, ed. by J. Kestens (Springer, Berlin, Heidelberg, New York 1975) pp. 102–114
12.22 P. S. Theocaris: Int. J. Mech. Sci. **18**, 171 (1976)
12.23 F. Zandman, M. Watter, S. S. Redner: Exp. Mech. **2**, 215 (1962)
12.24 P. S. Theocaris: J. Appl. Mech. **29**, 735 (1962)
12.25 P. S. Theocaris, E. Marketos: J. Mech. Phys. Sol. **11**, 411 (1963)
12.26 P. S. Theocaris, E. Marketos: J. Mech. Phys. Sol. **12**, 377 (1964)
12.27 P. S. Theocaris: J. Franklin Inst. **279**, 22 (1965)
12.28 F. Zandman: Welding J. **39**, 191 (1960)
12.29 W. W. Gerberich: Exp. Mech. **4**, Part I, II, 335, 345 (1964)
12.30 G. U. Oppel, P. W. Hill; Exp. Mech. **4**, 206 (1964)
12.31 J. W. Dally, I. Alfirevich: Exp. Mech. **9**, 97 (1969)
12.32 P. Bohler: Photoelasticity Lab. Swiss Fed. Inst. of Techn., Zurich, Switzerland (1967)
12.33 J. D. Hovanesian, G. Eggenberger, Y. Y. Hung: Exp. Mech. **12**, 196 (1972)
12.34 G. U. Oppel: Z. VDI **101**, 809 (1959)
12.35 F. Zandman, S. Redner, J. W. Dally: *Photoelastic Coatings* (Soc. Exp. Stress Analysis, Iowa State University Press, Iowa 1977)
12.36 R. B. Heywood: *Photoelasticity for Designers*, Vol. 2 (Pergamon Press, London 1969), Chap. 10, pp. 277–293
12.37 J. Javornicky: *Photoplasticity*, Pt. 4 (Czechoslovak Academy Sci., Prague 1974) Chaps. 15, 16, pp. 262–280
12.38 G. S. Holister: *Experimental Stress Analysis* (Cambridge, at Univ. Press 1967) Chap. 5, pp. 210–238

12.39 A. Kuske, G. Robertson: *Photoelastic Stress Analysis* (Wiley and Sons, New York 1974) Chap. 11, pp. 263–274

12.40 P. S. Theocaris: J. Appl. Pol. Sci. **8**, 399 (1964)

12.41 P. S. Theocaris: Exp. Mech. **5**, 105 (1965)

12.42 P. S. Theocaris: In *Proc. IUTAM Symp. on the Photoelastic Effect and Its Applications*, ed. by J. Kestens (Springer, Berlin, Heidelberg, New York 1975) pp. 146–230

12.43 P. S. Theocaris, K. Dafermos: J. Appl. Mech. **31**, 714 (1964)

12.44 P. S. Theocaris: Unpublished work

12.45 E. Mönch: Schweiz. Bauztg. **84**, 840 (1966)

12.46 F. Zandman: Exp. Mech. **2**, 225 (1962)

13. Graphical and Numerical Methods in Polarization Optics, Based on the Poincaré Sphere and the Jones Calculus

13.1 Introduction

The necessity of solving polarization-optics problems that involve the passage of polarized light through a series of optical elements, in a rapid, unique, and elegant manner, led to the introduction of the modern methods in polarization optics based on the Poincaré sphere and the Mueller and Jones calculi. These methods for describing the various forms of elliptical polarization were presented in Chapter 3; the corresponding calculi for predicting the polarization form that emerges from an optical element were described in Chapter 4. In those two chapters were also presented the relations that exist between the Poincaré-sphere representation of a state of polarization and the Stokes and Jones vectors, as well as between the related transformations on the sphere and their corresponding analytical representations through the Mueller and Jones matrices.

These three methods, based on the Poincaré sphere and the Mueller and Jones calculi, constitute the main modern methods for solving problems of polarization optics; they have been widely used by many researchers. The Poincaré-sphere method is graphical; the other two are analytical methods. In the method of the Poincaré sphere, the state of polarization and all transformations for predicting the polarization of light that emanates from an optical element are represented on the surface of a unit-radius sphere. This three-dimensional model for handling problems of polarization optics introduces many difficulties for quantitative determination of the various polarization states. For this reason, the Poincaré-sphere method is usually used only for qualitative formulation and solution of problems. In this case, the solution deduced by use of the sphere is obtained in an easy and rapid manner; a problem can be formulated and solved by consideration of only the sphere. Many examples of use of the Poincaré sphere in optics, particularly in photoelasticity, were presented in the previous chapters.

To facilitate the solution of problems by the Poincaré-sphere method and to reduce the work to a plane instead of a spherical surface, various methods of projection of a sphere on a plane can be used. Of all of these methods, those based on stereographic and orthographic projections are of particular importance.

Stereographic projections of a sphere on a plane were introduced for solution of navigation problems in the 1850s. Some years later, stereographic projections were adopted in problems of crystallography and petrology mainly in the U.S.A., as well as in Russia. For a brief description of the use of these methods in solving such problems, the reader is referred to [13.1].

From the various types of stereographic projection, the Wulff net is of particular importance and has been widely used in navigation, crystallography, and petrography, as well as in optics. In this type of projection, one pole of the sphere is chosen as the projecting center and a great circle of the sphere is taken as the plane of projection. Thus, the sphere is transformed into a plane by an inversion process with a pole that lies on the sphere and an inversion power equal to $2R^2$, where R is the radius of the sphere. In this case, the hemisphere that contains the pole is projected outside of the great circle, whereas the other hemisphere is projected inside that circle.

In the Wulff net type of projection, the parallel circles of the sphere are projected as circles, whereas the equator and the principal meridian are projected as mutually perpendicular diameters if either of sections C and C_0 of principal meridian and the equator is considered as the projecting center (Fig. 13.1). The Wulff-type stereographic projection, besides projecting the characteristic circles of the sphere as circles, preserves angles unaltered in the projection. Thus, the angle between any two characteristic circles of the sphere is the same as the angle between their projections.

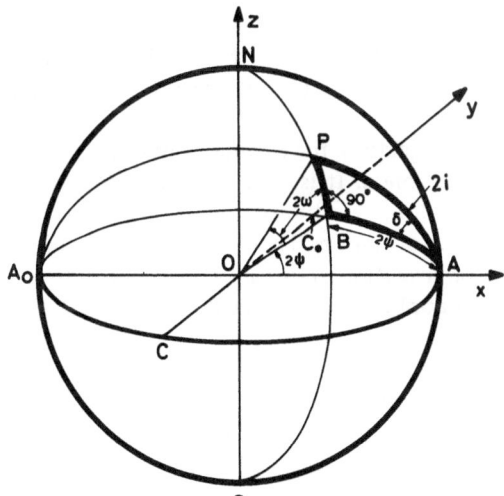

Fig. 13.1. Representation of an elliptically polarized light by a point P on the Poincaré sphere, and the presentation of azimuth ψ, ellipticity ω, amplitude ratio i, and the phase retardation δ of the corresponding elliptical vibration on the sphere

A comprehensive analysis of the use of the Wulff net in problems of polarization optics and photoelasticity was provided by *Wright* [13.2] as early as 1930. By starting with the relevant equations of elliptical vibration, *Wright* solved many problems with the Wulff net. These problems included the state of a plane polarized light that passes through a single birefringent plate or a system of two birefringent plates, determination of the intensity of a beam that passes through an analyzer, and analysis of an elliptical vibration by determining the azimuth and the ellipticity of its corresponding light ellipse by using the Friedel, Stokes, Brace, and Tool methods. Finally, *Wright* analyzed the optical transformations that occur in a typical plane polariscope. This paper by *Wright* included the first authoritative presentation of the use of the Wulff net in problems of optics. Although the Wulff-net method for solving problems of polarization optics presents many advantages, it has received little attention in the literature. After *Wright* in 1930 the Wulff net was used by *Aben* [13.3] in Russia to solve problems of polarization optics and photoelasticity. *Schwieger* [13.4] also used the Wulff net to determine the optical-equivalent plate of a system of two optical elements and also gave some numerical examples of its use.

Besides the stereographic projection of the Poincaré sphere, its parallel (orthographic) projection on the equatorial plane was also used to study photoelastic problems. This type of projection was first introduced in photoelasticity by *Menges* [13.5] and was subsequently developed by *Kuske*, who called it the *j*-circle method. *Kuske*, in a series of publications [13.6–11], gave the relevant features of the method and presented its use in problems of two- and three-dimensional photoelasticity. Although the concept of parallel projection is simpler than that of the stereographic projection, in stereographic projection all relevant circles of the Poincaré sphere are projected in a unified manner as circles or straight lines, whereas in the parallel projection they are projected as circles, ellipses, or straight lines. Also, in the Wulff net, angles are preserved along both families of lines in the net, whereas this does not occur in the parallel projection.

Finally, another type of geometric projection, the Schmidt net or otherwise called the Lambert net, should be mentioned. It is also used for geographic maps and in crystallography. The Schmidt net has the advantage of conserving, instead of angles, areas during the transformation. Therefore, it is convenient for statistical representation of the polarization states in various substances. However, this type of net has not been applied to photoelasticity as yet. Therefore, it will be not presented here.

Both the Wulff-net and *j*-circle methods are graphical. They result from stereographic and parallel projections of the Poincaré sphere, respectively. Use of them facilitates solution of polarization problems because they provide a two-dimensional model, with which to work, instead of a three-dimensional model, such as the Poincaré sphere.

Among the analytical methods, the method of quaternions, which is based on the Jones calculus, is also useful in the solution of a variety of problems in photoelasticity. The use of quaternions in photoelasticity was introduced by *Richartz* and *Hsu* [13.12], who remarked that the general two-by-two complex Jones matrix can be expressed as a linear combination of the four Pauli fundamental matrices. Therefore, the coefficients of the quaternion completely determine the corresponding optical element. The method of quaternions was widely used by *Cernosek* [13.13–16] to solve problems of polarization optics and photoelasticity. *Cernosek* [13.16] also presented a graphical method, that is based on quaternions, for the study of polarization-optics problems.

In the present chapter, the graphical methods of the Wulff net and the *j* circle, and the analytical method of quaternions will be presented. Particular emphasis will be given to the use of these methods for the solution of problems of photoelasticity.

13.2 The Wulff-Net Method

13.2.1 The Poincaré Sphere

Before proceeding to the development of the Wulff-net method, we shall review some basic results of the composition of two mutually perpendicular sinusoidal vibrations that result in an elliptical vibration, for which the Poincaré sphere may be applied.

As developed in Chapter 3, p. 26 the composition of two mutually perpendicular sinusoidal vibrations

$$
\begin{aligned}
\alpha_x &= A_x \cos(v + \delta_x) \\
\alpha_y &= A_y \cos(v + \delta_y)
\end{aligned}
\tag{13.1}
$$

leads to an elliptical vibration, which can be represented by the light ellipse. The characteristic elements of the light ellipse (Fig. 3.4), that is the inclination ψ of its major axis with the Ox axis, called the azimuth, and its ellipticity ω, defined by $\tan \omega = (b/a)$, where a and b are the lengths of the principal semi-axes of the ellipse, are related through (3.27, 31)

$$\tan 2\psi = \tan 2i \cos \delta, \tag{13.2}$$

$$\sin 2\omega = \sin 2i \sin \delta, \tag{13.3}$$

where $\delta = (\delta_y - \delta_x)$ is the phase difference between the two vibrations and i is defined by $\tan i = (A_y/A_x)$.

By mutually subtracting (3.21) it follows that

$$a^2 - b^2 = (A_x^2 - A_y^2)(\cos^2 \psi - \sin^2 \psi) + 2 A_x A_y \sin 2\psi \cos \delta.$$

By taking into account (3.22) and (13.2) and introducing the angles i and δ, we obtain

$$\cos 2i = \cos 2\omega \cos 2\psi. \tag{13.4}$$

Equations (13.2) to (13.4) constitute the fundamental relations that connect the four quantities ψ, ω, i, and δ.

According to the Poincaré-sphere method (Sect. 3.4.2) each point P of the sphere represents an elliptical polarization whose azimuth 2ψ and ellipticity 2ω are defined by the corresponding spherical angular coordinates shown in Fig. 13.1. From the right-angle spherical triangle ABP and by using the well-known relations of spherical triangles, we obtain (13.4) and (13.3), which means that the arc AP is equal to $2i$, and the angle BAP is δ.

13.2.2 The Wulff Net

The position of each point P on the Poincaré sphere is defined by the two spherical angular coordinates, namely the longitude 2ψ and latitude 2ω (Fig. 13.2a). Let us trace on the sphere equidistant meridians C_1 (constant

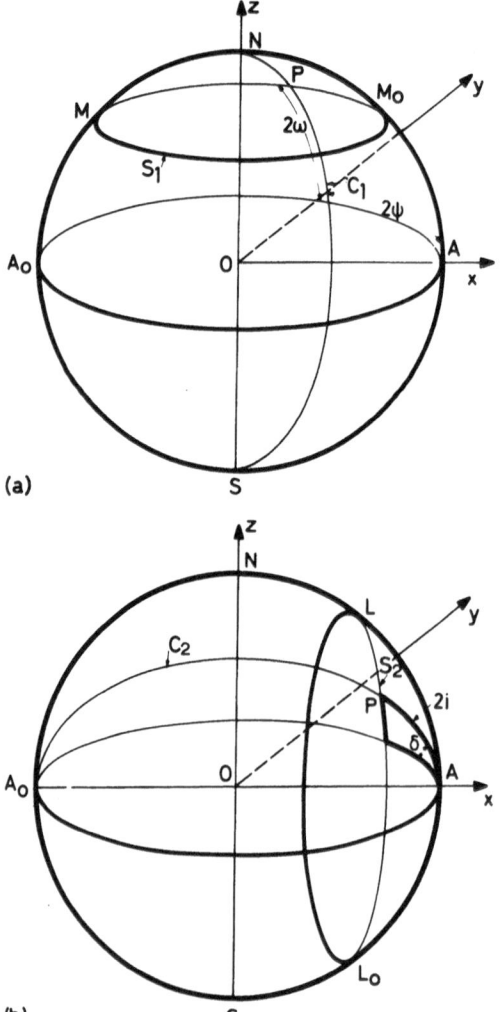

Fig. 13.2 a and b. Definition of a point P on the Poincaré sphere by the meridian C_1 and parallel S_1, corresponding to the azimuth ψ and ellipticity ω (a) or by the parallel to the principal meridian S_2 and the great circle through the Ox axis C_2 that correspond to the amplitude ratio i and the phase retardation δ (b)

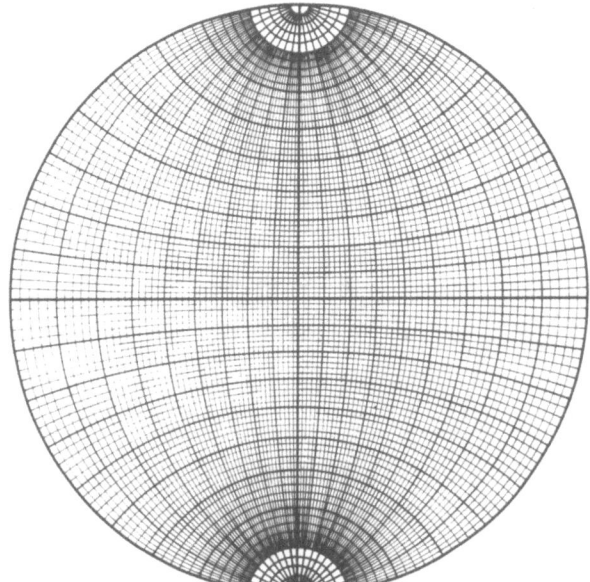

Fig. 13.3. The Wulff net used in stereographic projections of parallels and meridians with intervals of 2°

longitude 2ψ) and parallel circles S_1 (constant latitude 2ω). The stereographic projection of the system of meridians and parallels on the plane of the principal meridian $O\,x\,z$ yields the Wulff net, shown in Fig. 13.3. In this figure, meridians and parallels are traced at intervals of 2°. The center of the stereographic projection was either of the section points of the normal at the center of the principal meridian and the equator of the sphere. Thus, the sphere is transformed into a plane by inversion with a pole lying on the sphere and an inversion power equal to $2R^2$, where R is the radius of the sphere. By this type of stereographic projection, the hemisphere containing the center of projection is projected outside the outer circle of the Wulff net, whereas the other hemisphere is projected on the interior of this circle.

The fundamental curves of the Wulff net have the following properties:

I) The projection S_1' of each parallel circle S_1 of Fig. 13.2a is also a circle; the latitude 2ω of the circle S_1 is equal to the angle $M_0'\,O'\,A'$ in the projection formed by the $O'x$ horizontal axis and the radius to the point M_0' of the intersection of the circle S_1' with the outer circle of the net. For the complete determination of the circle S_1', a third point M_0'' is defined at the intersection of the straight line $A_0'\,M_0'$ with the Oz axis (Fig. 13.4), where line $A_0'\,M_0'$ subtends an angle ω with the $x'\,O'\,x$ axis.

II) The projection of the equator of the sphere is represented by the $x'\,O'x$ axis; the projection of the principal meridian is represented by the perpendicular $z'\,O'z$ axis.

III) The projection C_1' of each meridian C_1 of Fig. 12.2a is also a circle that passes through the points N' and S' of the net. A third point K' of this circle on the $x'O'x$ axis is defined by transferring the intersections of the diameter $z'O'z$ with the circles S_1' to corresponding points on the diameter $x'O'x$. Thus, if the longitude of the meridian NPS is 2ψ, the length $O'K'$ is equal to the length defined on the $z'O'z$ axis by the circle S_1' with $2\omega = 2\psi$ (Fig. 13.4).

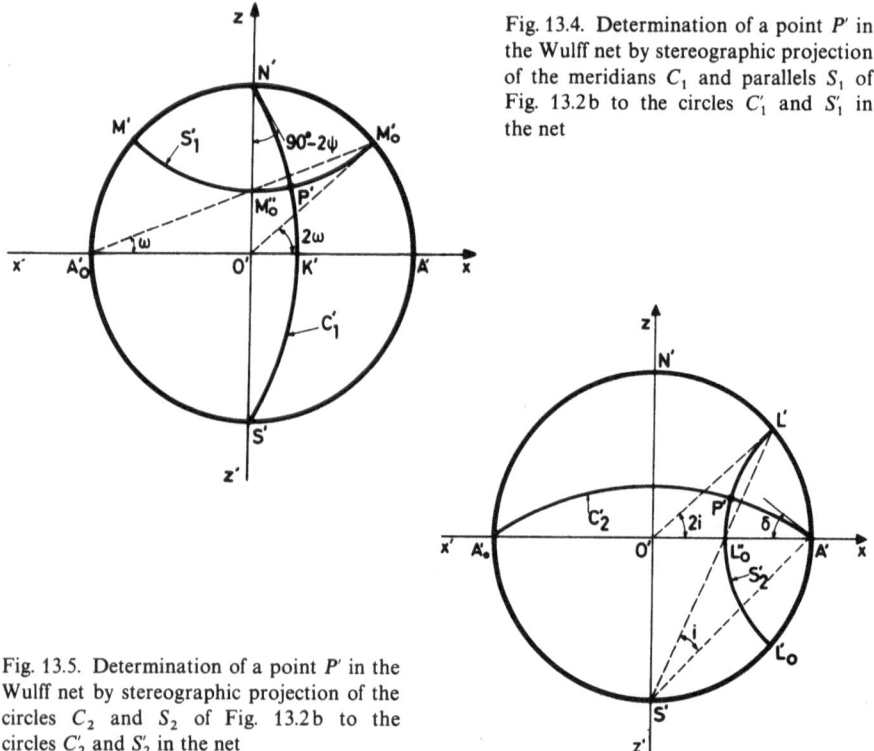

Fig. 13.4. Determination of a point P' in the Wulff net by stereographic projection of the meridians C_1 and parallels S_1 of Fig. 13.2b to the circles C_1' and S_1' in the net

Fig. 13.5. Determination of a point P' in the Wulff net by stereographic projection of the circles C_2 and S_2 of Fig. 13.2b to the circles C_2' and S_2' in the net

IV) The angles subtended between characteristic circles of the sphere are maintained in the projection. Thus, the projections of the parallel circles and the meridians constitute orthogonal pairs of curves. This property can be used to define the tangents of the circle C_1' at points N' and S' that subtend angles $(90° - 2\psi)$ with the $z'O'z$ axis.

Let us now consider that the position of each point on the Poincaré sphere is defined by the angle δ subtended between the equator and the great circle C_2 that passes through the $x'O'x$ axis and the angle $2i$ of the arc AP on the circle C_2 (Fig. 13.2b). If we trace on the sphere the corresponding circles C_2 and S_2 through the $x'O'x$ axis and parallel to the principal meridian, respectively, and project this system of circles stereographically, we obtain another Wulff net similar to the previous one, but angularly displaced 90° relative to the first net. A schematic diagram of this Wulff net, showing the meaning of its relevant quantities, is shown in Fig. 13.5.

13.2.3 Applications of Wulff Net to Problems of Polarization Optics

As developed in Section 13.2.1, when two mutually perpendicular linear vibrations with amplitude ratio $\tan i = (A_y/A_x)$ and phase difference δ combine, an elliptical vibration with an azimuth ψ and an ellipticity ω is formed. The four quantities i, δ, ψ, and ω are related through (13.2) to (13.4). From the development of the definition and properties of the Wulff net, the fundamental results concerning quantities i, δ, ψ, and ω of an elliptical vibration follow (see Fig. 13.4 and 13.5):

I) When the ellipticity ω of the light ellipse is constant, while its azimuth ψ varies, that is the ellipse is referred to various coordinate systems, the representative point P' of the light ellipse in the Wulff net moves along the circle $S_1' \equiv M'P'M_0'$ (Fig. 13.4). The angle of movement of the point P' along the circle S_1' is equal to the variation of ψ from one to the other coordinate system.

II) When the azimuth ψ of the light ellipse is constant, while its ellipticity ω varies, the representative point P' of the light ellipse in the Wulff net moves along the circle $C_1' \equiv N'P'S'$ by an angle equal to the variation of the ellipticity ω (Fig. 13.4).

III) When the amplitude ratio $\tan i$ of the two vibrations represented in (13.1) is kept constant, while the phase difference δ varies, the representative point P' of the corresponding elliptical vibration in the Wulff net moves along the circle $S_2' \equiv L'P'L_0'$ by an angle equal to the variation of δ (Fig. 13.5).

IV) When the phase difference δ is kept constant, while the amplitude ratio $\tan i$ varies, the representative point P' of the corresponding elliptical vibration in the Wulff net moves along the circle $C_2' \equiv A'P'A_0'$ by an angle equal to the variation of i (Fig. 13.5).

When a point P' is given in the Wulff net, the characteristic quantities ω, ψ, i, and δ of the corresponding light ellipse can be determined by tracing the four circles S_1', C_1', S_2', and C_2' in the net. Conversely, given an elliptical vibration, defined by the quantities i and δ, or the azimuth ψ and ellipticity ω of the corresponding light ellipse, the representative point P of the elliptical vibration in the Wulff net is defined and the other two of the four quantities, ψ, ω, i, and δ are determined.

We shall now proceed to solve the two basic problems of polarization optics that involve the passage of an elliptically polarized light through a birefringent plate and a polarizer. These problems will be solved on the basis of the above fundamental properties of the quantities ω, ψ, i, and δ in the Wulff net.

Passage of Polarized Light Through a Retarder (Birefringent Plate)

Let us consider the point P that represents the elliptically polarized light in the Wulff net (Fig. 13.6). We shall find the form of the polarization of the light that comes out of the linear retarder (birefringent plate) by two steps. We first

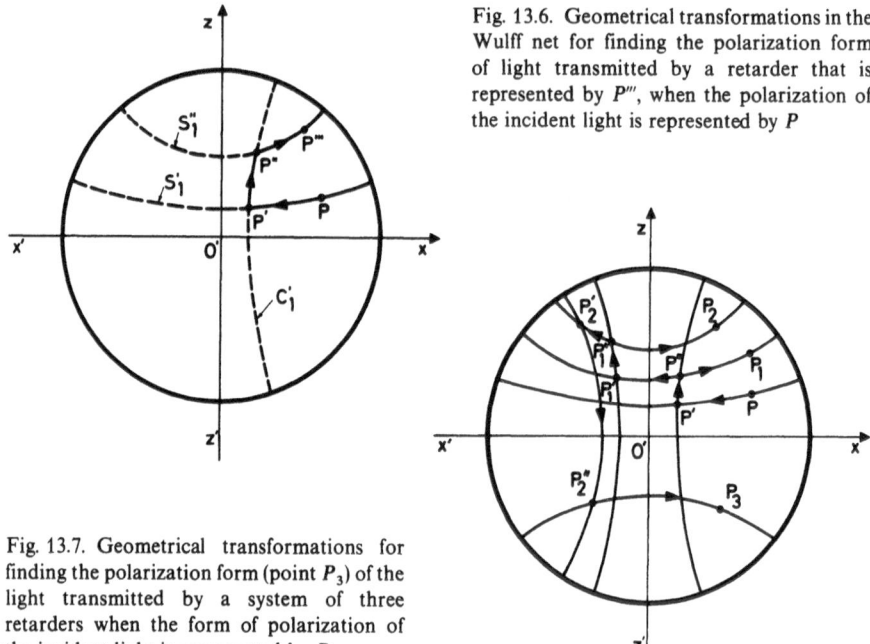

Fig. 13.6. Geometrical transformations in the Wulff net for finding the polarization form of light transmitted by a retarder that is represented by P''', when the polarization of the incident light is represented by P

Fig. 13.7. Geometrical transformations for finding the polarization form (point P_3) of the light transmitted by a system of three retarders when the form of polarization of the incident light is represented by P

refer the principal axes of the incident-light ellipse to the principal axes of the retarder. To find the corresponding point P' in the Wulff net, according to the property I) above, the point P must move along the circle S_1' by an angle equal to the difference of the orientations of the fast axis of the retarder and the azimuth of the light ellipse. The incident-light ellipse, represented by point P', has the same orientation as the principal axes of the retarder; to find the final form of polarization of the light we must add to the phase difference δ of the light ellipse the corresponding phase difference δ_0 introduced by the retarder. To do this, according to the property III) above, the point P' must be moved to the point P'' along the circle C_1' by an angle equal to δ_0. Point P'' represents in the Wulff net the final polarization form from the retarder referred to the principal axes of the plate. If we desire to refer the final form of polarization of the light to the axes of the incident-light ellipse, point P'' must be moved along the circle S_1'' to the point P''' by an angle opposite to the angle subtended when point P was first moved to point P'.

The foregoing procedure for finding the final form of polarization of light from a given retarder can be applied when the light beam encounters a pile of retarders, each of which is specified by the orientation of its principal axes and its retardance. This is of primary importance in problems of three-dimensional photoelasticity, where the photoelastic model can be considered as a pile of optical retarders. Figure 13.7 shows the graphical operations for finding the final form of polarization (point P_3) from a system of three retarders, when the incident light is given (point P).

Passage of Polarized Light Through a Polarizer

Consider a unit-intensity elliptically polarized light whose corresponding light ellipse has azimuth 2ψ and ellipticity 2ω, represented by point P in the Wulff net with $A'P' = 2\psi$ and $PP' = 2\omega$ and a linear polarizer whose pass axis has an azimuth 2θ and is represented in the Wulff net by point A (Fig. 13.8). According to the analysis of Section 4.3.5 the intensity I of the light transmitted through the analyzer is expressed by (4.49). In order to simplify this expression, we consider in the net the right-angle triangle $MA'M'$ ($\overline{MM'}A' = 90°$) in which $A'M' = 2(\theta - \psi)$ and $MM' = PP' = A'M'' = 2\omega$. From this triangle, we find $\cos A'M = \cos 2\omega \cos 2(\theta - \psi)$, and from (4.49) we find

$$I = \cos^2 \frac{A'M}{2}.$$ (13.5)

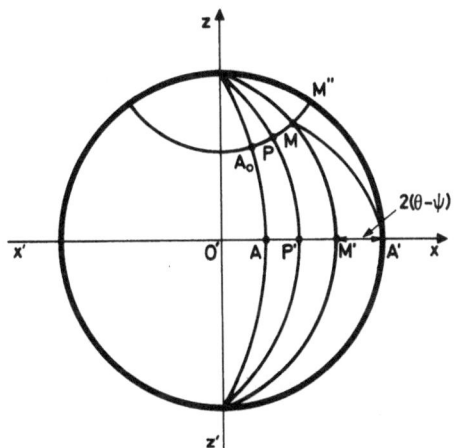

Fig. 13.8. Determination in the Wulff net of the intensity of a light beam (point P) that passes through a linear polarizer (point A). The arc $A'M$ is the independent variable in (13.5)

Thus, to find the light intensity transmitted by the polarizer, we form the triangle $MA'M'$ by taking $AM' = 2\psi$ and $M'M = 2\omega$ and apply relation (13.5) where $A'M$ is the hypotenuse of the triangle $MA'M'$.

From (13.5), we may conclude that for a given elliptical polarization and in order that the intensity transmitted through a polarizer be maximum, the hypotenuse $A'M$ must be minimum; that is, the polarizer must have the same azimuth as the incident-light ellipse ($\theta = \psi$).

Applications

Using these geometrical operations in the Wulff net for the fundamental cases of the passage of polarized light through a retarder or a polarizer, we shall now consider some relevant problems of polarization optics, which were solved previously by use of the Poincaré-sphere method or the Mueller and Jones calculi:

I) The Senarmont principle, treated in Chapter 5, p. 26 may be readily proved by the Wulff-net graphical method. According to this principle, when a quarter-wave plate is interposed in a beam of elliptically polarized light with its principal axes parallel to the principal axes of the light ellipse (with ellipticity ω) then the light that comes out is linearly polarized at an azimuth ω with respect to the slow axis of the quarter-wave plate. In order to prove the Senarmont principle with the Wulff-net graphical method, we proceed as follows (see Fig. 13.9): Let point P represent the elliptical vibration of the light incident on the quarter-wave plate. In order to find the final polarization form, the incident vibration must be referred to the coincident principal axes of its light ellipse and the quarter-wave plate; then the 90° retardation introduced by the quarter-wave plate must be added. These two operations are executed by rotating the point P along the circle S_1' until it reaches the position P' on the outer principal circle (for which $\delta = 90°$) and then by rotating point P' along the circle C_1' by 90°. The final position P'' of point P lies on the horizontal diameter of the Wulff net; therefore $A'P'' = A'P' = \omega$. That is, the final light is linearly polarized at an angle ω to the slow axis of the quarter-wave plate.

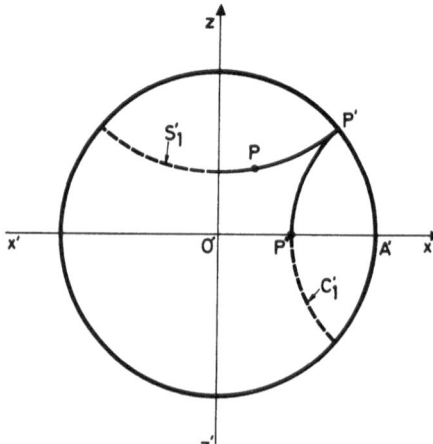

Fig. 13.9. Representation of the Senarmont principle in the Wulff net

II) When a retarder is interposed in elliptically polarized light, we can determine the position of the retarder for which the final light is linearly polarized.

In order to solve this problem graphically let point P represent the elliptically polarized light and point Q the retarder in the Wulff net (Fig. 13.10). The emerging light from the retarder, is found by rotating point P successively along two circles whose centers lie on the vertical and horizontal diameters of the net, respectively. The second rotation, about the horizontal axis, must be made through an angle δ equal to the retardation of the retarder; the final position P'' of P must therefore bring point P to the horizontal $A_0' A'$ axis.

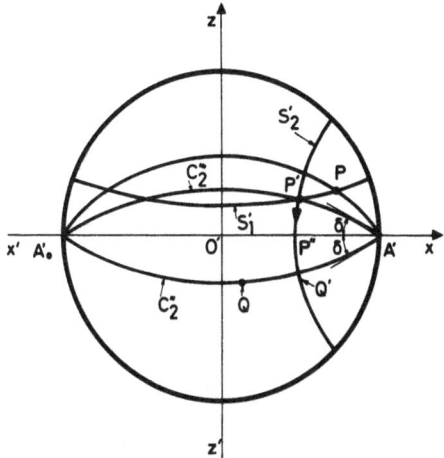

Fig. 13.10. Graphical solution of the problem of finding the orientation of a retarder (point Q) for which the transmitted light is linearly polarized, when the incident light is P

Thus, the point P must be rotated along the circle S_1' whose center is on the vertical axis to the intersection point P' of this circle with the circle C_2''' that is symmetric to the circle C_2'', with respect to the horizontal $A_0' A'$ axis. Then, point P' is rotated along the circle S_2' through an angle equal to $(-\delta)$ until it is brought to the diameter $A_0' A'$; that is, the final light is linearly polarized. The arc $A' P''$ on $A' A_0'$ determines the required position Q' of the retarder.

13.2.4 Applications of Wulff Net to Problems of Two-Dimensional Photoelasticity

We shall now apply the Wulff-net graphical method to some problems of two-dimensional photoelasticity that were treated in Chapter 7 by use of the method of the Poincaré sphere and the Mueller and Jones calculi.

I) *The Circular Polariscope.* First, we shall determine the optical transformations in the Wulff net that take place in a circular polariscope (Fig. 7.2) and lead to an isochromatic fringe pattern. In Fig. 13.11, the point N' that lies at the intersection of the vertical diameter and the outer principal circle of the net represents the circularly polarized light that emerges from the first quarter-wave plate. Let us consider all points on the specimen that have the same retardation δ but different principal axes. To find the final polarization form for all of these points, point N' must be rotated along the circle S_1' through angles that vary between $-90°$ and $+90°$; then, all points of this circle should rotate along the horizontal axis $A_0' A'$ through an angle δ equal to the common retardation of all points considered ($N' N'' = \delta$). Thus, all forms of polarization of the light emerging from the specimen are represented by the points of the circle S_1''. To find the form of polarization of the light that comes from the second quarter-wave plate, circle S_1'' is rotated about the horizontal axis $A_0' A'$ through an angle equal to $90°$ until it reaches the circle S_1'''. To refer all forms of polarization of the light that emerges from the second quarter-wave plate to the axis of the analyzer, circle S_1''' is rotated about the vertical axis $N' S'$

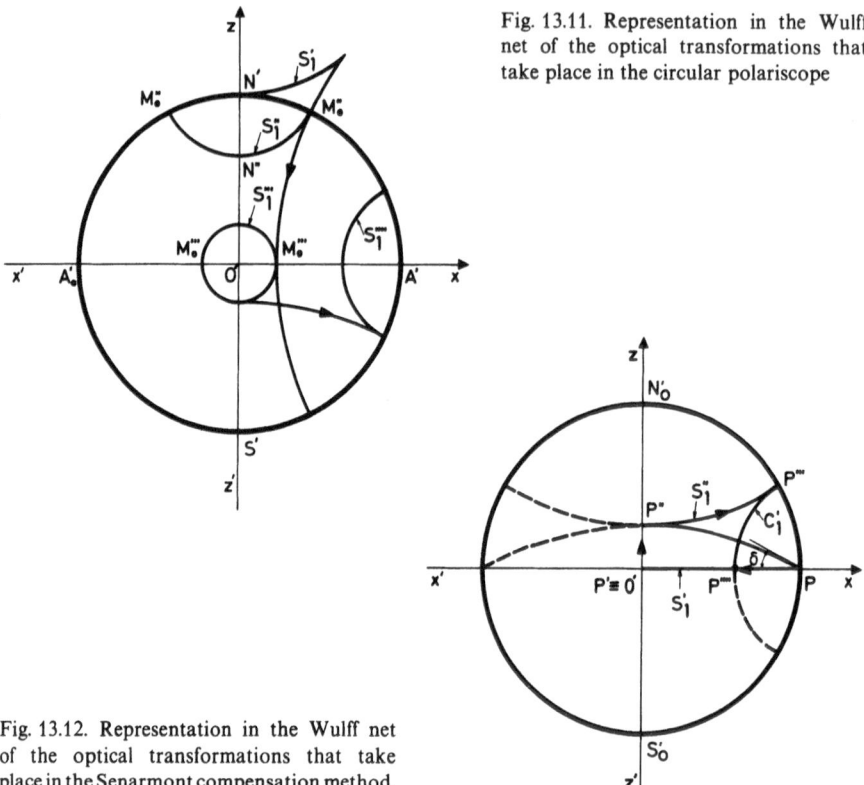

Fig. 13.11. Representation in the Wulff net of the optical transformations that take place in the circular polariscope

Fig. 13.12. Representation in the Wulff net of the optical transformations that take place in the Senarmont compensation method

through an angle equal to 90° and in this way the circle S_1'''' is derived. All points of this circle S_1'''' have the same intensity and correspond to the iso-chromatic fringe with retardation equal to δ.

II) *The Senarmont Compensation Method.* The optical arrangement for the Senarmont compensation method is shown in Fig. 7.4. The linearly polarized light that emerges from the polarizer is represented by point P (Fig. 13.12). (Figure 7.4 has been rotated through an angle of 90°.) To find the final form of polarization from the specimen, oriented with its principal birefringent axes at the point considered at an angle of 45° with respect to the axis of the polarizer, point P is rotated along the horizontal diameter S_1' through 90° and then is moved along the vertical axis $N_0' S_0'$ of the net through an angle δ equal to the stress-optical retardation δ at the point of the specimen. The polarization of the light emerging from the quarter-wave plate that follows the specimen is found by rotating point P'' along the circle S_1'' through 90°, so that the point P''' on the outer principal circle of the net is obtained, and by then rotating point P''' along the circle C_1' through 90° to point P''''. This point represents linearly polarized light at an angle equal to $(\delta/2)$ with respect to the axis of the polarizer $(P P'''' = P P''' = P' P'' = \delta)$.

13.3 The *j*-Circle Method

13.3.1 The *j* Circle

The *j* circle is a unit-radius circle in the Oxy plane, which results from parallel projection of the Poincaré sphere on this plane (Fig. 13.13). As parallel projection can be considered a special case of stereographic projection with the pole at infinity, the *j* circle can be deduced as a special case of the Wulff net. Each point of the circle defined by a vector with origin at the center of the circle, the so-called *j* vector, represents a different polarization form and especially that, whose azimuth is equal to half of the angle between the *j* vector and the Ox axis, and with ellipticity ω given by (see triangle OPP' in Fig. 13.13a)

$$\tan^2 \omega = \frac{1 - |j|}{1 + |j|},\qquad (13.6)$$

where $|j|$ is the length of the *j* vector.

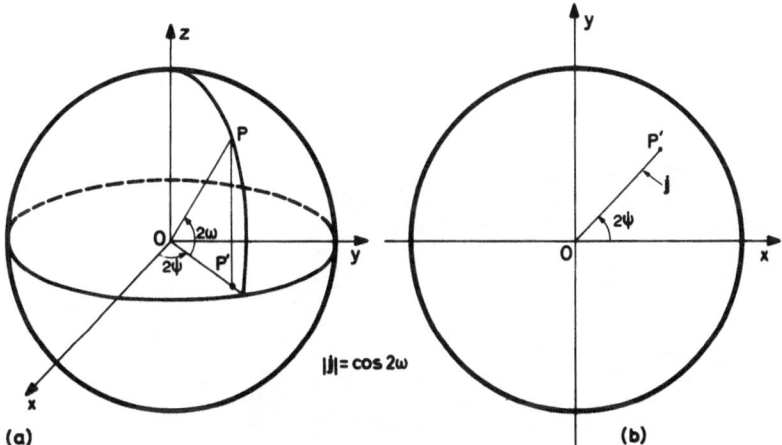

Fig. 13.13a and b. The Poincaré sphere (a) and its parallel projection (b), which gives the *j* circle

The fact that the tip of the *j* vector with the foregoing characteristics results from the parallel projection of the corresponding point on the Poincaré sphere with azimuth ψ and ellipticity ω was proved in Section 3.5.1.

From the definition of the *j* vector it follows that

I) All forms of linear polarization are represented by points on the circumference of the *j* circle.

II) Circularly polarized light is represented by the center of the *j* circle.

III) Any form of elliptical polarization is represented by a point inside the *j* circle and vice-versa.

Let us now consider the four fundamental circles through a point P of the Poincaré sphere: the circle S, parallel to the equator, the meridian C_1, the

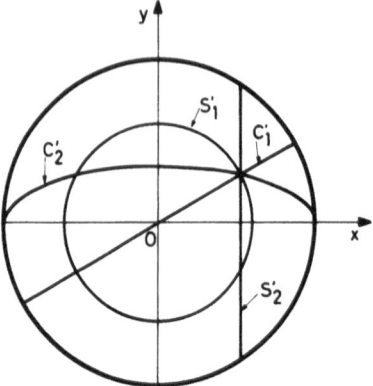

Fig. 13.14. Parallel projections of the fundamental circles C_1, S_1, C_2, S_2 of Fig. 13.2 to the straight line C_1', circle S_1', ellipse C_2', and straight line S_2' in the j circle

circle parallel to the principal meridian S_2, and the circle C_2 that passes through the Ox axis (Fig. 13.2). Along these circles S_1, C_1, S_2, C_2, the quantities 2ω, 2ψ, $2i$, and 2δ, respectively, are kept constant. In the j circle, the circles S_1 are transformed into circles S_1', the circles C_1 into straight lines C_1', the circles S_2 into straight lines S_2', and the circles C_2 into ellipses C_2' (Fig. 13.14). The

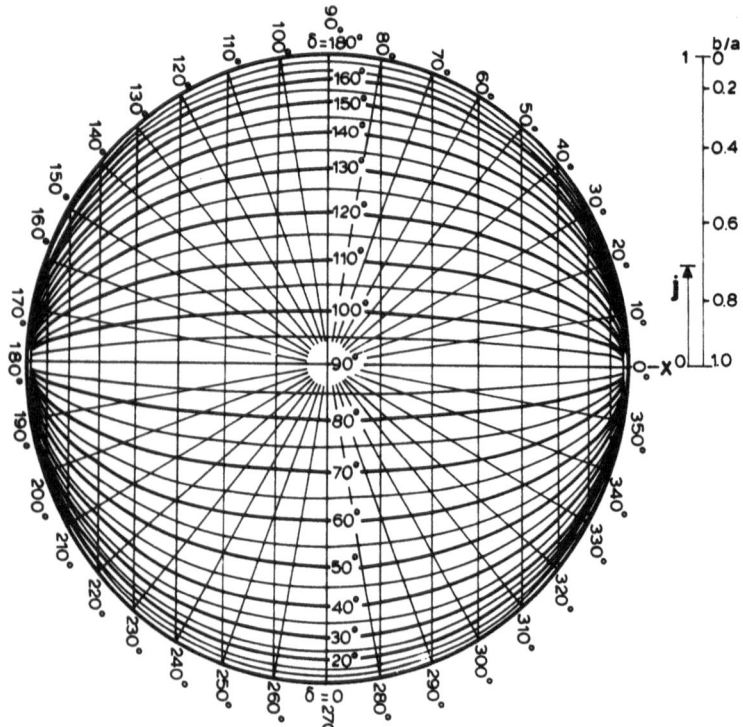

Fig. 13.15. Net of the characteristic curves C_1', S_2', and C_2', traced at intervals of $5°$ in the j circle

net of the characteristic curves C_1', S_2', and C_2' traced in the *j* circle at intervals of 5° is shown in Fig. 13.15. On the right-hand side of this figure, a scale has been added to facilitate the determination of the length of the *j* vector from the value of the ellipticity $\tan\omega = (b/a)$ of the corresponding light ellipse.

Comparing now the method of representation of elliptically polarized light by the *j* circle and the Wulff net, we can see that the Wulff-net method presents the following advantages:

I) All of the fundamental circles of the Poincaré sphere are projected into circles in the Wulff net.

II) All angles along the relevant circles of the net are preserved.

On the other hand, in the *j*-circle method, the fundamental circles of the Poincaré sphere are projected into circles, straight lines, or ellipses; angles are not preserved in the projection. However, the *j*-circle method has the advantage over the method of the Wulff net that it is based on parallel projection of the Poincaré sphere, which concept is simpler than the concept of stereographic projection upon which the Wulff-net method is based.

13.3.2 Applications to Problems of Polarization Optics

We shall now proceed to the solution of the two fundamental problems of polarization optics, which involve passage of elliptically polarized light through a retarder or a polarizer, by use of the method of the *j* circle. These problems are:

I) *Passage of Polarized Light Through a Retarder*. Let us consider an elliptically polarized light, represented by point *P* in the *j* circle, and a linear retarder whose eigenvectors are represented by points *F* and *S* on the circumference of the circle, and whose retardation is equal to δ (Fig. 13.16). To find the transmitted polarization form *P'* in the *j* circle, we consider the corresponding transformations on the Poincaré sphere and transfer them into the circle. Thus, we consider the small circle *C'* in the sphere through the point *P*, which

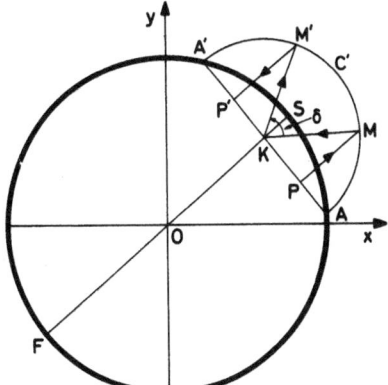

Fig. 13.16. Geometrical transformations in the *j* circle for finding the form of polarization (point *P'*) of the light transmitted by a retarder when the polarization form of the incident light is *P*. The retarder has principal directions represented by points *F* and *S*, and introduces a retardation δ

is perpendicular to the chord FS, and rotate it about its diameter AA' through $90°$, so as to lie in the plane of the j circle. Then we project point P to point M on the circle, take $MM' = \delta$ on the circle, and project point M' on the chord AA'. The projection P' of point M' on AA' represents the corresponding point of the transmitted polarization form in the j circle.

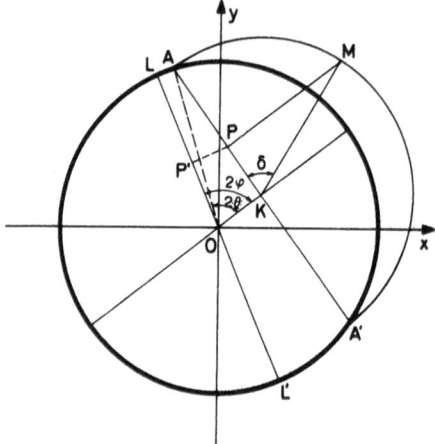

Fig. 13.17. Determination in the j circle of the intensity transmitted through a linear polarizer. Point P represents the incident light and L the pass axis of the polarizer. The transmitted intensity is represented by the length $P'L'$, where L' is diametrically opposite to L, and P' is the projection of P on LL'

II) *Passage of Polarized Light Through a Polarizer.* Let us consider an elliptically polarized light represented by point P in the j circle and a linear polarizer whose eigenvector is represented by point L. We consider that the elliptically polarized light is produced when a linearly polarized light passes through a linear retarder (Fig. 13.17). Let θ be the angle included between the linearly polarized light and the fast axis of the retarder, and φ the angle between this axis of the retarder and the pass axis of the polarizer. Then according to the results of Section 9.2 the light intensity I, passing through the polarizer, will be given by (9.5)

$$I = 1 + \cos 2\theta \cos 2\varphi + \sin 2\theta \sin 2\varphi \cos \delta . \tag{13.7}$$

From Fig. 13.17 it is concluded that

$$OP' = PK \cos(90 - 2\varphi) + OK \sin(90 - 2\varphi) ,$$

or

$$OP' = \sin 2\theta \cos \delta \sin 2\varphi + \cos 2\theta \cos 2\varphi . \tag{13.8}$$

Then, the intensity I is represented in the j circle by

$$I = 1 + OP' = OP' + OL' = P'L' . \tag{13.9}$$

Thus, if elliptically polarized light P passes through a polarizer L, to determine in the j circle the intensity of the transmitted light, we project the point P to the point P' on the diameter LL', where the point L represents the pass axis of the polarizer. The length of $P'L'$ represents the transmitted intensity.

From (13.9) the following basic results are derived. a) The locus in the j circle of the various polarization states for which the intensity of light that emerges from a polarizer is constant, is a chord normal to diameter of the circle defined by the eigenvector of the pass axis of the polarizer ($P'L'=$ constant). b) The intensity of the light that passes through a pair of crossed polarizers is equal to zero ($P'L'=0$). c) The intensity of the light transmitted through a linear polarizer with any azimuth, for circularly polarized incident light is half the intensity of the incident light ($P'L'=1$).

13.3.3 Applications to Photoelasticity

We apply now the above-proved geometrical operations on the j circle, involving the passage of elliptically polarized light through either a retarder, or a linear polarizer, to some basic problems of two-dimensional photoelasticity.

Circular Polariscope

The formation of isochromatics in the circular polariscope can be interpreted in the j circle as follows (Fig. 13.18). (The arrangement of the circular polariscope is shown in Fig. 7.2.)

Let P represent in the j circle the vertically polarized light, emitted from the polarizer. This light passes through the first quarter-wave plate, whose two

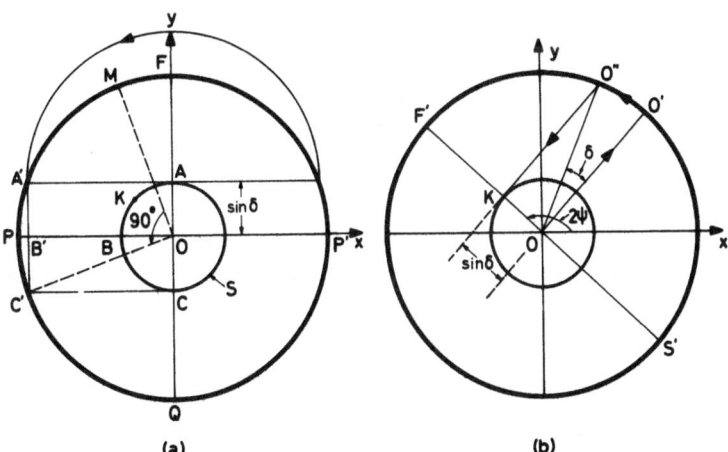

(a)   (b)

Fig. 13.18a and b. Representation in the j circle of the optical transformations that occur in the circular polariscope

eigenvectors are represented by points F and Q in the j circle. By rotating the point P along the circumference of the circle about the axis FQ by 90°, which is the angle that represents the phase difference introduced by the wave plate, we obtain point Q, whose projection on PO gives point O. This point represents the circularly polarized light that emerges from the first quarter-wave plate. Let us now consider all points on the specimen that have the same optical retardation δ, but different orientations ψ of the principal axes. To find the transmitted polarization form, by following the geometrical construction of Fig. 13.18b we obtain from point O the point K with $OK = \sin \delta$. Therefore, all points K that correspond to the polarization forms transmitted by the specimen lie on a small circle of radius $\sin \delta$, concentric with the j circle. To find the intensity of the light that emerges from the second quarter-wave plate, following the geometrical construction of Fig. 13.18b for points A and C, we obtain the chord $A'C'$ of the circle. Thus, point A is transformed to point A' and point C to C'. For the case of the dark-field circular polariscope, the analyzer is represented by point P'; therefore the intensity I of the transmitted light is equal to the length PB'. From the triangle $OA'B'$ it is deduced that $OB' = \cos \delta$, so that we have for PB'

$$I = PB' = 1 - OB' = 2 \sin^2 \frac{\delta}{2}. \tag{13.10}$$

Equation (13.10), which represents the intensity of light emitted from the dark-field circular polariscope, was found previously in Section 7.3 by using the Poincaré sphere and the Mueller and Jones calculi.

For the bright-field circular polariscope, the analyzer is represented by point P' and the transmitted-light intensity I' by the length $P'B'$. We have from Fig. 13.18a that

$$P'B' = 2 - PB' = 2 - 2\sin^2 \frac{\delta}{2} = 2 \cos^2 \frac{\delta}{2}. \tag{13.11}$$

The Senarmont Compensation Method

The optical transformations that take place in the Senarmont compensation method (Fig. 7.4) are shown in Fig. 13.19. The beam that emerges from the polarizer, whose pass axis is vertical, is represented in the j circle by the point P. By applying the geometrical operations described in Section 13.3.21 for the passage of light through the specimen, which introduces a retardation δ, we obtain point P'. This point, when the light passes through the quarter-wave plate, is transformed to point P'' on the circumference of the j circle; that is, the transmitted light is linearly polarized at an angle $\theta = PP''/2 = \delta/2$ with the axis of the polarizer.

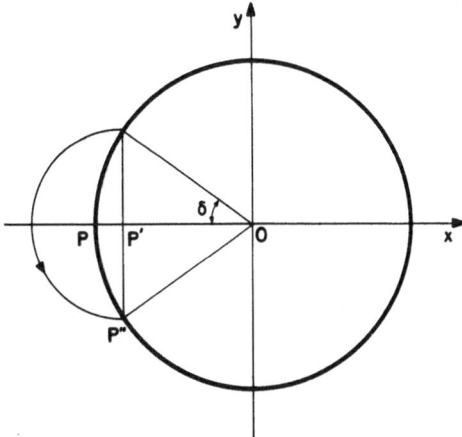

Fig. 13.19. Representation in the *j* circle of the optical transformations that occur in the Senarmont compensation method

The Tardy Compensation Method

The arrangement of the optical elements in the Tardy compensation method is shown in Fig. 7.6; the corresponding geometrical transformation in the *j* circle is shown in Fig. 13.20. Point *P* represents the linearly polarized light that emerges from the polarizer, and $P' \equiv O$ is the position of point *P* after the light passes through the first quarter-wave plate. As was shown in the first part of this section, the light that emerges from the specimen will be represented by point *P''* on the vertical diameter QQ' at a distance $P'P'' = \sin\delta$ from the center *O* of the *j* circle, where δ is the retardation introduced by the specimen. After the light passes through the second quarter-wave plate, point *P''* is transformed to point *P'''* on the circumference of the *j* circle by the geometrical construction shown in Fig. 13.20. Point *P'''* represents light linearly polarized at angle $\theta = PP'''/2 = \delta/2$ with respect to the axis of the polarizer.

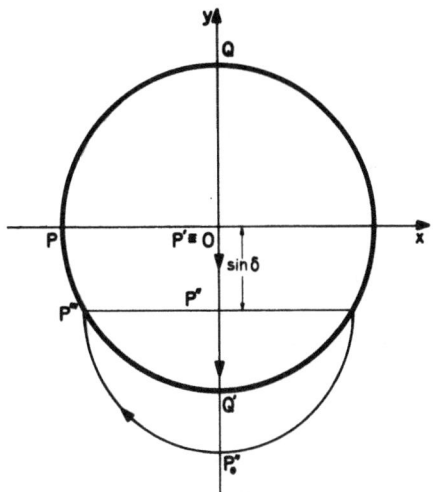

Fig. 13.20. Representation in the *j* circle of the optical transformations that occur in the Tardy compensation method

13.4 The Method of Quaternions

13.4.1 Definition of Quaternions

As was shown in Sections 4.3.2 and 4.3.3, every retarder can be represented by a unitary two-by-two complex matrix, called the Jones matrix, of the corresponding retarder. The method of quaternions is based on the fact that any unitary matrix can be represented by a quaternion of the four basic Pauli matrices [13.17].

Let us consider the more-general unitary matrix that can be expressed by

$$J = \begin{vmatrix} A_1 + iA_2 & A_3 + iA_4 \\ -A_3 + iA_4 & A_1 - iA_2 \end{vmatrix}, \tag{13.12}$$

where the real numbers A_1, A_2, A_3, and A_4 satisfy

$$A_1^2 + A_2^2 + A_3^2 + A_4^2 = 1, \tag{13.13}$$

and the four Pauli matrices I, i, j, and k are given by

$$I = \begin{vmatrix} 1 & 0 \\ 0 & 1 \end{vmatrix}, \quad i = \begin{vmatrix} i & 0 \\ 0 & -i \end{vmatrix}, \quad j = \begin{vmatrix} 0 & 1 \\ -1 & 0 \end{vmatrix}, \tag{13.14}$$

$$\text{and} \quad k = \begin{vmatrix} 0 & i \\ i & 0 \end{vmatrix}.$$

Matrix J can be written as a quaternion

$$J = A_1 I + A_2 i + A_3 j + A_4 k. \tag{13.15}$$

The coefficients A_1, A_2, A_3, and A_4 satisfy (13.13).

Multiplication of quaternions is greatly facilitated by use of the basic properties satisfied by the matrices i, j, and k,

$$i^2 = j^2 = k^2 = -I,$$

$$jk = -kj = i, \quad ki = -ik = j, \quad ij = -ji = k. \tag{13.16}$$

The Hermitian conjugate $J*$ of the quaternion expressed by (13.15) is

$$J* = A_1 I - A_2 i - A_3 j - A_4 k. \tag{13.17}$$

From the definition (13.17) and the properties of the quaternion, the complete analogy between quaternions and complex numbers can be deduced.

13.4.2 Representation of a Linear Retarder by Quaternions

Let us consider a linear retarder with retardation δ, whose fast axis subtends an angle θ with the Ox axis. The Jones matrix J of this retarder is

$$J(\theta) = R(-\theta)J_0 R(\theta),$$
(13.18)

where

$$R(\theta) = \begin{vmatrix} \cos\theta & \sin\theta \\ -\sin\theta & \cos\theta \end{vmatrix}, \quad J_0 = \begin{vmatrix} e^{i\delta/2} & 0 \\ 0 & e^{-i\delta/2} \end{vmatrix}.$$
(13.19)

The rotation matrix $R(\theta)$ can be expressed by

$$R(\theta) = \cos\theta\, I - \sin\theta\, j,$$
(13.20)

and the matrix J_0 of the retarder referred to its principal axes is expressed by

$$J_0 = \cos\frac{\delta}{2} I + \sin\frac{\delta}{2} i.$$
(13.21)

Then, matrix $J(\theta)$ referred to an arbitrary system of axes subtending an angle θ with the system of principal axes is expressed as a quaternion,

$$J(\theta) = (\cos\theta\, I - \sin\theta\, j)\left(\cos\frac{\delta}{2} I + \sin\frac{\delta}{2} i\right)(\cos\theta\, I + \sin\theta\, j).$$
(13.22)

Executing the multiplications indicated in (13.22) and taking into account (13.16), we obtain for $J(\theta)$,

$$J(\theta) = \cos\frac{\delta}{2} I + \sin\frac{\delta}{2}\cos 2\theta\, i + \sin\frac{\delta}{2}\sin 2\theta\, k.$$
(13.23)

Equation (13.23) yields the quaternion expression of a linear retarder; (13.20) yields the corresponding expression for a pure rotator.

Let us consider that a series of optical elements is represented by the quaternion J of (13.15) and that after this series of elements a pure rotator is interposed, whose quaternion is given by (13.20). The quaternion J' of the system that consists of the series of optical elements and the rotator is

$$J' = R(\theta)J,$$

or

$$J' = (\cos\theta\, I - \sin\theta\, j)(A_1 I + A_2 i + A_3 j + A_4 k).$$
(13.24)

By performing the multiplications in the right-hand member of (13.24) and taking into account (13.16), we obtain

$$J' = A'_1 I + A'_2 i + A'_3 j + A'_4 k , \tag{13.25}$$

where

$$A'_1 = A_1 \cos\theta + A_3 \sin\theta ,$$
$$A'_2 = A_2 \cos\theta - A_4 \sin\theta ,$$
$$A'_3 = A_3 \cos\theta - A_1 \sin\theta ,$$
$$A'_4 = A_4 \cos\theta + A_2 \sin\theta .$$

Similarly, the quaternion J' of the system of a series of optical elements, represented by quaternion J, followed by a retarder of retardation δ referred to its principal axes, whose quaternion J_0 is expressed by (13.21), may be represented by

$$J' = J_0 J ,$$

or

$$J' = \left(\cos\frac{\delta}{2} I + \sin\frac{\delta}{2} i \right) (A_1 I + A_2 i + A_3 j + A_4 k) . \tag{13.26}$$

By performing the multiplications in the right-hand member of (13.26) we obtain

$$J' = A''_1 I + A''_2 i + A''_3 j + A''_4 k , \tag{13.27}$$

where

$$A''_1 = A_1 \cos\frac{\delta}{2} - A_2 \sin\frac{\delta}{2} ,$$

$$A''_2 = A_2 \cos\frac{\delta}{2} + A_1 \sin\frac{\delta}{2} ,$$

$$A''_3 = A_3 \cos\frac{\delta}{2} - A_4 \sin\frac{\delta}{2} ,$$

$$A''_4 = A_4 \cos\frac{\delta}{2} + A_3 \sin\frac{\delta}{2} .$$

When the retarder is referred to an arbitrary system of coordinates, in order to transform it into one referred to its principal axes we may apply the transformation relation (13.22).

13.4.3 Applications to Photoelasticity

We shall now apply the method of quaternions to the solution of some basic problems of two-dimensional photoelasticity. However, before we proceed to the solution of these problems, we must define first the light intensity that corresponds to a system of optical elements placed between either two crossed or two parallel polarizers.

Let us consider that the system of optical elements is represented by the quaternion of (13.15), whose corresponding Jones matrix is given by (13.12). If the two polarizers are crossed and the axis of the first is vertical, then the Jones vector a of the light transmitted through the second polarizer will be

$$a = \begin{bmatrix} 1 & 0 \\ 0 & 0 \end{bmatrix} \begin{bmatrix} A_1+iA_2 & A_3+iA_4 \\ -A_3+iA_4 & A_1-iA_2 \end{bmatrix} \begin{bmatrix} 0 \\ 1 \end{bmatrix} = \begin{bmatrix} A_3+iA_4 \\ 0 \end{bmatrix}. \tag{13.28}$$

The corresponding intensity I is

$$I = \tilde{a}\,a = A_3^2 + A_4^2. \tag{13.29}$$

Similarly, when the axes of the two polarizers are parallel, we obtain

$$I = A_1^2 + A_2^2. \tag{13.30}$$

We apply now (13.29) and (13.30) to the cases of plane and circular polariscopes.

The Plane Polariscope (Fig. 7.1)

In the plane polariscope, the specimen is placed between crossed or parallel polarizers for dark- or bright-fields, respectively. The specimen under stress behaves as a retarder, whose quarternion is given by (13.23). Applying (13.29) and (13.30), we obtain

$$I = \sin^2 \frac{\delta}{2} \sin^2 2\theta \tag{13.31}$$

for the dark-field arrangement, and

$$I = \cos^2 \frac{\delta}{2} + \sin^2 \frac{\delta}{2} \cos^2 2\theta = 1 - \sin^2 \frac{\delta}{2} \sin^2 2\theta \tag{13.32}$$

for the bright-field plane polariscope. Relations (13.31) and (13.32) were derived in Section 7.2.

The Circular Polariscope (Fig. 7.2)

In the circular polariscope, two crossed quarter-wave plates $Q_{45°}$ and $Q_{-45°}$, together with the specimen located between the plates, are placed between two crossed (dark-field) or parallel (bright-field) polarizers. The axes of the quarter-wave plates subtend angles of $45°$ and $-45°$ with the respective axes of the polarizers. Quaternions Q_{45} and Q_{-45} are

$$Q_{45} = \frac{\sqrt{2}}{2}(I+k) \quad Q_{-45} = \frac{\sqrt{2}}{2}(I-k).$$ (13.33)

The quaternion of the specimen is given by (13.23). Thus, we obtain for the quaternion J' of the system of the two quarter-wave plates and the specimen

$$J' = Q_{45°} J(\theta) Q_{-45°},$$ (13.34)

or

$$J' = \frac{\sqrt{2}}{2}(I+k)\left(\cos\frac{\delta}{2}I + \sin\frac{\delta}{2}\cos 2\theta i + \sin\frac{\delta}{2}\sin 2\theta k\right)\frac{\sqrt{2}}{2}(I-k).$$ (13.35)

By performing the multiplications in the right-hand member of (13.35), we obtain for J'

$$J' = \cos\frac{\delta}{2}I + \sin\frac{\delta}{2}\cos 2\theta j + \sin\frac{\delta}{2}\sin 2\theta k.$$ (13.36)

Applying (13.29) and (13.30) to the quaternion J', we obtain for the intensity

$$I = \sin^2\frac{\delta}{2}\cos^2 2\theta + \sin^2\frac{\delta}{2}\sin^2 2\theta = \sin^2\frac{\delta}{2}$$ (13.37)

for the dark-field circular polariscope, and

$$I = \cos^2\frac{\delta}{2}$$ (13.38)

for the bright-field circular polariscope.

Again, (13.37) and (13.38) were derived in Section 7.3.

13.4.4 A Geometric Method Based on Quaternions

A simple geometric method, based on quaternions, was developed by *Cernosek* [13.16]. According to this method, a quaternion, which is completely defined by its four components A_1, A_2, A_3, and A_4 [(13.15)], is geometrically represented by four points K_1, K_2, K_3, K_4 with coordinates (A_1, A_2), (A_3, A_1), (A_3, A_4), (A_4, A_2) in the systems OQ_1Q_2, OQ_3Q_1, OQ_3Q_4, OQ_4Q_2, respectively (Fig. 13.21). Given that the light may be represented by a quaternion defined by points K_1, K_2, K_3, K_4, any light through either a pure rotator or a retarder (a birefringent plate) can be readily deduced from (13.25) and (13.27), respectively. Thus, (13.25) suggests that the effect of a pure rotator is to rotate points K_2 and K_4 in the clockwise direction through an angle equal to θ, whereas (13.27) shows that the effect of a retarder is to rotate point K_1 clockwise and point K_3 anticlockwise through an angle $\delta/2$. The geometric operations for finding the new set K_1', K_2', K_3', K_4' of points K_1, K_2, K_3, K_4 that corresponds to passage of light through a pure rotator or a retarder are shown in Figs. 13.22 and 13.23. By using these fundamental transformations, we can find the geometrical representation of a given quaternion after passage of the corresponding light through a series of optical elements.

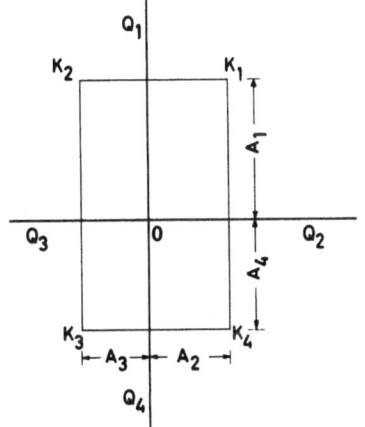

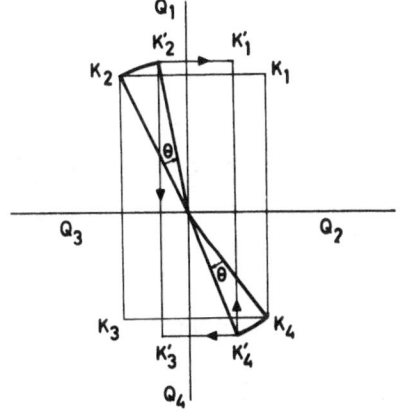

Fig. 13.21. Geometrical representation of a quaternion by the points K_1, K_2, K_3, K_4 in the systems OQ_1Q_2, OQ_3Q_1, OQ_3Q_4, OQ_4Q_2

Fig. 13.22. Geometrical transformations for finding the points K_1', K_2', K_3', K_4' that represent the quaternion of the light that emerges from a pure rotator when the quaternion of the incident light is represented by points K_1, K_2, K_3, K_4

Furthermore, for a given quaternion J, we can find the characteristic quantities of the corresponding pure rotator and retarder, according to the equivalence theorem developed in Section 4.5. Let the pure rotator and the retarder be represented by the quaternions $R(\theta)$ and $J(\varphi)$ given by (13.20) and (13.23), respectively. We then have

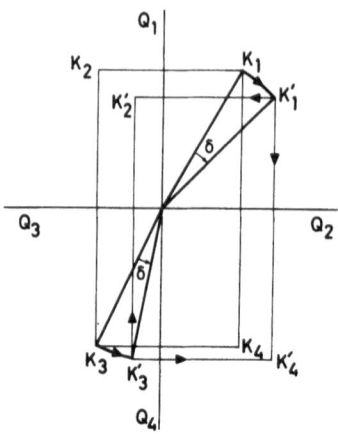

Fig. 13.23. As in Fig. 13.22, for a linear retarder referred to its principal axes

$$J = R(\theta) J(\varphi) , \tag{13.39}$$

and if

$$J = A_1 I + A_2 i + A_3 j + A_4 k ,$$

we have

$$A_1 = \cos \theta \cos \frac{\delta}{2}, \quad A_2 = \sin \frac{\delta}{2} \cos(\theta + 2\varphi) ,$$

$$A_3 = -\sin \theta \cos \frac{\delta}{2}, \quad A_4 = \sin \frac{\delta}{2} \sin(\theta + 2\varphi) . \tag{13.40}$$

From (13.40) we conclude that

$$\tan \theta = -\frac{A_3}{A_1} ,$$

$$\tan(\theta + 2\varphi) = \frac{A_4}{A_2} , \tag{13.41}$$

$$\tan \frac{\delta}{2} = \left(\frac{A_2^2 + A_4^2}{A_1^2 + A_3^2} \right)^{1/2} .$$

By the geometrical construction of Fig. 13.24 for a quaternion expressed by points K_1, K_2, K_3, K_4, the corresponding quantities θ, $(\theta + 2\varphi)$, and $\delta/2$ can be determined. By the inverse procedure, if these quantities are known, points K_1, K_2, K_3, K_4 can be determined.

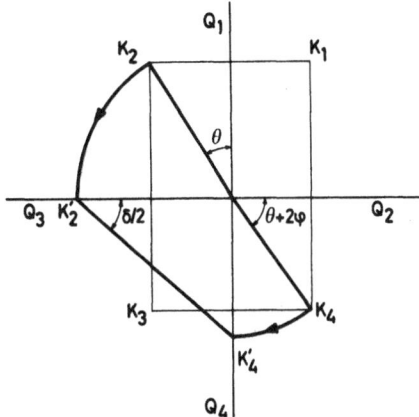

Fig. 13.24. Geometrical construction for finding the characteristic quantities θ, $(\theta + 2\varphi)$ and $\delta/2$ that correspond to a quaternion that is represented by points K_1, K_2, K_3, K_4

References

13.1 P. Badgley: *Structural Methods for the Exploration Geologist* (Peter C. Badgley, New York 1959) Chap. 8, pp. 187–242

13.2 F. E. Wright: J. Opt. Soc. Am. **20**, 529 (1930)

13.3 H. K. Aben: *Integrated Photoelasticity* (Valgus, Tallin, USSR 1975) pp. 70–75

13.4 H. Schwieger: Exp. Mech. **9**, 67 (1969)

13.5 H. J. Menges: Z. Angew. Math. Mech. **20**, 210 (1940)

13.6 A. Kuske: *Einführung in die Spannungsoptik* (Wissenschaftliche Verlagsgesellschaft, Stuttgart 1959)

13.7 A. Kuske: Optik **19**, 261 (1962)

13.8 A. Kuske: Rev. Franc. Méc. **9**, 49 (1964)

13.9 A. Kuske: Int. Spannungsopt. Symp. Berlin, 11.–15. 4. 1961 (Akademie Verlag, Berlin 1962) pp. 115–126

13.10 A. Kuske: Exp. Mech. **6**, 218 (1966)

13.11 A. Kuske, G. Robertson: *Photoelastic Stress Analysis* (Wiley and Sons, New York 1974) Chap. 13, pp. 304–328

13.12 M. Richartz, H. Y. Hsü: J. Opt. Soc. Am. **39**, 136 (1949)

13.13 J. Cernosek: J. Opt. Soc. Am. **61**, 324 (1971)

13.14 J. Cernosek: Exp. Mech. **13**, 83 (1973)

13.15 J. Cernosek: Exp. Mech. **13**, 273 (1973)

13.16 J. Cernosek: Exp. Mech. **15**, 354 (1975)

13.17 C. Whitney: J. Opt. Soc. Am. **61**, 1207 (1971)

Bibliography

2. Electromagnetic Theory of Light

Books

Born, M., Wolf, E.: *Principles of Optics*, 5th. ed. (Pergamon Press, London 1975) Chap.14, pp.1-18, 665-690

Ramachandran, G.N., Ramaseshan, S.: "Crystal Optics", in *Crystal Optics Diffraction*, ed. by S. Flügge, Encyclopedia of Physics, Vol. 25/1 (Springer, Berlin, Göttingen, Heidelberg 1961) pp.54-85

Rossi, B.: *Optics*, 3rd ed. (Addison-Wesley, Reading, Mass. 1965) Chaps.6,7, pp.266-347

3. Description of Polarized Light

4. Passage of Polarized Light Through Optical Elements

Books

Aben, H.K.: *Integrated Photoelasticity* (Valgus Tallin, USSR 1975)

Born, M., Wolf, E.: *Principles of Optics*, 5rd ed. (Pergamon Press, London 1975) Chap.10, pp.544-555

Chandrasekhar, S.: *Radiative Transfer* (Oxford University Press, London 1950) pp.24-35

Clarke, D., Grainger, J.F.: *Polarized Light and Optical Measurement* (Pergamon Press, Oxford 1971)

Gerrard, A., Burch, J.M.: *Introduction to Matrix Methods in Optics* (Wiley and Sons, London 1975)

Hartshorne, N.H., Stuart, A.: *Crystals and the Polarizing Microscope*, 2nd ed. (Arnold Press, London 1950)

Pockels, F.: *Lehrbuch der Kristalloptik*, Vol. 11-13 (Teubner, Leipzig 1906) pp.267-283

Poincaré, H.: *Théorie Mathématique de la Lumière*, Vol.2 (Gauthiers-Villars, Paris 1892) Chap.12

Ramachandran, G.N., Ramaseshan, S.: "Crystal Optics", in *Crystal Optics Diffraction*, ed. by S. Flügge, Encyclopedia of Physics, Vol.25/1 (Springer, Berlin, Göttingen, Heidelberg 1961) pp.1-217

Shurcliff, W.A.: *Polarized Light, Production and Use* (Harvard University Press, Cambridge, Mass. 1962)

Shurcliff, W.A., Ballard, S.S.: *Polarized Light* (Van Nostrand, Princeton, N.J. 1964)

Simmons, J.W., Gutmann, M.J.: *States, Waves and Photons: A Modern Introduction to Light* (Addison-Wesley, Reading, Mass. 1970)

Walker, J.: *Analytical Theory of Light* (Cambridge, at the University Press 1904)

Papers

Aben, H.K.:"On the Matrix Representation of Optical Phenomena in Three-Dimensional Photoelasticity",Proc. Conf. on Exp. Methods of Investigating Stress and Strain in Structures, Prague (Building Research Institute of the Technical University of Prague, Prague 1965) pp.33-42

Aben, H.K.: Magnetophotoelasticity-photoelasticity in a magnetic field. Exp. Mech. *10*, 97-105 (1970)

Aben, H.K.: "Principles of Magnetophotoelasticity", Proc. 4th Int. Conf. Exp. Stress Analysis, Cambridge, 1970, ed. by M.L. Meyer (The Institution of Mechanical Engineers, London 1971) pp.175-182

Aben, H.K.: Discussion of "Towards the achromatic quarterwave plate". Exp. Mech. *14*, 249-250 (1974)

Anile, A.M., Breuer, R.A.: Gravitational Stokes parameters. Astrophys. J. *189*, 39-49 (1974)

Aspnes, D.E.: Measurement and correction of first-order errors in ellipsometry. J. Opt. Soc. Am. *61*, 1077-1085 (1971)

Azzam, R.M.A., Bashara, N.M.: Unified analysis of ellipsometry errors due to imperfect components, cell-window birefringence, and incorrect azimuth angles. J. Opt. Soc. Am. *61*, 600-607 (1971)

Azzam, R.M.A., Bashara, N.M.: General treatment of the effect of cell windows in ellipsometry. J. Opt. Soc. Am. *61*, 773-776 (1971)

Azzam, R.M.A., Bashara, N.M.: Ellipsometry with imperfect components including incoherent effects. J. Opt. Soc. Am. *61*, 1380-1391 (1971)

Azzam, R.M.A.: Polarization orthogonalization properties of optical systems. Appl. Phys. *13*, 281-285 (1977)

Beardsley, G.F.: Mueller scattering matrix of sea water. J. Opt. Soc. Am. *58*, 52-57 (1968)

Becquerel, J.: The existence in a mono-axial crystal of two different values for the magnetic rotation of polarisation in directions parallel with the axis and perpendicular to it. Commun. Phys. Lab. Univ. Leiden *191* C, 19-34 (1928)

Becquerel, J.: Pouvoir rotatoire magnétique d'un cristal uniaxe suivant les directions obliques sur l'axe; détermination de la rotation de la tysonite suivant une direction voisine d'un axe binaire, à la température du nitrogène liquide. Commun. Phys. Lab. Univ. Leiden *211* A, 1-18 (1930)

Becquerel, J., De Haas, W.J.: Détermination du pouvoire rotatoire paramagnétique d'un cristal de tysonite suivant une direction normale a l'axe optique, aux températures réalisables avec l'hydrogène liquide. Commun. Phys. Lab. Univ. Leiden *211* A, 19-34 (1930)

Becquerel, J., De Haas, W.J.: Anisotropic magnétooptique dans un plan normal a l'axe optique d'un cristal hexagonal. Pouvoirs rotatoires paramagnétiques dans les directions voisines des axes binaires, aux très basses températures. Commun. Phys. Lab. Univ. Leiden *211* A, 35-51 (1930)

Bigelow, J.E., Kashnow, R.A.: Poincaré sphere analysis of liquid crystal optics. Appl. Opt. *16*, 2090-2096 (1977)

Billings, B.H., Land, E.H.: A comparative survey of some possible systems of polarized headlights. J. Opt. Soc. Am. *38*, 819-829 (1948)

Billings, B.H.: A monochromatic depolarizer. J. Opt. Soc. Am. *41*, 966-975 (1951)

Billings, B.H.: The electro-optic effect in uniaxial crystals of the dihydrogen phosphate (XH_2PO_4) Type. IV. Angular field of the electro-optic shutter. J. Opt. Soc. Am. *42*, 12-20 (1952)

Björnstahl, Y.: Über die Bracesche Halbschattenmethode. Physik. Zeitschr. *40*, 437-443 (1939)

Björnstahl, Y.: Über die Methode von Brace mit variablem Halbschatten. Z. Instrumentenkd. *59*, 425-430 (1939)

Bolinder, E.F.: Geometric-analytic theory of noisy two-port networks. Proc. IRE *46*, 1959-1960 (1958)

Bolinder, E.F.: Theory of noisy two-port networks. J. Franklin Inst. *267*, 1-23 (1959)

Bolinder, E.F.: Geometric analysis of partially polarized electromagnetic waves. IEEE Trans. AP-*15*, 37-40 (1967)

Bolinder, E.F.: Comments on "Poincaré sphere representation of partially polarized fields". IEEE Trans. AP-*23*, 747-748 (1975)

Bourret, R.: The depolarization of electromagnetic radiation in a random medium: Evolution of the Stokes parameters. Opt. Acta *21*, 721-735 (1974)

Bradshaw, J.A.: Discussion of "The Stokes parameters of a beam in a plasma". Am. J. Phys. *37*, 934-936 (1969)

Bruhat, G., Grivet, P.: Le pouvoir rotatoire du quartz pour des rayons perpendiculaires à l'axe optique et sa dispersion dans l'ultra-violet. J. de Physique *6* (1), 12-26 (1935)

Cernosek, J.: Simple geometrical method for analysis of elliptical polarization. J. Opt. Soc. Am. *61*, 324-327 (1971)

Cernosek, J.: Towards the achromatic quarterwave plate. Exp. Mech. *13*, 83-85 (1973)

Cernosek, J., Perla, M.: A new method of determining the isoclinic parameter. J. Strain. Anal. *5*, 263-268 (1970)

Chaumont, M.L.: Sur la théorie des appareils servant à l'étude de la lumière polarisée elliptiquement. C.R. Acad. Sci. *150*, 1604-1606 (1913)

Chaumont, M.L.: Recherches expérimentales sur le phénomène electro-optique de Kerr et sur les méthodes servant à l'étude de la lumière polarisée elliptiquement. Ann. Phys. (Leipzig) *4*, 101-206 (1915)

Cocke, W.J., Holm, D.A.: Lorentz transformation properties of the Stokes parameters. Nature (London) Phys. Sci. *240*, 161-162 (1972)

Collett, E.: The description of polarization in classical physics. Am. J. Phys. *36*, 713-725 (1968)

Collett, E.: Stokes parameters for quantum systems. Am. J. Phys. *38*, 563-574 (1970)

Collett, E.: Discussion of "Stokes matrices for light beams in a double rhomb". Am. J. Phys. *39*, 225-226 (1971)

Collett, E.: Mueller-Stokes matrix formulation of Fresnel's equations. Am. J. Phys. *39*, 517-528 (1971)

Collett, E.: Mathematical formulation of the Interference laws of Fresnel and Arago. Am. J. Phys. *39*, 1483-1495 (1971)

Dawson, E.F., Young, N.O.: Helical Kerr cell. J. Opt. Soc. Am. *50*, 170-171 (1960)

Degl'innocenti, E.L., Degl'innocenti, M.L.: A perturbative solution of the transfer equations for the Stokes parameters in a magnetic field. Solar Phys. *31*, 299-305 (1973)

Deschamps, G.A.: Geometrical representation of the polarization of a plane electromagnetic wave. Proc. Inst. Radio Eng. *39*, 540-544 (1951)

Deschamps, G.A.: New chart for the solution of transmission-line and polarization problems. Electrical Commun. *28*, 247-254 (1953)

Deschamps, G.A., Mast, P.E.: Poincaré sphere representation of partially polarized fields. IEEE Trans. AP-*21*, 474-478 (1973)

Domanski, A.: The Mueller matrix formalism in diffraction phenomena. Opt. Commun. *14*, 281-286 (1975)

Evans, J.W.: Solc birefringent filter. J. Opt. Soc. Am. *48*, 142-145 (1958)

Falkoff, D.L., MacDonald, J.E.: On the Stokes parameters for polarized radiation. J. Opt. Soc. Am. *41*, 861-862 (1951)

Fano, U.: Remarks on the classical and quantum-mechanical treatment of partial polarization. J. Opt. Soc. Am. *39*, 859-863 (1949)

Fano, U.: A Stokes-parameter technique for the treatment of polarization in quantum mechanics. Phys. Rev. *93*, 121-123 (1954)

Fano, U.: Description of states in quantum mechanics by density matrix and operator techniques. Rev. Mod. Phys. *29*, 74-93 (1957)

Grechushnikov, B.N., Konstantinova, A.F.: Müller matrices for optically active crystals. Sov. Phys.-Crystallogr. *16*, 378-379 (1971)

Hagyard, M.J.: Analytic solutions to the Unno transfer equations for the Stokes parameters in a Milne-Eddington atmosphere. Solar Phys. *16*, 286-287 (1971)

Hammerschlag, R.H.: Jones-Vektoren geschrieben in rechts- und linkszirkularen Komponenten. Optik *34*, 595-597 (1972)

Hecht, E.: Note on "An operational definition of the Stokes parameters". Am. J. Phys. *38*, 1156-1158 (1970)

Hege, G., Leonhardt, K.: Polarisationseigenschaften von Tripelprismen-Angabe experimentell bestätigter Bauteilmatrizen. Optik *47*, 167-184 (1977)

Hewitt, M.H.: Polarization, cross section, and the Stokes parameters. Proc. IEEE *53*, 1143-1144 (1965)

Hillion, P.: Paramètres de Stokes et neutrino. Ann. Inst. Henri Poincaré, Sect. A *13*, 253-261 (1970)

Hodgdon, E.B.: Theory, design, and calibration of a uv spectrophotopolarimeter. Appl. Opt. *4*, 1479-1483 (1965)

Holoubek, J.: The use of Mueller matrices in the intensity methods of birefringence measurements. Czech. J. Phys. B *24*, 1162-1167 (1974)

Hsü, H.Y., Richartz, M., Liang, Y.K.: A generalized intensity formula for a system of retardation plates. J. Opt. Soc. Am. *37*, 99-106 (1947)

Hurwitz, H., Jones, R.C.: A new calculus for the treatment of optical systems. II. Proof of three general equivalence theorems. J. Opt. Soc. Am. *31*, 493-499 (1941)

Jerrard, H.G.: Transmission of light through birefringent and optically active media: The Poincaré sphere. J. Opt. Soc. Am. *44*, 634-640 (1954)

Jones, R.C.: A new calculus for the treatment of optical systems. I. Description and discussion of the calculus. J. Opt. Soc. Am. *31*, 488-493 (1941)

Jones, R.C.: A new calculus for the treatment of optical systems. III. The Sohncke theory of optical activity. J. Opt. Soc. Am. *31*, 500-503 (1941)

Jones, R.C.: A new calculus for the treatment of optical systems. IV. J. Opt. Soc. Am. *32*, 486-493 (1942)

Jones, R.C.: A new calculus for the treatment of optical systems. V. A more general formulation, and description of another calculus. J. Opt. Soc. Am. *37*, 107-110 (1947)

Jones, R.C.: A new calculus for the treatment of optical systems. VI. Experimental determination of the matrix. J. Opt. Soc. Am. *37*, 110-112 (1947)

Jones, R.C.: A new calculus for the treatment of optical systems. VII. Properties of the N-matrices. J. Opt. Soc. Am. *38*, 671-685 (1948)

Jones, R.C.: New calculus for the treatment of optical systems. VIII. Electromagnetic theory. J. Opt. Soc. Am. *46*, 126-131 (1956)

Jones, R.C.: Transmittance of a train of three polarizers. J. Opt. Soc. Am. *46*, 528-533 (1956)

Kastner, S.O.: Polarization in multiple scattering using random Stokes vectors. J. Quant. Spectrosc. Radiat. Transfer. *6*, 317-324 (1966)

Ko, H.C.: The use of the statistical matrix and the Stokes vector in formulating the effective aperture of antennas. IRE Trans. AP-*9*, 581-582 (1961)

Koester, C.J.: Achromatic combinations of half-wave plates. J. Opt. Soc. Am. *49*, 405-409 (1959)

Kuske, A.: "Beiträge zur spannungsoptischen Untersuchung von Flächentragwerken". Int. Spannungsopt. Symp., Berlin, April 1961 (Akademie Verlag Berlin 1962) pp.115-126

Kuske, A.: Die Gesetzmässigkeiten der Doppelbrechung. Optik *19*, 261-272 (1962)

Kuske, A.: L'analyse des phénomènes optiques en photoélasticité à trois dimensions par la méthode du cercle de "J". Rev. Fr. Méc. *9*, 49-58 (1964)

Kuske, A.: The J-circle method. Exp. Mech. *6*, 218-224 (1966)

De Lang, H.: Flow lines on the Poincaré sphere as an aid to the study of mode polarization in lasers. IEEE J. QE-*7*, 441-444 (1971)

Lenhardt, K., Burckhardt, C.: Die Jonesmatrix eines Tripelstreifens. Optik *47*, 215-222 (1977)

Mark, R.: A simple geometric method for analyzing polarization states in photoelasticity. AIAA J. *2*, 150-152 (1964)

McDonough, R.: Comment on "Use of Mueller matrices for determination of the actual polarization of light". J. Opt. Soc. Am. *59*, 1004 (1969)

McMaster, W.H.: Polarization and the Stokes parameters. Am. J. Phys. *22*, 351-362 (1954)

McMaster, W.H.: Matrix representation of polarization. Rev. Mod. Phys. *33*, 8-28 (1961)

Meeks, M.L., Ball, J.A., Carter, J.C., Ingalls, R.P.: Stokes parameters for 1665-megacycles-per-second emission from OH near source W3. Science *153*, 978-981 (1966)

Mueller, H.: "Momorandum on the polarization optics of the photoelastic shutter". Rpt. *2*, OSRD Project OEMsr-576 (Nov. 1943)

Mueller, H.: The foundation of optics. J. Opt. Soc. Am. *38*, 661 (1948)

O'Handley, R.C.: Modified Jones calculus for the analysis of errors in polarization-modulation ellipsometry. J. Opt. Soc. Am. *63*, 523-528 (1973)

Pancharatnam, S.: Achromatic combinations of birefringent plates. Pt. I. An achromatic circular polarizer. Proc. Indian Acad. Sci. A *41*, 130-136 (1955)

Pancharatnam, S.: Achromatic combinations of birefringent plates. Pt. II. An achromatic quarter-wave plate. Proc. Indian Acad. Sci. A *41*, 137-144 (1955)

Pancharatnam, S.: Generalized theory of interference, and its applications. Pt. I. Coherent pencils. Proc. Indian Acad. Sci. A *44*, 247-262 (1956)

Pancharatnam, S.: Generalized theory of interference, and its applications. Pt. II. Partially coherent pencils. Proc. Indian Acad. Sci. A *44*, 398-417 (1956)

Parke, N.G.: "Matrix Optics"; Ph. D. Thesis, Dept. of Physics (MIT, Cambridge, Mass. 1948) pp.181

Parke, N.G.: "Matrix Algebra of Electromagnetic Waves", Techn. Rpt. 70, Res. Lab. of Electr. (MIT, Cambridge, Mass. 1948) pp.28

Parke, N.G.: "Statistical Optics: I. Radiation", Techn. Rpt. No. 95, Res. Lab. of Electr. (MIT, Cambridge, Mass 1949) pp. 15

Parke, N.G.: "Statistical Optics: II. Mueller phenomenological algebra", Techn. Rpt. No. 119, Res. Lab. of Electr. (MIT, Cambridge, Mass. 1949) pp. 19

Parrent, G.B., Roman, P.: On the matrix formulation of the theory of partial polarization in terms of observables. Nuovo Cimento *15*, 370-388 (1960)

Pascu, M.L.: Semnificaţia şi ecuaţille de transformare ale parametrilor Stokes. Stud. Cercet. Fiz. *24*, 253-264 (1972)

Perrin, F.: Polarization of light scattered by isotropic opalescent media. J. Chem. Phys. *10*, 415-427 (1942)

Plechata, R.: A geometric interpretation of double refraction in photoelasticity. Acta Tech. Prague 2, 230-261 (1957)

Pomraning, G.C.: The Stokes parameters for light arising from induced processes. Astrophys. J. 191, 183-189 (1974)

Pospergelis, M.M.: Measurement and computation of the instrumental Stokes vector. Sov. Astron.-AJ 12, 512-521 (1968)

Priebe, J.R.: Operational form of the Mueller matrices. J. Opt. Soc. Am. 59, 176-180 (1969)

Pritchard, B.S., Elliott, W.G.: Two instruments for atmospheric optics measurement. J. Opt. Soc. Am. 50, 191-202 (1960)

Raimond, E., Eliasson, B.: Positions and Stokes parameters of seven OH-emission sources. Astrophys. J. 155, 817-830 (1969)

Ramachandran, G.N., Chandrasekharan, V.: Photo-elastic constants of sodium chlorate. Proc. Indian Acad. Sci. A 33, 199-215 (1951)

Ramachandran, G.N., Ramaseshan, S.: Magneto-optic rotation in birefringent media-application of the Poincaré sphere. J. Opt. Soc. Am. 42, 49-56 (1952)

Ramaseshan, S.: Faraday effect and birefringence-I. Proc. Indian Acad. Sci. A 34, 32-40 (1951)

Ramaseshan, S., Chandrasekharan, V.: Faraday effect and birefringence. Curr. Sci. (India) 20, 150-151 (1951)

Richartz, M., Hsü, H.Y.: Analysis of elliptical polarization. J. Opt. Soc. Am. 39, 136-157 (1949)

Riera, J.D., Mark, R.: The optical-rotation effect in photoelastic shell analysis. Exp. Mech. 9, 9-16 (1969)

Robert, A.J.: New methods in photoelasticity. Exp. Mech. 7, 224-232 (1967)

Robert, A.: Application de la sphère de Poincaré à l'ellipsométrie de précision. Bull. Soc. Fr. Minéral. Cristallogr. 91, 415-421 (1968)

Robert, A.: The application of Poincaré's sphere to photoelasticity. Int. J. Solids Struct. 6, 423-432 (1970)

Robert, A.: La sphère de Poincaré et ses applications à la mesure des formes de lumière et à la photoélasticimétrie classique. Sci. et techniques de l'Armement. 45, 309-379 (1971)

Robert, A.: Polarimétrie et photoélasticimétrie. Serv. Techn. Const. Armes Navales, Paris, Chap. 7, 145-171 (1972)

Rodimov, A.P., Potekhin, V.A.: Distribution of the probabilities of the position of the polarization point of a partially polarized wave on a Poincaré sphere. Radio Eng. 12, 2038-2039 (1967)

Roman, P.: Generalized Stokes parameters for waves with arbitrary form. Nuovo Cimento 13, 974-982 (1959)

Schmieder, R.W.: Stokes-algebra formalism. J. Opt. Soc. Am. 59, 297-302 (1969)

Schwieger, H.: Graphical methods for determining the resulting photoelastic effect of compound states of stress. Exp. Mech. 9, 67-74 (1969)

Skinner, C.A.: A universal polarimeter. J. Opt. Soc. Am. 10, 491-520 (1925)

Smirnov, V.S.: Relativistic transformation of the Stokes parameters and the coherency matrix. Opt. Spectrosc. USSR 28, 557-558 (1970)

Snellman, O., Björnstahl, Y.: Einige Untersuchungen über Strömungsdoppelbrechung. Kolloid Beih. 52, 403-466 (1941)

Soleillet, P.: Sur les paramètres charactérisant la polarisation partielle de la lumière dans les phénomènes de fluorescence. Ann. Phys.(Paris) 12, 23-97 (1929)

Stokes, G.G.: On the composition and resolution of streams of polarized light from different sources. Trans. Cambridge Philos. Soc. 9, 399-416 (1852)

Tewarson, S.P.: Reciprocity equations for isotropic opalescent scattering media. Indian J. Phys. 40, 281-293 (1966)

Tewarson, S.P.: Orientation effect on the reciprocity relation for Stokes vectors of scattered light. Kolloid z. u. Z. Polymere *225*, 52-54 (1968)

Tewarson, S.P.: Matrix Optics in dipolar scattering. Kolloid Z. u. Z. Polymere *225*, 69-71 (1968)

Tewarson, S.P., Vachaspati: Reciprocity relation for Stokes vectors of scattered light. Kolloid Z. u. Z. Polymere *213*, 131-134 (1966)

Tronko, V.D., Golovach, G.P.: Jones and Mueller matrices of a phase-shifting slab with a rotating axis of the highest velocity. Sov. Phys.-Crystallogr. *18*, 291-294 (1973)

Troshin, B.I., Bagaev, S.N.: Use of the Poincaré sphere method for the analysis of the polarization characteristics of a laser with an anisotropic element. Opt. Spectrosc. USSR *23*, 424-425 (1967)

Uri, J.B.: Polarization and interference in optics, Pt. I: The transfer function-OTF. Optik *47*, 337-350 (1977)

Uri, J.B.: Polarization and interference in optics, Pt. II: Interference, coherency matrix and degree of polarization. Optik *47*, 405-420 (1977)

Uri, J.B.: Polarization and interference in optics, Pt. III: Reflection, refraction, and diffraction. Optik *48*, 1-22 (1977)

Vasil'yev, B.I.: Use of Mueller matrices for determination of the actual polarization of light. Sov. J. Opt. Technol. *35*, 87-89 (1968)

Walker, M.J.: Matrix calculus and the Stokes parameters of polarized radiation. Am. J. Phys. *22*, 170-174 (1954)

Weeks, D.W.: A study of sixteen coherency matrices. J. Math. Phys. (N.Y.) *13*, 380-386 (1957)

Whitney, C.: Pauli-algebraic operators in polarization optics. J. Opt. Soc. Am. *61*, 1207-1213 (1971)

Wittmann, A.: Computation and observation of Zeeman multiplet polarization in Fraunhofer lines, Pt. I: Photographic measurement of Stokes parameters. Sol. Phys. *33*, 107-118 (1973)

Wittmann, A.: Computation and observation of Zeeman multiplet polarization in Fraunhofer lines, Pt. II: Computation of Stokes parameter profiles. Sol. Phys. *35*, 11-29 (1974)

Wolf, E.: Optics in terms of observable quantities. Nuovo Cimento *12*, 884-888 (1954)

Woods, H.J.: The optical properties of twisted fibres. J. Text. Inst. *55*, 243-250 (1964)

Wright, F.E.: A spherical projection chart for use in the study of elliptically polarized light. J. Opt. Soc. Am. *20*, 529-564 (1930)

De Young, D.S.: Relativistic spinor formulation of Stokes parameters with application to the inverse compton effect. J. Math. Phys. *7*, 1916-1923 (1966)

Zamkov, V.A., Kondratyev, A.S., Kuchma, A.Y.: Jones and Mueller matrices for a uniaxial Plane-parallel plate. Opt. Spectrosc. USSR *38*, 592-593 (1975)

5. Measurement of Elliptically Polarized Light

Books

Clarke, D., Grainger, J.F.: *Polarized Light and Optical Measurement* (Pergamon Press, Oxford 1971) Chap.4, pp.118-154

MacCullagh, J.: Collected Works, pp.138, 230 (Dublin, London 1880)

Ramachandran, G.N., Ramaseshan, S.: "Crystal Optics", in *Crystal Optics, Diffraction*, ed. by S. Flügge, Encyclopedia of Physics, Vol.25/1 (Springer, Berlin, Göttingen, Heidelberg 1961) pp.34-53

Schönrock, O.: "Polarimetrie", in *Elektrische Leistungsphänomene I*, Handbuch der Physik, Hrsg. H. Geiger, K. Scheel, Bd. 19 (Springer, Berlin 1928) pp.705-776

Papers

Archard, J.F., Clegg, P.L., Taylor, A.M.: Photoelectric analysis of ellipti-cally polarized light. Proc. Phys. Soc. London B *65*, 758-768 (1952)

Azzam, R.M.A.: Alternate arrangement and analysis of systematic errors for dynamic photometric ellipsometers employing an oscillating-phase retarder. Optik *45*, 209-218 (1976)

Bergholm, C., Björnstahl, Y.: Elektrische Doppelbrechung in Kolloiden. Physik. Zeitschr. *21*, 137-141 (1920)

Bertrand, E.: De l'application du microscope à l'étude de la minéralogie. Bull. Soc. Mineral. *1*, 22-28 (1878)

Budde, W.: Photoelectric analysis of polarized light. Appl. Opt. *1*, 201-205 (1962)

Chaumont, M.L.: Recherches expérimentales sur le phénomène électrooptique de Kerr et sur les méthodes servant à l'étude de la lumière polarisée elliptiquement. Ann. de Phys. Paris *4*, 175-206 (1915)

Chauvin, M.: Polarisation rotatoire magnétique dans le spath d'Islande. Ann. de Toulouse *3*, 30-49 (1889)

Cheng, J.C.: Polarization scrambling using a photoelastic modulator: Appli-cation to linear dichroism measurement. Rev. Sci. Instrum. *48*, 1086-1089 (1977)

Cornu, M.A.: Appareil destiné aux mesures des pouvoirs rotatoires. Bull. Soc. Chim. *14*, 140-142 (1870)

Ferre, M.: Description et élaboration d'un ellipsomètre dynamique. Appli-cation à la photoélasticimétrie dynamique. Ecole Nationale Superieure de Techniques Avancées, Rpt. 073, Paris (1976) pp. 84

Françon, M., Sergent, B.: Compensateur biréfringent à grand champ. Opt. Acta *2*, 182-184 (1955)

Holoubek, J.: The use of mueller matrices in the intensity methods of bire-fringence measurements. Czech. J. Phys. B *24*, 1162-1167 (1974)

Hyde, W.L., Tubbs, E.F., Koester, C.J.: An automatic photoelectric polari-meter. J. Opt. Soc. Am. *49*, 513 (1959)

Ioshpa, B.A., Obridko, V.N.: Photoelectric analysis of polarized light. Opt. Spectrosc. USSR *15*, 60-62 (1963)

Jellett, J.H.: On a new instrument for determining the plane of polarization. Rpt. Br. Assoc. *30*, 13 (1860)

Jerrard, H.G.: Optical compensators for measurement of elliptical polariz-ation. J. Opt. Soc. Am. *38*, 35-59 (1948)

Jerrard, H.G.: Accurate adjustment of the wedges of a Babinet compensator. J. Sci. Instrum. *26*, 353-357 (1949)

Jerrard, H.G.: Examination and calibration of a Babinet compensator. J. Sci. Instrum. *27*, 62-66 (1950)

Jerrard, H.G.: Accurate adjustment of the wedges of a Soleil compensator. J. Sci. Instrum. *27*, 164-167 (1950)

Jerrard, H.G.: Use of a half-shadow plate with uniform field compensators. J. Sci. Instrum. *28*, 10-14 (1951)

Jerrard, H.G.: The examination and calibration of Soleil compensators. J. Sci. Instrum. *30*, 65-70 (1953)

Kent, C.V., Lawson, J.: A photoelectric method for the determination of the parameters of elliptically polarized light. J. Opt. Soc. Am. *27*, 117-119 (1937)

Lagarde, A., Oheix, P.: "Méthodes ponctuelles statiques et dynamiques de photoélasticimétrie pour des problèmes bidimensionnels"; Proc. 5th Int. Conf. Exp. Stress Analysis, Udine, 1974, ed. by G. Bartolozzi (Tecnoprint-Pitagora, Bologna 1974) pp.1.21-1.28

Lagarde, A., Oheix, P.: "Static and Dynamic Ponctual Method for the Determination of the Neutral Axis and the Algebraic Value of the Birefringence for a Plane Plate", in *Photoelastic Effect and Its Applications* (IUTAM Symposium) ed. by J. Kestens (Springer, Berlin, Heidelberg, New York 1975) pp.525-531

Lippich, F.: Über polaristrobometrische Methoden, insbesondere über Halbschattenapparate. Wien. Ber. *91*, 1059-1096 (1885)

Lipskii, Y.N., Pospergelis, M.M.: Some results of measurements of the total Stokes vector for details of the lunar surface. Sov. Astron.-AJ *11*, 324-326 (1967)

Nakamura, S.: Neue Instrumente und Beobachtungsmethoden. Zentralblatt f. Min. 267-279 (1905)

Oheix, P., Lagarde, A.: Sur une conception des mesures en photoélasticimétrie bidimensionelle. Rev. Fr. de Méc. 63-70 (1973)

Oheix, P.: Méthode globale de détermination de l'ordre de frange en valeur algébrique d'un modèle photoélastique bidimensional et visualisation des franges de quart d'onde. C.R. Acad. Sci. B *279*, 361-363 (1974)

Pospergelis, M.M.: The "Taimyr" electronic polarimeter. Sov. Astron.-AJ *9*, 313-321 (1965)

Pospergelis, M.M.: Measurement and computation of the instrumental Stokes vector. Sov. Astron.-AJ *12*, 512-521 (1968)

Pospergelis, M.M.: Spectroscopic measurements of the four Stokes parameters for light scattered by natural objects. Sov. Phys.-Astron. *12*, 973-977 (1969)

Richartz, M.: Einfache Halbschattenvorrichtung für den Viertelwellenlängenkompensator. Z. Instrumentenkd. *60*, 357-360 (1940)

Richartz, M., Hsü, H.Y.: Analysis of elliptical polarization. J. Opt. Soc. Am. *39*, 136-157 (1949)

Robert, A.: "New Methods in Three-Dimensional Photoelasticity", Proc. 2nd SESA Int. Congress on Exp. Mech., Washington, D.C. Sept. 1965, ed. by B.E. Rossi (1966) pp.123-131

Robert, A.J.: New methods in photoelasticity. Exp. Mech. *7*, 224-232 (1967)

Robert, A.: Application de la sphère de Poincaré à l'ellipsométrie de précision. Bull. Soc. Fr. Minéral. Cristallogr. *91*, 415-421 (1968)

Robert, A.: The application of Poincaré's sphère to photoelasticity. Int. J. Solids Struct. *6*, 423-432 (1970)

Robert, A.: La sphère de Poincaré et ses applications à la mesure des formes de lumière et à la photoélasticimétrie classique. Sci. et techniques de l'Armement *45*, 309-379 (1971)

Robert, A.: "Polarimétrie et photoélasticimétrie". Serv. Techn. Const. Armes Navales, Paris (1972)

Robert, A., Bourdon, C., Le Goer, J.L.: Principe et description d'un photoélasticimètre tridimensionel à lumière diffusée. Rev. Fr. Méc. *24*, 93-116 (1967)

Robert, A., Dore, A.: "Méthode de photoélasticimétrie dynamique", Proc. 5th Int. Conf. Exp. Stress Analysis, Udine, 1974, ed. by G. Bartolozzi (Tecnoprint-Pitagora, Bologna 1974) pp.1.38-1.42

Robert, A., Ferre, M.: Principe et description d'un photoélasticimètre
    automatique application à l'étude d'une tranche de pétrolier. Bull.
    Association Technique Maritime et Aéronautique *69*, Memoire No. 1549,
    1-15 (1969)
Robert, A., Guillemet, E.: Nouvelle méthode d'utilisation de la lumière
    diffusée en photoélasticimétrie à trois dimensions. II. Mise au point
    pratique. Rev. Fr. Méc. *5-6*, 147-157; *7-8*, 39-46 (1963)
Seielstad, G.A., Berge, G.L.: Time dependence of the integrated Stokes para-
    meters of compact radio sources at 5 GHz. Astron. J. *80*, 271-281 (1975)
Sekera, Z.: Recent developments in the study of the polarization of sky light.
    Adv. Geophys. *3*, 43-104 (1956)
Skinner, C.A.: A universal polarimeter. J. Opt. Soc. Am. *10*, 491-520 (1925)
Som, S.C., Chowdhury, C.: New ellipsometric method for the determination of
    the optical constants of thin films and surfaces. J. Opt. Soc. Am. *62*,
    10-15 (1972)
Stokes, G.G.: On a new elliptic analyser. Math. and Phys. Papers, Cambridge,
    3, 197-202 (1901)
Szivessy, G.: Über den Gebrauch des Braceschen Halbschattenkompensators bei
    der gleichzeitigen Messung von Azimut und Elliptizität der Schwingungs-
    ellipse. Z. Instrumentenkd. *47*, 148-154 (1927)
Takasaki, H., Isobe, M., Masaki, T., Konda, A., Agatsuma, T., Watanabe, Y.:
    An automatic retardation meter for automatic polarimetry by means of an
    ADP polarization modulator. Appl. Opt. *3*, 345-350 (1964)
Takasaki, H., Okazaki, N., Kida, K.: An automatic polarimeter, Pt. II. Auto-
    matic polarimetry by means of an ADP polarization modulator. Appl. Opt.
    3, 833-837 (1964)
Tardy, M.H.L.: Méthode pratique d'éxamen et de mesure de la biréfringence des
    verres d'optique. Rev. Opt. *8*, 59-69 (1929)
Tool, A.Q.: A method for measuring ellipticity and the determination of opti-
    cal constants of metals. Phys. Rev. *31*, 1-25 (1910)

6. The Photoelastic Phenomenon

Books

Coker, E.G., Filon, L.N.G.: *A Treatise on Photoelasticity*, 2nd ed. (Cambridge
    University Press, Cambridge 1957) Chap.3, pp.181-295
Ramachandran, G.N., Ramaseshan, S.: "Crystal Optics", in *Crystal Optics Dif-
    fraction*, ed. by S. Flügge, Encyclopedia of Physics, Vol. 25/1 (Springer,
    Berlin, Göttingen, Heidelberg 1961) pp.1-217
Stein, R.S.: *Rheo-Optics of Polymers*, J. Pol. Sci., Pt. C., Polymer Symposia
    (Wiley-Interscience, New York 1964)

Papers

Amba-Rao, C.L.: "Stress-Strain-Time-Birefringence Relations in Photoelastic
    Plastics with Creep", "Reo-Optics of Polymers". J. Polymer Sci., Pt. C.,
    Pol. Symp. *5*, 75-86 (1964)
Arakawa, I.: On the determination of the stress-optical coefficient of
    bakelite with initial stress. Proc. Phys.-Math. Soc. Jpn. 117-136 (1923)
Arthur, R.E.: Theory of photoelasticity and its application to problems of
    stress analysis. J. Roy. Aero. Soc. *47*, 263-272 (1943)

Baud, R.V.: Contribution to study of the effect of elliptical polarization upon energy transmission. J. Opt. Soc. Am. *21*, 119-123 (1931)

Bayoumi, S.E.A., Frankl, E.K.: Fundamental relations in photoplasticity. Br. J. Appl. Phys. *4*, 306-310 (1953)

Braybon, J.E.H.: A new method of measurement of the variation with wavelength of the refractive index and absolute stress optical coefficients of amorphous solids. Proc. Phys. Soc. London *63*, 446-451 (1950)

Brewster, D.: Additional observations on the optical properties and structure of heated glass and annealed glass drops. Phil. Trans. R. Soc. London *105*, 1-8 (1815)

Brewster, D.: On the effect of simple pressure in producing that species of crystallization which forms two oppositely polarised images and exhibits the complementary colours by polarised light. Phil. Trans. R. Soc. London *105*, 60-64 (1815)

Brewster, D.: On the laws which regulate the polarisation of light by reflexion from transparent bodies. Phil. Trans. R. Soc. London *105*, 125-159 (1815)

Brewster, D.: On the communication of the structure of doubly refracting crystals to glass, muriate of soda, fluor spar and other substances, by mechanical compression and dilatation. Phil. Trans. R. Soc. London *106*, 156-178 (1816)

Brewster, D.: On the effects of compression and dilatation in altering the polarising structure of doubly refracting crystals. Phil. Trans. R. Soc. Edinburgh *8*, 281-286 (1818)

Brewster, D.: On the laws which regulate the distribution of the polarising force in plates, tubes, and cylinders of glass, that have received the polarising structure. Phil. Trans. R. Soc. Edinburgh *8*, 353-371 (1818)

Brewster, D.: On the production of regular double diffraction in the molecules of bodies by simple pressure with observations on the origin of the doubly refracting structure. Phil. Trans. R. Soc. London *120*, 87-97 (1830)

Brewster, D.: On the production of crystalline structure in crystallised powders, by compression and traction. Phil. Trans. R. Soc. Edinburgh *20*, 555-559 (1853)

Brinson, H.F.: Mechanical and optical viscoelastic characterization of hysol 4290. Exp. Mech. *8*, 561-566 (1968)

Canit, J.C., Berger, D., Billardon, M.: Mesure de variations d'indice par polarimétrie. Opt. Acta *13*, 255-270 (1966)

Coleman, B.D., Dill, E.H.: Theory of induced birefringence in materials with memory. J. Mech. Phys. Solids *19*, 215-243 (1971)

Coleman, B.D., Dill, E.H.: "Photoviscoelasticity: Theory and Practice", in *Photoelastic Effect and Its Applications* (IUTAM Symposium) ed. by J. Kestens (Springer, Berlin, Heidelberg, New York 1975) pp.455-505

Daniel, I.M.: Quasi-static properties of a photoviscoelastic material. Exp. Mech. *5*, 83-89 (1965)

De Candia, F., Russo, R., Vittoria, V.: Mechanical and photoelastic properties of ehtylene-propylene copolymers related to chain microstructure. J. Phys. Chem. *80*, 2961-2966 (1976)

Dill, E.H.: "A Theory of Photothermoviscoelasticity", Proc. 4th Int. Congr. Rheology, Pt. 2 (1963) pp.51-59

Dill, E.H.: " A Theory of Photothermoviscoelasticity", Proc. 4th Int. Congr. Rheologoy, Pt. 2 (1963) pp.51-59

Dill, E.H.: On Phenomenological rheo-optic constitutive relations. J. Poly. Sci., Pt. C *5*, 67-74 (1964)

Dill, E.H.: "Photoviscoelasticity", Proc. 4th Symp. Naval Structural Mechanics: Mechanics and Chemistry of Solid Propellants, ed. by A.C. Eringen et al. (Pergamon Press, New York 1967) pp.443-461

Dill, E.H., Fowlkes, C.W.: Photoviscoelastic experiments I. Trend Eng. (Univ. Washington Publ.) *16*, 5-9 (1964)

Durelli, A.J.: Discussion of "The birefringent response of a potential
photoelastic material with variable elastic properties. Exp. Mech. *16*,
151-153 (1976)

Favre, H.: Sur l'application de la théorie des êrreurs à la résolution d'un
systême d'êquations utilisê en photoêlasticitê. Rev. Opt. *34*, 305-322
(1955)

Feldman, A., Horowitz, D., Waxler, R.M.: Photoelastic constants of potassium
chloride at 10.6 µm. J. Opt. Soc. Am. *67*, 254 (1977)

Filon, L.N.G.: On the dispersion in artificial double refraction. Phil.
Trans. R. Soc. London A *207*, 263-306 (1907)

Filon, L.N.G.: On the temperature variation of the photoelastic effect in
strained glass. Proc. R. Soc. London A *89*, 587-593 (1912)

Filon, L.N.G., Jessop, H.T.: On the stress optical effect in transparent
solid strained beyond the elastic limit. Phil. Trans. R. Soc. London
A *223*, 89-125 (1922-1923)

Frocht, M.M., Cheng, Y.F.: "An Experimental Study of the Laws of Double
Refraction in the Plastic State in Cellulose Nitrate-Foundations for
Three-Dimensional Photoplasticity", Proc. Int. Symp. on Photoelasticity,
Illinois Inst. of Technology, Chicago, 1961, ed. by M.M. Frocht (Pergamon
Press, New York 1963) pp.195-216

Frocht, M.M., Thomson, R.A.: Experiments in mechanical and optical coinci-
dence in photoplasticity. Exp. Mech. *1*, 43-47 (1961)

Ghaswala, S.K.: Elements of the theory of photoelasticity. Civ. Eng. Public
Works Rev. *45*, 237-238 (1950)

Grishchenko, A.E., Myznikov, V.G., Afonin, S.N., Skazka, V.S.: Study of the
photoelasticity of polymers in an oscillating mechanical field. Vysoko-
molekulyarnye Soedineniya, Seriya B-Kratkie Soobshcheniya *19*, 504-506
(1977) (in Russian)

Hauret, G., Cao, X.A.: Determination of photoelastic constants in thiourea
by brillouin scattering. Rev. Phys. Appl. (in French) *12*, 995-997 (1977)

Heller, W., Quimfe, G.: On the validity of the Langevin theory of orientation and the possibility of distinguishing between inherent and photoelastic anisotropy. Phys. Rev. *61*, 382 (1942)

Janeschitz-Kriegl, H., Wales, J.L.S.: "Streaming Double Refraction, a
General Report", in *Photoelastic Effect and Its Applications* (IUTAM
Symposium), ed. by J. Kestens (Springer, Berlin, Heidelberg, New York
1975) pp.271-293

Javornicky, J.: Concerning the double nature of birefringence in plastics.
Exp. Mech. *3*, 175-176 (1963)

Kerr, J.: Experiments on the birefringent action of strained glass. Phil.
Mag. *26*, 321-342 (1888)

Kiesling, E.:"Non Linear, Quasi Static Behavior of Some Photoelastic and
Mechanical Model Materials"; Ph. D. Thesis, Mich. State Univ., Ann Arbor,
Michigan (1966)

Kuno, J.: Effect of the age of phenolite on the coefficient of photoelastic
extinction. Phil. Mag. *23*, 63-64 (1937)

Lee, G.H.: "The Theory of the Photoelastic Effect"; Ph. D. Thesis, Cornell
Univ. (1937)

Mach, E.: Über die temporäre Doppelbrechung der Körper durch einseitigen
Druck. Ann. Phys. Chem., Ser. II *146*, 313-316 (1872)

Maxwell, J.C.: On double refraction in a viscous fluid in motion. Proc. R.
Soc. London *22*, 46-47 (1873)

Mindlin, R.D.: Analysis of doubly refracting materials with circularly and
elliptically polarized light. J. Opt. Soc. Am. *27*, 288-291 (1937)

Mindlin, R.D.: A mathematical theory of photoviscoelasticity. J. Appl. Phys.
20, 206-216 (1949)

Mindlin, R.D.: "Rotation-Dependence of the Photoelastic Effect", in *Photo-elastic Effect and Its Applications* (IUTAM Symposium) ed. by J. Kestens (Springer, Berlin, Heidelberg, New York 1975) pp.376-388

Mueller, H.: Theory of photoelasticity in amorphous solids. Physics *6*, 179-184 (1935)

Mueller, H.: The theory of photoelasticity. J. Am. Ceram. Soc. *21*, 27-33 (1938)

Nelson, D.F., Lax, M., Lazay, P.D.: "Electrodynamics of Photoelastic Media: Theory and Experiment", in *Photoelastic Effect and Its Applications* (IUTAM Symposium) ed. by J. Kestens (Springer, Berlin, Heidelberg, New York 1975) pp.316-350

Neumann, F.E.: Die Gesetze der Doppelbrechung des Lichtes in komprimierten und ungleichförmig erwärmten unkristallinischen Körpern. Abh. d. königl. Akad. d. Wissensch. Berlin, Pt. II, 1-254 (1841)

Nishitani, T.: Birefringence of non-linear viscoelastic-plastic polymers. J. Phys. D: Appl. Phys. *10*, 791-797 (1977)

Noblet, A., Sylin, G., Vandaele-Dossche, M., Van Geen, R.: La birêfringence mêcanique diffêrêe de signe nêgatif et sa reprêsentation analogique. Bull. Cl. Sci. Acad. R. Belg.,Ser. 5e, Vol. L, 534-561 (1964)

Pauthier-Camier, S.: "The Available Experimental Methods for Measuring Optical Properties of Matter", in *Photoelastic Effect and Its Applications* (IUTAM Symposium) ed. by J. Kestens (Springer, Berlin, Heidelberg, New York 1975) pp.231-256

Philippoff, W.:"The Present Stand of Rheo-Optics of Polymer Solutions", Proc. 5th Congr. Rheology, Kyoto, 1968, ed. by S. Onogi, Vol. *4* (Univ. of Tokyo Press, Tokyo 1970) pp.3-12

Pindera, J.T.: Remarks on properties of photoviscoelastic model materials. Exp. Mech. *6*, 375-380 (1966)

Pindera, J.T., Cloud, G.: On Dispersion of birefringence of photoelastic materials. Exp. Mech. *6*, 470-480 (1966)

Richard, T.G., Young, W.C.: The birefringent response of a potential photo-elastic material with variable elastic properties. Exp. Mech. *15*, 226-229 (1975)

Rivlin, R.S., Smith, G.F.: Birefringence in viscoelastic materials. ZAMP *22*, 325-339 (1971)

Sapriel, J., Vacher, R.: Photoelastic tensor components of $Gd_2(MoO_4)_3$. J. Appl. Phys. *48*, 1191-1194 (1977)

Shanker, J., Sharma, H.P., Sharma, O.P., Sharma, J.C.: Theory of the photo-elastic effect in ZnO, ZnS, and CdS crystals. Solid State Commun. *22*, 401-404 (1977)

Shanker, J., Bakhshi, P.S., Sharma, O.P.: Theory of the photoelasticity of cuprous halides. Solid State Commun. *24*, 217-219 (1977)

Sharma, H.P., Shanker, J., Verma, M.P.: Effect of Lundqvist's many-body potential on the photoelastic behavior of ionic crystals. Phys. Rev. B: Solid State *15*, 4100-4102 (1977)

Sharma, J.C.,Sharma, H.P., Shanker, J.: Effect of the exchange charge polarization on the photoelastic behavior of diatomic crystals. Can. J. Phys. *55*, 1510-1511 (1977)

Soroka, V.V.: Dispersion of the photoelasticity. Sov. Phys.-Solid State *18*, 366-367 (1976)

Sugie, M., Tada, K.: A theory of the photoelastic coefficients and unclamped values of the linear electrooptic coefficients. Jpn. J. Appl. Phys. *15*, 257-263 (1976)

Tada, K., Kikuchi, K., Sato, K.: Dispersion of photoelastic coefficients in ZnSe. Jpn. J. Appl. Phys. *16*, 757-760 (1977)

Theocaris, P.S.: "Viscoelastic Properties of Epoxy Resins Derived from Creep and Relaxation Tests at Different Temperatures", Rpt. NSF-G 8188/3, Brown Univ. (1960)

Theocaris, P.S.: A review of the rheo-optical properties of linear high polymers. Exp. Mech. *5*, 105-114 (1965)

Theocaris, P.S.: Dependence of the stress-optical coefficients on the mechanical and optical properties of polymers. J. Strain Anal. *8*, 267-276 (1973)

Theocaris, P.S.: "Phenomenological Analysis of Mechanical and Optical Behaviour of Rheo-Optically Simple Materials", in *Photoelastic Effect and Its Applications* (IUTAM Symposium), ed. by J. Kestens (Springer, Berlin, Heidelberg, New York 1975), pp.146-230

Tokuoka, T.: Theoretical investigation of birefringence deformation relations in photo-elasto-plasticity. Int. J. Solids Struct. *1*, 343-350 (1965)

Tokuoka, T.: Mechanical foundations of birefringence of photo-elastoplastic media. Int. J. Solids Struct. *2*, 49-58 (1966)

Tokuoka, T.: Mechanical foundations of birefringence of elastic media and viscous media. Int. J. Eng. Sci. *4*, 23-40 (1966)

Tokuoka, T.: A mathematical theory of birefringence of viscoelastic media. J. Phys. Soc. Jpn. *23* (2), 430-435 (1967)

Tokuoka, T., Miyakawa, M.:"Birefringent Formula in Elastic-Plastic Deformation Region of High Polymer Solids", Proc. 14th Jpn. Natl. Congr. Appl. Mech. (1964)pp.100-105

Van Geen, R.E.: Non-coïncidence des singularités mécaniques et optiques. C.R. Acad. Sci. Paris *258*, 5164-5166 (1964)

Van Geen, R.E.: Deux phénomènes dus à la non-linéarité des relations contrainte-biréfringence. C.R. Acad. Sci. Paris *259*, 999-1001 (1964)

Van Geen, R.E.: Contribution à l'étude du mécanisme de la biréfringence mécanique. Bull. Cl. Sci. Acad. R. Belg. Ser. 5e, Vol. L, 869-892 (1964)

Van Geen, R.E.: Effect photoélastique et matériaux adéquats en photoélasticimétrie. Sci. Techniques de l'Armement *45*, 381-476 (1971)

Vandaele-Dossche, M., Van Geen, R.E.: La biréfringence mécanique en lumière ultra-violette et ses applications. Bull. Cl. Sci. Acad. R. Belg., Ser. 5e, Vol. L, 125-141 (1964)

Waxler, R.M., Horowitz, D., Feldman, A.: Precision interferometer for measuring photoelastic constants. Appl. Opt. *16*, 20-22 (1977)

Wertheim, M.G.: Memoire sur la polarisation chromatique produite par le verre comprimé. C.R. Acad. Sci. *32*, 289-292 (1851). Note sur la double réfraction artificiellement produite dans des cristaux du systême régulier. ibid. *33*, 576-579 (1851). Deuxième note sur la double réfraction artificiellement produite dans des cristaux du systême régulier. ibid. *35*, 276-278 (1852)

Wertheim, M.G.: Mémoire sur la double réfraction temporairement produite dans les corps isotropes et sur la rélation entre l'élasticité mécanique et l'élasticité optique. Ann. Chim. Phys. III, *40*, 156-221 (1854)

Williams, M.L., Arenz, R.J.: The engineering analysis of linear photoviscoelastic materials. Exp. Mech. *4*, 249-262 (1964)

Wilson, C.A.C.: The influence of surface-loading on the flexure of beams. Phil. Mag. *32*, 481-503 (1891)

7. Two-Dimensional Photoelasticity

Books

Le Boiteux, H., Boussard, R.: *Elasticité et Photoélasticimétrie* (Hermann Editor, Paris 1940)
Brcic, V.: "Photoelasticity in Theory and Practice", (Course Held at the Department for Mechanics of Deformable Bodies, Sept.-Oct. 1970). CISM International Centre for Mechanical Sciences, Vol. 59 (Springer, Berlin, Heidelberg, New York 1975)

Bricas, M.: *La Théorie de l'Elasticité Bidimensionnelle* (Pyrsos, Athens 1936)
Coker, E.G., Filon, L.N.G.: *A Treatise on Photoelasticity*, 2nd ed. (Cambridge University Press, Cambridge 1957)
Dally, J.W., Riley, W.F.: *Experimental Stress Analysis* (McGraw-Hill, New York 1965)
Durelli, A.J.: *Applied Stress Analysis* (Prentice-Hall, Englewood Cliffs N.J. 1967)
Durelli, A.J., Phillips, E.A., Tsao, C.H.: *Introduction to the Theoretical and Experimental Analysis of Stress and Strain* (McGraw-Hill, New York 1958)
Filon, L.N.G.: *A Manual of Photoelasticity for Engineers* (Cambridge University Press, Cambridge 1936)
Föppl, L., Mönch, E.: *Praktische Spannungsoptik* (Springer, Berlin, Göttingen, Heidelberg 1950)
Föppl, L., Neuber, H.: *Festigkeitslehre Mittels Spannungsoptik* (R. Oldenbourg, Munich, Berlin 1935)
Frocht, M.M.: *Photoelasticity*, Vol.I, Vol.II (Wiley and Sons, New York 1941, 1948)
Hendry, A.W.: *Photoelastic Analysis* (Pergamon Press, London 1966)
Hetényi, M.: *Handbook of Experimental Stress Analysis* (Wiley and Sons, New York 1950)
Heywood, R.B.: *Designing by Photoelasticity* (Chapman and Hall, London 1952)
Holister, G.S.: *Experimental Stress Analysis* (Cambridge University Press, Cambridge 1967)
Jessop, H.T., Harris, F.C.: *Photoelasticity, Principles and Methods* (Cleaver-Hume Press, London 1949)
Kammerer, A.: *Recherches sur la Photoélasticimétrie* (Paris 1944)
Kuske, A.: *Verfahren der Spannungsoptik* (VDI Edition, Düsseldorf 1951)
Lee, G.H.: *An Introduction to Experimental Stress Analysis* (Wiley and Sons, New York 1950)
Pirard, A.: *La Photoélasticité* (H. Vaillant-Carmanne, Liège 1947)
Robert, A.: *Polarimétrie et Photoélasticimétrie* (Serv. Techn. Const. Armes Navales, Paris 1972)
Wolf, H.: *Spannungsoptik* (Springer, Berlin, Göttingen, Heidelberg 1961)

8. Three-Dimensional Photoelasticity

Books

Coker, E.G., Filon, L.N.G.: *A Treatise on Photoelasticity,* 2nd ed. (Cambridge University Press, Cambridge 1957) Chap.3, pp.253-258
Dally, J.W., Riley, W.F.: *Experimental Stress Analysis* (McGraw-Hill, New York 1965) Chap.10, pp.252-279
Durelli, A.J., Riley, W.F.: *Introduction to Photomechanics* (Prentice-Hall, Englewood Cliffs N.J. 1965) Chap.7, pp.173-179, Chap.10, pp.254-291
Frocht, M.M.: *Photoelasticity,* Vol. I, Vol. II (Wiley and Sons, New York 1941, 1948) Chap.10, pp.334-344; Chap.10, pp.333-465
Hetényi, M.: *Handbook of Experimental Stress Analysis* (Wiley and Sons, New York 1950) Chap.17, pp.924-965
Heywood, R.B.: *Photoelasticity for Designers* (Pergamon Press, Oxford 1959) Chap.8, pp.209-276
Holister, G.S.: *Experimental Stress Analysis* (Cambridge University Press, Cambridge 1967) pp.239-248
Kuske, A., Robertson, G.: *Photoelastic Stress Analysis* (Wiley and Sons, New York 1974) Chap.14, pp.329-390

Papers

Aben, H.K.: Optical phenomena in photoelastic models by the rotation of principal axes. Exp. Mech. *6*, 13-22 (1966)
Aben, H.K.: The optical investigation of twisted fibres. J. Text. Inst. *59*, 523-527 (1968)
Aben, H.K.: Optical theory of the multilayer-reflection technique for three-dimensional photoelastic studies. Exp. Mech. *9*, 25-30 (1969)
Aben, H.K.: Magnetophotoelasticity-photoelasticity in a magnetic field. Exp. Mech. *10*, 97-105 (1970)
Aben, H.K.: "Principles of Magneto-Photoelasticity", Proc. 4th Int. Conf. Exp. Stress Analysis, Cambridge, 1970, ed. by M.L. Meyer (The Institution of Mechanical Engineers, London 1971) pp.175-182
Aben, H.K.: "Additional Physical Fields in Photoelasticity", in *Photoelastic Effect and Its Applications* (IUTAM Symposium), ed. by J. Kestens (Springer, Berlin, Heidelberg, New York 1975) pp.294-315
Aben, H.K., Idnurm, S.: "Stress Concentration in Bent Plates by Magneto-photoelasticity", Proc. 5th Int. Conf. Exp. Stress Analysis, Udine, 1974, ed. by G. Bartolozzi (Tecnoprint-Pitagora, Bologna 1974) pp.4.5-4.10
Aben, H.K., Idnurm, S.J., Klabunovskii, E.I., Uffert, M.M.: Photoelasticity with additional physical fields and optically active models. Exp. Mech. *14*, 361-366 (1974)
Allison, I.M.: Oblique incidence observations as an aid to three-dimensional stress separation. J. Strain Anal. *1*, 322-330 (1966)
Allison, I.M.: Analysis of photoelastic data for three-dimensional stress separation. Strain *6*, 177-181 (1970)
Ansevin, R.W.: The non destructive measurement of surface stresses in glass. ISA Trans. *4*, 339-343 (1965)
Arfaei, A.H., Lang, B.R.: Three-dimensional photoelastic stress analysis of a free-end partial denture with parallel precision attachment. J. Dent. Res. *56*, A187 (1977)

Becker, H., Colao, A.: A bonded grid polariscope for photoviscoelasticity. J. Appl. Mech. *32*, 702-703 (1965)

Brock, J.S.: The determination of effective stress and maximum shear stress by means of small cubes taken from photoelastic models. Proc. Soc. Exp. Stress Anal. *16*, 1-8 (1958)

Cernosek, J., Perla, M.:"Composite Model Technique for Three-Dimensional Photoelastic Stress Analysis", Proc. 4th Int. Conf. Exp. Stress Analysis, Cambridge, 1970, ed. by M.L. Meyer (The Institution of Mechanical Engineers, London 1971) pp.189-197

Chandrashekhara, K., Jacob, K.A.: A method for separation of stresses in two- and three-dimensional photoelasticity. J. Indian Inst. Sci. *58*, 331-344 (1976)

Chandrashekhara, K., Jacob, K.A.: Photoelastic stress analysis of a composite cylinder subjected to mechanical loading. AIAA J. *15*, 1432-1435 (1977)

Cheng, Y.F.: Stress at notch root of shafts under axially symmetric loading. Exp. Mech. *10*, 534-536 (1970)

Chiba, M., Shimada, H.: "Remarkable Optical Effects by the Rotation of Secondary Principal Stress Axes in Three-Dimensional Photoelasticity", Proc. Ann. Meeting of Jpn. Soc. Mech. Eng., Tokyo 1964, No. 740-1, pp.169-172

Chiba, M., Shimada, H.: A numerical method for determining membrane and bending stresses in photoelastic models of plates and shells. Exp. Mech. *15*, 142-147 (1975)

Cook, R.D.: "Correlation of Variables in Photoelastic Stress Freezing", Univ of Ill., Dept. Theor. Appl. Mech., Rpt. 162 (1960)

Craig, R.J., Gutzwiller, M.J., Lee, R.H., Stitz, E.O.: A composite three-dimensional photoelastic method. Exp. Mech. *17*, 433-438 (1977)

Dally, J.W., Riley, W.F.: Initial studies in three-dimensional dynamic photoelasticity. J. Appl. Mech. *34*, 405-410 (1967)

Dally, J.W., Durelli, A.J., Riley, W.F.: A new method to "Lock-in" elastic effects for experimental stress analysis. J. Appl. Mech. *25*, 189-195 (1958)

Dascola, G., Giori, D.C., Varacca, V.: On the measurements of Faraday effect, associated with EPR, in a birefringent medium. Nuovo Cimento *33*, 702-708 (1964)

Davidson, G.B.: A method for three dimensional photoelastic stress analysis of the remaining tooth structure associated with various cavity preparations. J. Dent. Res. *43*, 912 (1964)

Desai, J.N., Patel, R.M.: Optics of a twisted birefringent monofilament. Indian J. Pure Appl. Phys. *4*, 436-438 (1966)

Desailly, R., Lagarde, A.: Application des propriétés des champs de granularité à la photo-élasticimétrie tridimensionnelle. C.R. Acad. Sci. B *284*, 13-16 (1977)

Desailly, R., Lagarde, A.: Rectilinear and circular analysis of a plane slice optically isolated in a three dimensional photoelastic model. Mech. Res. Commun. *4*, 99-107 (1977)

Drucker, D.C.: The photoelastic analysis of transverse bending of plates in the standard transmission polariscope. J. Appl. Mech. *9*, A161-164 (1942)

Drucker, D.C., Mindlin, R.D.: Stress Analysis by three-dimensional photoelastic methods. J. Appl. Phys. *11*, 724-732 (1940)

Drucker, D.C., Woodward, W.B.: Interpretation of photoelastic transmission patterns for a three-dimensional model. J. Appl. Phys. *25*, 510-512 (1954)

Durelli, A.J.: Complete experimental solution of three-dimensional elastic problems. J. Strain Anal. *10*, 42-52 (1975)

Durelli, A.J., Lake, R.L.: Three-dimensional photoelasticity. Mach. Des. *22*, 122-125 (1950)

Durelli, A.J., Hasseem, H.M., Parks, V.J.: New experimental method in three-dimensional elastostatics. Acta Mech. *24*, 219-238 (1976)

Favre, H.: Sur une mēthode optique de détermination des tensions intērieures dans les solides à trois dimensions. C.R. Acad. Sci. *190*, 1182-1184 (1930)

Fessler, H.: "Frozen Stress" phenomenon in photoelasticity. Proc. Inst. Mech. Eng. *1B*, 613-619 (1952)

Fessler, H., Mansell, D.O.: Photoelastic study of stresses near cracks in thick plates. J. Mech. Eng. Sci. *4*, 213-225 (1962)

Fessler, H., Litewka, A., Perla, M.: Repeated use of frozen stress photoelastic models. J. Strain Anal. *11*, 186-190 (1976)

Fisher, W.A.P.: Basic physical properties relied upon in the frozen stress technique. Proc. Inst. Mech. Eng. *158*, 230-235 (1948)

Friedrich, G.: Superposition of photoelastic data. Exp. Mech. *17*, 33-36 (1977)

Frocht, M.M.: Studies in three-dimensional photoelasticity: Stress concentrations in shafts with transverse circular holes in tension. Relation between two- and three-dimensional factors. J. Appl. Phys. *15*, 72-88 (1944)

Frocht, M.M.: Studies in three-dimensional photoelasticity. J. Appl. Mech. *11*, A10-16 (1944)

Frocht, M.M.: Studies in three-dimensional photoelasticity torsional stresses by oblique incidence. J. Appl. Mech. *11*, A 229-234 (1944)

Frocht, M.M.: The growth and present state of three-dimensional photoelasticity. Appl. Mech. Rev. *5*, 337-340 (1952)

Frocht, M.M.: Discussion of "The elastic sphere under concentrated loads". J. Appl. Mech. *20*, 304-307 (1953)

Frocht, M.M.: Final report on the three-dimensional photoelastic investigation of the principal stresses and maximum shears in the head of a railroad rail. Rpt. Com. for Rail, Am. Rly Eng. Assoc. Bull. *55*, (514) (1954)

Frocht, M.M., Guernsey, R.: "Three-Dimensional Photoelasticity - The Application of the Shear Difference Method to the General Space Problem", Proc. 1st U.S. Natl. Congr. Appl. Mech. (Edwards Brothers, Ann. Arbor, Mich. 1952) pp.301-307

Frocht, M.M., Guernsey, R.: "The Solution of the General Three-Dimensional Photoelastic Problem", Proc. 8th Int. Congr. Appl. Mech.(1952)

Frocht, M.M., Guernsey, R.: A special investigation to develop a general method for three-dimensional photoelastic stress analysis. NACA Rpt. 1148 (1953)

Frocht, M.M., Guernsey, R.: Further work on the general three-dimensional photoelastic problem. J. Appl. Mech. *22*, 183-189 (1955)

Frocht, M.M., Leven, M.M.: On the state of stress in thick bars. J. Appl. Phys. *13*, 308-313 (1942)

Frocht, M.M., Pih, H., Landsberg, D.: The use of photometric devices in the solution of the general three-dimensional photoelastic problem. Proc. Soc. Exp. Stress Anal. *12*, 181-190 (1954)

Frocht, M.M., Srinath, L.S.: "A Non-Destructive Method for Three-Dimensional Photoelasticity", Proc. 3rd U.S. Natl. Congr. Appl. Mech., Providence, R.I. 1958 (ASME, New York 1958) pp.329-337

Frocht, M.M., Wang, B.C.: "A Three-Dimensional Photoelastic Study of Interior Stresses in the Head of a Railroad Rail in the Region Under a Wheel", Proc. 4th U.S. Natl. Congr. Appl. Mech., Berkeley, Calif.1962, Vol. 1 (ASME, New York 1962) pp.603-609

Frocht, M.M., Wang, B.C.: "A Three-Dimensional Photoelastic Investigation of a Propeller Blade Retention", Proc. Int. Symp. on Photoelasticity, Illinois Institute of Technology, Chicago, 1961, ed. by M.M. Frocht (Pergamon Press, London 1963) pp.123-144

Froehly, C., Desailly, R.: Polychromatic speckle technique for three-dimensional non-destructive photoelasticimetry. Opt. Commun. 21, 258-262 (1977)

Gaudfernau, C.L.: Photoelasticité tridimensionnelle: Aspects théoriques et experimentaux. Publ. Sci. Minist. Air Fr. 330, 1-86 (1957)

Ginsburg, V.L.: On the influence of the terrestrial magnetic field on the reflection of radio waves from the ionosphere. J. Phys. 7, 289-304 (1943)

Goodier, J.N., Lee, G.H.: An extension of the photoelastic method of stress measurement to plates in transverse bending. J. Appl. Mech. 8, A27-29 (1941)

Gorman, D., Evan-Iwanowski, R.M.: Photoelastic analysis of prebuckling deformations of cylindrical shells, AIAA J. 3, 1956-1958 (1965)

Gross-Petersen, J.F.: The gamma-ray-irradiation method applied to three-dimensional thermal photoelasticity. Exp. Mech. 12, 414-419 (1972)

Guernsey, R.: "Extension of the Shear-Difference Method in Photoelasticity to General Space Problems"; Doctoral Thesis, Ill. Inst. of Technology (1952)

Gurtman, G.A., Colao, A.A.: Photothermoelastic investigation of stresses around a hole, in a plate subjected to thermal shock. Exp. Mech. 5, 97-104 (1965)

Hetényi, M.: "Photoelastic Studies of Three-Dimensional Stress Problems", Proc. 5th Int. Congr. Appl. Mech. (ASME, New York 1966) pp.208-212

Hetényi, M.: Photoelastic stress analysis made in three dimensions. Mach. Des. 10, 40-41 (1938)

Hetényi, M.: The fundamentals of three-dimensional photoelasticity. J. Appl. Mech. 5, A149-155 (1938)

Hetényi, M.: The application of hardening resins in three-dimensional photoelastic studies. J. Appl. Phys. 10, 295-300 (1939)

Hetényi, M.: A photoelastic study of bolt and nut fastenings. J. Appl. Mech. 10, A93-100 (1943)

Heywood, R.B.: Modern applications of photoelasticity. Proc. Inst. Mech. Eng. London 158, 235-240 (1948)

Hickson, V.M.: Photoelastic determination of free boundary stresses on "Frozen Stress" models by an oblique incidence method. Br. J. Appl. Phys. 2, 261-269 (1951)

Hickson, V.M.: Errors in stress determination at the free boundaries of "Frozen Stress" photoelastic models. Br. J. Appl. Phys. 3, 176-181 (1952)

Hung, Y.Y., Durelli, A.J.: An optical slicing method for three-dimensional photoelasticity. Mech. Res. Commun. 4, 265-270 (1977)

Huszár, I., Csizmadia, L.: Three-dimensional photoelastic investigation of a twisted bar. Acta Tech. Acad. Sci. Hung. 51, 381-401 (1965)

Jessop, H.T.: The determination of the separate stresses in three-dimensional stress investigations by the "Frozen Stress" method. J. Sci. Instrum. and Phys. Ind. 26, 27-31 (1949)

Jessop, H.T.: A tilting stage method for three-dimensional photoelastic investigations. Br. J. Appl. Phys. 8, 30-32 (1957)

Jessop, H.T., Stableford, W.H.: A method of correcting for initial stresses in frozen-stress observations. Br. J. Appl. Phys. 4, 281-283 (1953)

Jessop, H.T., Wells, M.K.: The determination of the principal stress differences at a point in a three-dimensional photoelastic model. Br. J. Appl. Phys. 1, 184-189 (1950)

Johnson, E.W.:"A Three Dimensional Photoelastic Investigation of Stress Concentrations in Operably Deformed Human Teeth"; Thesis, Univ. Alberta, Edmonton, Canada (1965)

Johnson, E.W., Castaldi, C.R., Gau, D.J., Wysocki, G.P.: Stress pattern variations in operatively prepared human teeth, studied by three-dimensional photoelasticity. J. Dent. Res. *47*, 548-558 (1968)

Jullien, J.F., Duhau, R., Cubaud, J.C.: Applications of photoelasticity to structural analysis. Proc. Am. Soc. Civ. Eng. (EM5) *102*, 867-882 (1976)

Kaliski, S.: Wave Equations of thermo-electro-magnetoelasticity. Proc. Vibration Problems, Polish Acad. Sci. *6*, 231-265 (1965)

Kammerer, M.A.: Photoelasticity in three dimensions. Ann. Ponts Chaussées *121*, 71-104 (1951)

Kenny, B., Hugill, J.W.: Comparison of fringe multiplication and tardy compensation methods for a frozen stress investigation. Strain *3*, 3-7 (1967)

Kuske, A.: Das Kunstharz Phenolformaldehyd in der Spannungsoptik. Forsch. Geb. Ingenieurwes. *9*, 139-149 (1938)

Kuske, A.: Vereinfachte Auswerteverfahren räumlicher spannungsoptischer Versuche. Z. VDI *86*, 541-544 (1942)

Kuske, A.: Frozen-stress techniques: Three-dimensional photoelastic stress analysis *183*, 397-402 (1957)

Kuske, A.: L'analyse des phénomènes optiques en photoélasticité à trois dimensions par la méthode du cercle de "j". Rev. Fr. de Méc. *9*, 49-58 (1964)

Lee, L.H.N.: Effects of rotation of principal stresses on photoelastic retardation. Exp. Mech. *4*, 306-312 (1964)

Lerchenthal, Ch.H., Betser, A.A.: " A New Two-Layer Technique for the Photoelastic Analysis of Loaded Plates", Proc. Int. Symp. on Photoelasticity, Illinois Institute of Technology, Chicago, 1961, ed. by M.M. Frocht (Pergamon Press, London 1963) pp.97-107

Leven, M.M.: A new material for three-dimensional photoelasticity. Proc. Soc. Exp. Stress Anal. *6*, 19-28 (1948)

Leven, M.M.: Quantitative three-dimensional photoelasticity. Proc. Soc. Exp. Stress Anal. *12*, 157-171 (1955)

Leven, M.M., Sampson, R.C.: Photoelastic stress and deformation analysis of nuclear reactor components. Proc. Soc. Exp. Stress Anal. *17*, 161-180 (1959)

Leven, M.M., Wahl, A.M.: Three-dimensional photoelasticity and its application in machine design. Trans. ASME, *80*, 1683-1694 (1958)

Lingaiah, K., Gargesa, G.: Photoelastic investigation of a shrink-fit problem. Exp. Mech. *16*, 75-80 (1976)

Lingaiah, K., Ramachandra, K.: Three-dimensional photoelastic study of the load-carrying capacity/face width ratio of Wildhaber-Novikov gears for automotive applications. Exp. Mech. *17*, 392-397 (1977)

Mark, R.: Photoelastic analysis of microelectric-component thermal stresses. Exp. Mech. *17*, 121-127 (1977)

Marloff, R.H., Daniel, I.M.: Three-dimensional photoelastic analysis of a fiber-reinforced composite model. Exp. Mech. *9*, 156-162 (1969)

Masaki, J.I.: "A Simple Method for Measuring the Stress in Shell-Type Glass-to-Metal-Seals", Proc. 1st Jpn. Congr. Test. Mat., Tokyo (1958) pp.159-163

Masaki, J.: "On the Problem of Rotation of Polarizing Axes in Stress Measurement of Shell-Type Glass-to-Metal Seals", Proc. 2nd. Jpn. Congr. Test. Mat., Kyoto (1959) pp.189-191

Matsumoto, E., Sumi, S., Sekiya, T.: A photothermoelastic investigation of transient thermal stresses in wing ribs. J. Strain Anal. *7*, 117-124 (1972)

Mehrotra, C.L., Meyer, M.L.: Self-calibration of fractional fringe order lines obtained in photoelastic Frozen-stress slices by controlled density techniques. Int. J. Mech. Sci. *18*, 547-553 (1976)

Michalski, B.: Photoelastic investigations of a three-dimensional state of stress on models with a perforated optically sensitive layer. Rozpr. Inz. (in Polish) *24*, 71-91 (1976)

Mindlin, R.D.: A review of the photoelastic method of stress analysis, Pt.
  I,II. J. Appl. Phys. *10*, 222-241, 273-294 (1939)
Mindlin, R.D.: Discussion of"An extension of the photoelastic method of
  stress measurement to plates in transverse bending". J. Appl. Mech. *8*,
  A187-189 (1941)
Mindlin, R.D.: Optical aspects of three-dimensional photoelasticity. J.
  Franklin Inst. *233*, 349-364 (1942)
Mindlin, R.D.: The present status of three-dimensional photoelasticity. Civ.
  Eng. (ASCE) *12*, 255-258 (1942)
Mindlin, R.D.: A mathematical theory of photo-viscoelasticity. J. Appl. Phys.
  20, 206-216 (1949)
Mindlin, R.D., Goodman, L.E.: The optical equations of three-dimensional
  photoelasticity. J. Appl. Phys. *20*, 89-95 (1949)
Miyazono, S.: Fixation photoelastic fringe patterns by gamma rays. J. Appl.
  Mech. *34*, 2319-2323 (1967)
Miyazono, S.: Some applications of fixation of photoelastic-fringe patterns
  by gamma rays. Exp. Mech. *9*, 473-477 (1969)
Mönch, E.: "Frozen stress" photoelasticity using phenol resins wrapped in
  protecting foils. Eng. Dig. *9*, 180-182 (1948)
Mönch, E.: "Photoelastic Investigation of a Model in Whose Middle Surface
  a Semi-Transparent Mirror Layer is Embedded", Actes IX Congr. Int. de
  Mécanique appliquée,Univ. des Bruxelles, Bruxelles, 1957, Vol. 8, pp.384-
  394
Mylonas, C., Drucker, D.C.: Twisting stresses in tape. Exp. Mech. *1*, 23-32
  (1961)
Nelson, J.M., Cook, W.A., Stibor, G.S.: Three-dimensional photoelastic and
  finite-element analysis of a propellant grain. Exp. Mech. *12*, 436-440
  (1972)
Neumann, F.: "Die Gesetze der Doppelbrechung des Lichtes in komprimierten
  oder ungleichförmig erwärmten unkristallinischen Körpern", Abh. d.
  königl. Akad. d. Wissensch. Berlin,Pt.II, 1-254 (1841)
Nisida, M.: "New Photoelastic Methods for Torsion Problems", Proc. Int.
  Symp. on Photoelasticity, Illinois Institute of Technology, Chicago, 1961,
  ed. by M.M. Frocht (Pergamon Press, London 1963) pp.109-121
Oppel, G.: Photoelastic investigation of three-dimensional stress and strain
  conditions. Forsch. Geb. Ingenieurwes. *7*, 240-248 (1936)NACA Techn. Memo 824
  (1937)
O'Rourke, R.C., Saenz, A.W.: Quenching stresses in transparent isotropic
  media and the photoelastic method. Q. Appl. Math. *8*, 303-311 (1950)
O'Rourke, R.C.: Three-dimensional photoelasticity. J. Appl. Phys. *22*,
  872-878 (1951)
Papirno, R., Becker, H.: A multilayer-reflection technique for three-dimensional photoelastic studies of perforated plates in bending. Exp.
  Mech. *7*, 361-371 (1967)
Parks, V.J., Durelli, A.J., Chandrashekhara, K., Chen, T.L.: Stress distribution around a circular bar, with flat and spherical ends, embedded
  in a matrix in a triaxial stress field. J. Appl. Mech. *37*, Trans. ASME *92*,
  578-586 (1970)
Pih, H.: Three-dimensional photoelastic investigations of circular cylinders
  with spherical cavities in axial loading. Exp. Mech. *5*, 90-96 (1965)
Pih, H., Vanderveldt, H.: Stresses in spheres with concentric spherical
  cavities under diametral compression by three-dimensional photoelasticity.
  Exp. Mech. *6*, 244-250 (1966)
Plechata, R.: Determination of resultant parameters in a three dimensional
  elastically loaded continuum. Acta Tech. ČSAV *9*, 432-439 (1964)
Plechata, R.: Elliptical polarization at a threedimensionally loaded continuum. Acta Tech. ČSAV *9*, 43-50 (1964)

Plechata, R.: About azimuts of polarization directions in a three-dimensional continuum. Acta Tech. ČSAV *9*, 250-254 (1964)

Plechata, R.: Photoelasticity of continuum. Acta Tech. ČSAV *10*, 270-281 (1965)

Post, D.: Fringe multiplication in three-dimensional photoelasticity. J. Strain Anal. *1*, 380-388 (1966)

Post, D.: Photoelastic-fringe multiplication-for tenfold increase in sensitivity. Exp. Mech. *10*, 305-312 (1970)

Prigorovsky, N.I., Preiss, A.K.: Study of the state of stress in transparent three-dimensional models by means of a beam of parallel polarised light. Izv. Akad. USSR, Ser. Tekh. Nauk. *5*, 686-700 (1949)

Read, W.T.: An optical method for measuring the stress in glass bulbs. J. Appl. Phys. *21*,250-257 (1950)

Reissner, E.: "On the Calculation of Three-Dimensional Corrections for the Two-Dimensional Theory of Plane Stress", Proc. 15th Semi-Annual Eastern Photoelasticity Conf. 1942 (Addison-Wesley Reading, Cambridge, Mass. 1942) pp.23-31

Riera, J.D., Mark, R.: The optical-rotation effect in photoelastic shell analysis. Exp. Mech. *9*, 9-16 (1969)

Riney, T.D.: Photoelastic determination of the residual stress in the dome of electron tube envelopes. Proc. Soc. Exp. Stress Anal. *15*, 161-170 (1957)

Robert, A., Royer, J.:"Photoêlasticimêtrie Tridimensionnelle", DGRST-7470721, CNRS Paris (1976) p.62

Rubayi, N.A., Ved, R.: Photoelastic analysis of a thick square plate containing central crack and loaded by pure bending. Int. J. Fracture *12*, 435-451 (1976)

Rubayi, N.A., Yadava, V.: Three-dimensional photoelastic analysis in axially loaded thick plate with hole. J. Strain Anal. *8*, 220-227 (1973)

Sampson, R.C.: A three-dimensional photoelastic method for analysis of differential-contraction stresses. Exp. Mech. *3*, 225-237 (1963)

Sanford, R.J.: The validity of three-dimensional photoelastic analysis of non-homogeneous elastic field problems. Br. J. Appl. Phys. *17*, 99-108 (1966)

Sanford, R.J., Beaubien, L.A.: Stress analysis of a complex part: Photoelasticity vs. finite elements. Exp. Mech. *17*, 441-448 (1977)

Sarma, A.V.S.S.S.R., Srinath, L.S.: Photoelastic analysis with Stokes vector and new methods for the determination of characteristic parameters in three-dimensional photoelasticity. J. Aero. Soc. Ind. *24*, 300-306 (1972)

Smith, C.W.: Use of three-dimensional photoelasticity in fracture mechanics. Exp. Mech. *13*, 539-544 (1973)

Smith, D.G., Smith, C.W.: A photoelastic evaluation of the influence of closure and other effects upon the local bending stresses in cracked plates. Int. J. Fract. Mech. *6*, 305-318 (1970)

Smith, D.G., Smith, C.W.: Influence of precatastrophic extention and other effects on local stresses in cracked plates under bending fields. Exp. Mech. *11*, 394-401 (1971)

Srinath, L.S.: Some observations on three-dimensional photoelasticity. Br. J. Appl. Phys. *18*, 225-231 (1967)

Srinath, L.S.: Principal-stress differences in transverse planes of symmetry. Exp. Mech. *11*, 130-132 (1971)

Srinath, L.S.,Bhave, S.K.: A new nondestructive method for three-dimensional photoelasticity. Exp. Mech. *14*, 367-372 (1974)

Srinath, L.S., Sarma, A.V.S.S.R.: Effects of stress-induced optical activity in photoelasticity. J. Phys. D: Appl. Phys. *5*, 883-895 (1972)

Srinath, L.S., Sarma, A.V.S.S.R.: Determination of the optically equivalent model in three-dimensional photoelasticity. Exp. Mech. *14*, 118-122 (1974)

Taylor, C.E.,Schweiker, J.W.: A three-dimensional photoelastic investigation of the stresses near a reinforced opening in a reactor pressure vessel. Proc. Soc. Exp. Stress Anal. *17*, 25-36 (1959)

Tennyson, R.: An experimental investigation of the buckling of circular cylindrical shells in axial compression using the photoelastic technique. UTIAS Rpt. 102, Inst. Aerosp. Stud., Univ. Toronto (1964)

Theocaris, P.S.: Discussion of "Extended frozen stress method". Proc. Am. Soc. Civ. Eng. (EM4) *89*, 73-77 (1963)

Theocaris, P.S.: Discussion of "Composite Model Technique for Three-Dimensional Photoelastic Stress Analysis", Proc. 4th Int. Conf. Exp. Stress Analysis, Cambridge, 1970, ed. by M.L.Meyer (The Institution of Mechanical Engineers, London 1971) pp.200, 204

Theocaris, P.S., Gdoutos, E.E.: "On a Novel Matrix Method for Three-Dimensional Photoelasticity", Proc. 6th Int. Conf. Exp. Stress Analysis, Munich, 1978, ed. by VDI (VDI-Verlag, Düsseldorf 1978) pp.599-605

Theocaris, P.S., Paipetis, S.A.: Shrinkage stresses in three-dimensional two-phase systems. J. Strain Anal. *8*, 286-293 (1973); *9*, 264-267 (1974)

Thomson, R.A.: The use of digital computers in the photoelastic solution of the general space problem by the Shear-difference method. Proc. Soc. Exp. Stress Anal. *16*, 11-16 (1959)

Tramposch, H., Gerard, G.: An exploratory study of three-dimensional photothermoelasticity. J. Appl. Mech. *28*, 35-40 (1961)

Tuzi, Z.: Photoelastic study of stress on a specimen of three-dimensional form. Sci. Pap. Inst. Phys. Chem. Res. Jpn. *7* (113), 97-103 (1927)

Young, W.C.: An automated process for three-dimensional photoelastic analysis. Exp. Mech. *9*, 275-280 (1969)

9. Scattered-Light Photoelasticity

Books

Bokstein, M.F.: "Scattered–Light Stress Analysis", in *Polarization–Optical Method of Stress Analysis*, ed. by N.I. Prigorovski (Acad. Sci., USSR 1956) pp.182-213

Dally, J.W., Riley, W.F.: *Experimental Stress Analysis* (McGraw-Hill, New York 1965) Chap.11, pp.295-306

Frocht, M.M.: *Photoelasticity*, Vol.II (Wiley and Sons, New York 1948) Chap.14, pp.466-482

Kuske, A., Robertson, G.: *Photoelastic Stress Analysis* (Wiley and Sons, New York 1974) Chap.16, pp.391-417

Papers

Aderholdt, R.W., McKinney, J.M., Ranson, W.F., Swinson, W.F.: Effect of rotating secondary principal axes in scattered-light photoelasticity. Exp. Mech. *10*, 160-165 (1970)

Aderholdt, R.W., Ranson, W.F., Swinson, W.F.: "Stress Distributions in Thin Wall Pressure Vessels by Scattered-Light Photoelasticity", Proc. 4th Int. Conf. Exp. Stress Analysis, Cambridge, 1970, ed. by M.L. Meyer (The Institution of Mechanical Engineers, London 1971) pp.318-323

Aderholdt, R.W., Swinson, W.F.: Establishing the boundary retardation with respect to the observed fringes in scattered-light photoelasticity. Exp. Mech. *11*, 521-523 (1971)

Barker, D.B., Fourney, M.E.: Displacement measurements in the interior of 3-D bodies using scattered-light Speckle Patterns. Exp. Mech. *16*, 209-214 (1976)

Berghaus, D.G., Aderholdt, R.W., Buban, J.P., Womack, D.R.: Instrumentation improvements for scattered-light photoelasticity. Exp. Mech. *14*, 505-506 (1974)

Berghaus, D.G., Aderholdt, R.W.: Photoelastic analysis of interlaminar matrix stresses in fibrous composite models. Exp. Mech. *15*, 409-417 (1975)

Braswell, D.W., Ranson, W.F., Swinson, W.F.: Scattered-light photoelastic thermal stress analysis of a solid-propellant rocket motor. J. Spacecr. Rockets *5*, 1411-1416 (1968)

Bokstein, M.F.: "Methods of Stress Analysis in Scattered Light", Proc. 4th All-Union Conf. Polarization-Optical Method for Stress Analysis, ed. by S.P. Shikhobalov (Leningrad University Press, Leningrad 1960) pp.94-102

Cernosek, J.: On the effect of rotating secondary principal stresses in scattered-light photoelasticity. Exp. Mech. *13*, 273-279 (1973)

Cheng, Y.F.: Some new techniques for scattered-light photoelasticity. Exp. Mech. *3*, 275-278 (1963)

Cheng, Y.F.: Scattered-light photoelastic polariscope for three-dimensional stress analysis. Rev. Sci. Instrum. *35*, 976-977 (1964)

Cheng, Y.F.: Some techniques for employing a continuous-wave gas laser as a light source in scattered-light static photoelasticity. Exp. Mech. *6*, 431-432 (1966)

Cheng, Y.F.: A dual-observation method for determining photoelastic parameters in scattered light. Exp. Mech. *7*, 140-144 (1967)

Cheng, Y.F.: Discussion of "Applications of scattered-light photoelasticity to doubly connected tapered torsion bars". Exp. Mech. *8*, 217 (1968)

Cheng, Y.F.: An automatic system for scattered-light photoelasticity. Exp. Mech. *9*, 407-412 (1969)

Cheng, Y.F.: Discussion of "A study of the scattered-light technique in two-dimensional problems". Exp. Mech. *12*, 434-435 (1972)

Chiang, F.P.: Discussion of "Displacement measurements in the interior of 3-D bodies using scattered-light Speckle patterns". Exp. Mech. *17*, 120 (1977)

Drucker, D.C., Mindlin, R.D.: Stress analysis by three-dimensional photo-elastic methods. J. Appl. Phys. *11*, 724-732 (1940)

Drucker, D.C., Frocht, M.M.: Equivalence of photoelastic scattering patterns and membrane contours for torsion. Proc. Soc. Exp. Stress Anal. *5*, 34-41 (1947)

Durelli, J.: Discussion of "Photoelastic analysis of interlaminar matrix stresses in fibrous composite models". Exp. Mech. *16*, 316-317 (1976)

Fishburn, J.D.: "Analysis of Thermal Stresses in Complex Models Using Scattered Light Photoelasticity", Proc. 5th Int. Conf. Exp. Stress Analysis, Udine, 1974, ed. by G. Bartolozzi (Tecnoprint-Pitagora, Bologna 1974) pp.1.10-1.20

Fishburn, J.D., Aderholdt, R.W., Ranson, W.F., Swinson, W.F.: Scattered-light rosette. Exp. Mech. *11*, 554-559 (1971)

Fried, B., Weller, R.: Photoelastic analysis of two- and three-dimensional stress systems. Ohio State Univ. The Engineering Experiment Station Bull., No. 106, 1-28, 1940

stress systems. Ohio State Univ. Exp. Bull., Vol.106, July 1940

Frigon, R.A.: "Some Three Dimensional Studies with Scattered Light", Proc. 15th Semi-Annual Eastern Photoelasticity Conf. (Addison-Wesley Press, Cambridge, Mass. 1942) pp.68-73

Frocht, M.M., Srinath, L.S.: "A Non-Destructive Method for Three-Dimensional Photoelasticity", Proc. 3rd U.S. Natl. Congr. Appl. Mech. Providence, R.I. (ASME, New York 1958) pp.329-337

Gaudfernau, C.L.: Photoelasticité tridimensionnelle: Aspects théoriques et experimentaux. Publ. Sci. Techn. Minist. Air, Fr. *330*. 1-86 (1957)

Gross-Petersen, J.F.: A scattered-light method in photoelasticity. Exp. Mech. *14*, 317-322 (1974)

Gross-Petersen, J.F.: "A Compensation Method in Scattered Light Photoelasticity", in *Photoelastic Effect and Its Applications* (IUTAM Symposium), ed. by J. Kestens (Springer, Berlin, Heidelberg, New York 1975) pp.547-572

Hemann, J.H., Becherer, R.J.: A study of the scattered-light technique in two-dimensional problems. Exp. Mech. *12*, 43-46 (1972)

Hemann, J.H., Becherer, R.J.: Stresses and strains in axisymmetric problems using scattered-light photoelasticity. Exp. Mech. *17*, 233-236 (1977)

Jenkins, D.R.: Analysis of behavior near a cylindrical glass inclusion by scattered-light photoelasticity. Exp. Mech. *8*, 467-473 (1968)

Jessop, H.T.: The scattered light method of exploration of stresses in two- and three-dimensional models. Br. J. Appl. Phys. *2*, 249-260 (1951)

Johnson, R.L.: Measurement of elastic-plastic stresses by scattered-light photomechanics. Exp. Mech. *16*, 201-208 (1976)

Kammerer, M.A.: Photoelasticity in three dimensions. Ann. Ponts Chaussées *121*, 71-104 (1951)

Leven, M.M.: Stresses in keyways by photoelastic methods and comparison with numerical solution. Proc. Soc. Exp. Stress Anal. *7*, 141-154 (1950)

McAfee, W.J., Pih, H.: A scattered-light polariscope for three-dimensional birefringent flow studies. Rev. Sci. Instrum. *42*, 221-223 (1971)

McAfee, W.J., Pih, H.: Scattered-light flow-optic relations adaptable to three-dimensional flow birefringence. Exp. Mech. *14*, 385-391 (1974)

McKinney, J.M., Swinson, W.F.: "Location of Maximum Secondary Principal Axis in Scattered-Light Photoelasticity", Proc. 4th Southeastern Conf. Theor. Appl. Mech. (Pergamon Press, New York 1970) pp.407-413

Menges, H.J.: Die experimentelle Ermittlung räumlicher Spannungszustände an durchsichtigen Modellen mit Hilfe des Tyndalleffektes. Z. Angew. Math. Mech. *20*, 210-217 (1940)

Pindera, J.T., Straka, P.: Response of the integrated polariscope. J. Strain Anal. *8*, 65-76 (1973)

Prigorovski, N.I.: "Stress Analysis on Three-Dimensional Models", Selected Papers on Stress Analysis, The Inst. of Phys. (Chapman and Hall, Ltd., London 1961) pp.77-82

Ranson, W.F., Swinson, W.F., Tucker, R.: "Stress Distributions in Composites by Scattered Light Photoelasticity", Developments in Theoretical and Applied Mechanics, Proc. of the 8th Southeastern Conf. on Theoretical and Applied Mechanics, Blacksburg, VA 1976, Vol.8 (1976) pp.213-214

Robert, A.J.: New methods in photoelasticity. Exp. Mech. *7*, 224-232 (1967)

Robert, A.J.: Application de la sphère de Poincaré à l'ellipsométrie de précision. Bull. Soc. fr. Minéral. Cristallogr. *91*, 415-421 (1968)

Robert, A.J.: The application of Poincaré's sphere to photoelasticity. Int. J. Solids Struct. *6*, 423-432 (1970)

Robert, A.J.: "Polarimétrie et Photoélasticimétrie", Serv. Techn. Const., Armes Navales Paris, 145-171 (1972)

Robert, A.J., Guillemet, E.: New scattered-light method in three-dimensional photoelasticity. Br. J. Appl. Phys. *15*, 567-578 (1964)

Rosenberg, P.R.: "Study of a Shrink Fit Model by the Scattered Light Method", Proc. 13th Semi-Annual Eastern Photoelasticity Conf. (Mass. Institute of Technology, Cambridge Mass. 1941) pp.99-103

Rowlands, R.E.: "A Sequentially Modulated Ruby Laser System for Transmitted- and Scattered-Light Dynamic Photoelasticity", Ph. D. Thesis (TAM Report 304), Dept. Theor. and Appl. Mech., Univ. of Illinois, Urbana, Ill. (1967)

Rowlands, R.E., Taylor, C.E., Daniel, I.M.: A multiple-pulse ruby-laser system for dynamic photomechanics: Applications to transmitted- and scattered-light photoelasticity. Exp. Mech. *9*, 385-393 (1969)

Saleme, E.M.: Three-dimensional photoelastic analysis by scattered light. Proc. Soc. Exp. Stress Anal. *5*, 49-55 (1948)

Shelson, W., Smith, I.W.: A photoelastic method employing scattered light
  for the solution of plane stress problems. Br. J. Appl. Phys. *7*, 436-439
  (1956)
Srinath, L.S.: "A Non-Destructive Method for Three-Dimensional Photoelasticity",
  Ph. D. Thesis, Ill. Inst. of Technology (1958)
Srinath, L.S.: Analysis of scattered-light methods in photoelasticity. Exp.
  Mech. *9*, 463-468 (1969)
Srinath, L.S., Frocht, M.M.: "Scattered Light in Photoelasticity-Basic
  Equipment and Techniques", Proc. 4th U.S. Natl. Congr. Appl. Mech.,
  Berkeley, Calif. 1962, ed. by ASME, Vol.2 (ASME, New York 1962) pp.775-781
Srinath, L.S., Frocht, M.M.: "The Potentialities of the Method of Scattered
  Light", Proc. Int. Symp. Photoelasticity, Illinois Institute of Techno-
  logy, Chicago, 1961, ed. by M.M. Frocht (Pergamon Press, London 1963)
  pp.277-292
Sutliff, D.R., Pih, H.: Three-dimensional scattered-light stress analysis
  of discontinuous fiber-reinforced composites. Exp. Mech. *13*, 294-298 (1973)
Swinson, W.F.: "Theoretical and Scattered Photoelastic Solutions of Doubly
  Connected Circular Shafts of Varying Diameter", Doctoral Thesis, Univ. of
  Illinois, Urbana, Ill., Dept. Theor. Appl. Mech. (1964)
Swinson, W.F., Bowman, C.E.: Application of scattered-light photoelasticity
  to doubly connected tapered torsion bars. Exp. Mech. *6*, 297-305 (1966)
Takada, T., Kuramoto, M., Kunio, T., Kuno, H.: A feasibility study of scat-
  tered-light photoelasticity in the determination of the side pressure
  distribution of the pressed powder bed. Powder Techn. *14*, 51-60 (1976)
Taylor, C.E., Hemann, J.H.: A photoelastic study of stress waves in solids
  using scattered light. Proc. JSME Semi-Int. Symp., Jpn. 19-20 (1967)
Weller, R.: A new method for photoelasticity in three dimensions. J. Appl.
  Phys. *10*, 266 (1939)
Weller, R.: "Three-Dimensional Photoelastic Analysis by Scattered Light",
  Proc. 9th Semi-Annual Eastern Photoelasticity Conf. (1939)
Weller, R.: "A Photoelastic Analysis of Three-Dimensional Stress Systems
  Using Scattered Light", NACA Technical Note 737 (1938)
Weller, R.: Three-dimensional photoelasticity using scattered light. J.
  Appl. Phys. *12*, 610-616 (1941)
Weller, R.: "Photoelastic Analysis of Three-Dimensional Stress Problems",
  Proc. 13th Semi-Annual Eastern Photoelasticity Conf. (Mass. Institute
  of Technology, Cambridge, Mass. 1941) pp.119-125
Weller, R., Shortley, G.H.: Calculation of stresses within the boundary of
  photoelastic models. J. Appl. Mech. *6*, A71-78 (1939)
Weller, R., Bussey, J.K.: Photoelastic analysis of three-dimensional stress
  systems using scattered light. J. R. Aeronaut. Soc. *44*,74-88 (1940)

10. Interferometric Photoelasticity

Books

Coker, E.G., Filon, L.N.G.: *A Treatise on Photoelasticity*, 2nd ed. (Cambridge
  University Press, London 1957) pp.181-295
Frocht, M.M.: *Photoelasticity*, Vol.II (Wiley and Sons, New York 1948)
  pp.202-237
Hetényi, M.: *Handbook of Experimental Stress Analysis*, 1st ed. (Wiley and
  Sons, New York 1950) pp.906-912
Kuske, A., Robertson, G.: *Photoelastic Stress Analysis* (Wiley and Sons, New
  York 1974) pp.240-256

Papers

Antropius, K., Lind, N.C.: A note on the relation between absolute photoelastic coefficients for incompressible media. Acta Techn. ČSAV *16*,617-620 (1971)

Azuma, K., Soezima, Y.: Immersion technique for interferometric-photoelastic stress analysis. Jpn. J. Appl. Phys. *5*, 259-260 (1966)

Azuma, K., Soezima, Y.: Interferometric measurements of absolute stress-optical coefficients in amorphous polymers I. Direct determination method of absolute stress-optical coefficients, employing immersion technique. Jpn. J. Appl. Phys. *6*, 909-914 (1967)

Azuma, K., Soezima, Y.: Interferometric measurements of absolute stress-optical coefficients in amorphous polymers II. "Direct" and "Transverse" stress-optical coefficients in photoviscoelastic effects of polyester mixture. Jpn. J. Appl. Phys. *6*, 915-920 (1967)

Azuma, K., Yamada, T., Soezima, Y.: A new method of photoelastic stress analysis by multiple-beam interferometry. Jpn. J. Appl. Phys. *6*, 791-792 (1967)

Balas, J.: Some applications of experimental analysis of models and structures. Exp. Mech. *7*, 127-139 (1967)

Barker, L.M.: Laser interferometry in shock-wave research. Exp. Mech. *12*, 209-215 (1972)

Barker, L.M., Hollenbach, R.E.: Interferometer technique for measuring the dynamic mechanical properties of materials. Rev. Sci. Instrum. *36*, 1617-1620 (1965)

Baud, R.V.: Entwicklung und heutiger Stand der Photoelastizität und der Photoplastizität im Rahmen der Gesamt-Experimentalelastizität. Schweizer Arch. *4* (1) 1-16, 48-53 (1938)

Bohler, P.: On the complete determination of dynamic states of stress at the surface of nontrasparent models or prototypes. Exp. Mech. *8*, 527-528 (1968)

Bohler, P., Schumann, W.: On the complete determination of dynamic states of stress. Exp. Mech. *8*, 115-121 (1968)

Bohler, P., Schumann, W.: On a multiplication technique applied to very thin photoelastic coatings. Exp. Mech. *11*, 289-295 (1971)

Bouricius, G.M.B., Clifford, S.F.: An optical interferometer using polarization coding to obtain quadrature phase components. Rev. Sci. Instrum. *41*, 1800-1803 (1970)

Brahtz, J.H.A., Bruggeman, J.R.: "The Bureau of Reclamation Photoelastic Laboratory", Proc. 13th Semi-Annual Eastern Photoelasticity Conf. (Mass. Institute of Technology, Cambridge Mass. 1941) pp.67-89

Brahtz, J.H.A., Soehrens, J.E.: Direct optical measurement of individual principal stresses. J. Appl. Phys. *10*, 242-247 (1939)

Bubb, F.W.: A complete photoelastic instrument. J. Opt. Soc. Am. *30*, 297-298 (1940)

Drouven, G.: "Measurement of the Sum of the Principal Stresses in Plane Problems of Elasticity by Interference", Doctorate Thesis, Washington Univ., St. Louis, Mont. (1952)

Drucker, D.C.: Photoelastic separation of principal stresses by oblique incidence. J. Appl. Mech. *10*, A156-160 (1943)

Fabry, C.: Sur une nouvelle méthode pour l'étude expérimentale des tensions élastiques. C.R. Acad. Sci. *190*, 457-460 (1930)

Favre, H.: Sur une nouvelle méthode optique de détermination des tensions intérieures. Rev. Opt. *8*, 193-213, 241-261, 289-307 (1929)

Favre, H.: Etude théorique de l'influence des réflexions intérieures sur la marche d'un rayon lumineux traversant une lame transparente soumise à des forces agissant dans son plan moyen. Ing. Arch. *28*, 39-52 (1959)

Favre, H., Schumann, W., Stromer, E.: Ein photoelektrisch-interferometrisches Verfahren zur vollständigen Bestimmung von ebenen Spannungszuständen. Schweiz. Bauz. *78* (36) 577-584 (1960)

Favre, H., Schumann, W.: "A Photoelectric-Interferometric Method to Determine Separately the Principal Stresses in Two-Dimensional States and Possible Applications to Surface and Thermal Stresses", Proc. Int. Symp. on Photoelasticity, Illinois Institute of Technology, Chicago, 1961, ed. by M.M. Frocht (Pergamon Press, London 1963) pp.3-25

Frappier, E.: Etude d'un interféromètre pour la détermination des lignes isopachiques. An. Contraintes *2* (8) 29-36 (1957)

Frocht, M.M.: On the application of interference fringes to stress analysis. J. Franklin Inst. *216*, 73-89 (1933)

Frocht, M.M.: Isopachic stress patterns. J. Appl. Phys. *10*, 248-257 (1939)

Grigull, U., Rottenkolber, H.: Two-beam interferometer using a laser. J. Opt. Soc. Am. *57*, 149-155 (1967)

Kestens, J., Vandaele-Dossche, M., Van Geen, R.: "Exemples d'application de techniques expérimentales avec problèmes de dimensionnement en construction civile", Proc. 4th Int. Conf. Exp. Stress Analysis, Cambridge, 1970, ed. by M.L. Meyer (The Institution of Mechanical Engineers, London 1971) pp.3-12

Khayyat, F.A., Stanley, P.: The integrated relative retardation in a photoelastic cylinder with a radial temperature gradient. J. Phys. D: Appl. Phys. *10*, 1389-1396 (1977)

Khesin, G.L., Sakharov, V.N.: "Methods of Strain Measurement on the Surface of Concrete and Reinforced Concrete Constructions by Means of Photoelastic Coatings", Proc. 4th Int. Conf. Exp. Stress Analysis, Cambridge, 1970, ed. by M.L. Meyer (The Institution of Mechanical Engineers, London 1971) pp.47-57

Leonhardt, K.: Gütezahlen und Gütefunktionen für Strahlenteiler und praktische Berechnung von Jonesmatrizen in Zweistrahlinterferometern. Optik *36*, 529-546 (1972)

Maury, V.M.R.: "Application de nouvelles méthodes interférométriques aux milieux stratifiés", Proc. 4th Int. Conf. Exp. Stress Analysis, Cambridge, 1970, ed. by M.L. Meyer (The Institution of Mechanical Engineers, London 1971) pp.257-264

Mehrotra, C.L.: A new interpretation of interferometric fringe patterns. Int. J. Mech. Sci. *15*, 227-235 (1973)

Mesmer, G.: The interference screen method for isopachic patterns (Moiré Method). Proc. Soc. Exp. Stress Anal. *13*, 21-26 (1956)

Nisida, M., Saito, H.: A new interferometric method of two-dimensional stress analysis. Exp. Mech. *4*, 366-376 (1964)

Nisida, M., Saito, H.: Application of an interferometric method to studies of contact problems. Sci. Pap. Inst. Phys. Chem. Res., Tokyo *59*, 112-123 (1965)

Nisida, M., Saito, H.: Stress distributions in a semi-infinite plate due to a pin determined by interferometric method. Exp. Mech. *6*, 273-279 (1966)

Parks, V.J., Sanford, R.J.: On the role of material and optical properties in complete photoelastic analysis. Exp. Mech. *16*, 441-447 (1976)

Post, D.: Characteristics of the series interferometer. J. Opt. Soc. Am. *44*, 243-249 (1954)

Post, D.: Photoelastic stress analysis for an edge crack in a tensile field. Proc. Soc. Exp. Stress Anal. *12*, 99-116 (1954)

Post, D.: A new photoelastic interferometer suitable for static and dynamic measurements. Proc. Soc. Exp. Stress Anal. *12*, 191-202 (1954)

Post, D.: Photoelastic evaluation of individual principal stresses by large field absolute retardation measurements. Proc. Soc. Exp. Stress Anal. *13*, 119-132 (1956)

Post D.: "Study of the Complete Solution of Three-Dimensional Photoelasticity Problems by Interferometry", Ph. D. Thesis, University of Illinois, Urbana, Ill. (1957)

Post, D.: Multiple-beam fringe sharpening with the series interferometer. J. Opt. Soc. Am. 48, 309-312 (1958)

Post, D.: The generic nature of the absolute-retardation method of photoelasticity. Exp. Mech. 7, 233-241 (1967)

Post, D.: On the interferometric method of stress analysis in three-dimensional photoelasticity. Exp. Mech. 10, 172-174 (1970)

Post, D.: Holography and interferometry in photoelasticity. Exp. Mech. 12, 113-123 (1972)

Primak, W., Post, D.: Photoelastic constants of vitreous silica and its elastic coefficient of refractive index. J. Appl. Phys. 30, 779-788 (1959)

Saenz, A.W.: Determination of residual stresses of quenching origin in solid and concentric hollow cylinders from interferometric observations. J. Appl. Phys. 21, 962-965 (1950)

Sanford, R.J., Parks, V.J.: On the limitations of interferometric methods in three-dimensional photoelasticity. Exp. Mech. 13, 464-471 (1973)

Schumann, W., Haenggi, H.: Über einen Versuch zur vollständigen Bestimmung zeitlich veränderlicher ebener Spannungszustände mittels eines photoelektrisch-interferometrischen Verfahrens. Forsch. Ingenieurwes. 30, 78-85 (1964)

Sinclair, D.: "A New Method for the Determination of the Principal Stress Sum", Proc. 10th Semi-Annual Eastern Photoelasticity Conf. (1940) pp.1-7

Sinclair, D.: Interferometer method of plane stress analysis. J. Opt. Soc. Am. 30, 511-513 (1940)

Sinclair, D., Bucky, P.B.: Photoelasticity and its application to mine-pillar and tunnel problems. Trans. Am. Inst. Mining Metall Pet. Enging. 139, 224-252 (1940)

Solakian, A.G.: New developments in photoelasticity a purely optical method of stress analysis. J. Opt. Soc. Am. 22, 293-306 (1932)

Srinath, L.S., Mehrotra, C.L.: On the interferometric method of stress analysis. Exp. Mech. 10, 170-171 (1970)

Stell, W.H.: Two-beam interferometry. Prog. Opt. 5, 147-197 (1966)

Theocaris, P.S.: Diffused light interferometry for measurement of isopachics. J. Mech. Phys. Solids 11, 181-195 (1963)

Theocaris, P.S.: Discussion of "Application de nouvelles méthodes interférometriques aux milieux stratifiés", Proc. 4th Int. Conf. Exp. Stress Analysis, Cambridge 1970, ed. by M.L. Meyer (The Institution of Mechanical Engineers, London 1971) pp.279-283

Theocaris, P.S.: Dependence of the stress-optical coefficients on the mechanical and optical properties of polymers. J. Strain Anal. 8, 267-276 (1973)

Theocaris, P.S., Gdoutos, E.E.: An interferometric method for the direct evaluation of principal stresses in plane-stress fields. J. Phys. D: Appl. Phys. 7, 472-482 (1974)

Theocaris, P.S., Gdoutos, E.E.: A unified interpretation of interferometric and holographic fringe patterns in photoelasticity. J. Strain Anal. 13 (2) 95-102 (1978)

Wells, A.A., Post, D.: The dynamic stress distribution surrounding a running crack - a photoelastic analysis. Proc. Soc. Exp. Stress Anal. 16, 69-92 (1958)

11. Holographic Photoelasticity

Books

Brcic, V.: *Application of Holography and Hologram Interferometry to Photo-elasticity*, 2nd ed. (Springer, Berlin, Heidelberg, New York 1975)
Butters, J.N.: *Holography and Its Technology* (Peter Peregrinus, London 1971)
Collier, R.J., Burckhardt, C.B., Lin, L.H.: *Optical Holography* (Academic Press, New York 1971)
DeVelis, J.B., Reynolds, G.O.: *Theory and Applications of Holography* (Addison-Wesley, Reading, Mass. 1967)
Françon, M.: *Holography* (Academic Press, New York 1974)
Goodman, J.W.: *Introduction to Fourier Optics* (McGraw-Hill, New York 1968)
Jones, T.: *Holography: A Source Book to Information* (New Dimensional Studio, Nashville, TN, 1977)
Smith, H.M.: *Principles of Holography* (Wiley-Interscience, New York 1969)
Stroke, G.W.: *An Introduction to Coherent Optics and Holography*, 2nd ed. (Academic Press, New York 1969)

Papers

Abramson, N.: The holo-diagram. I: A practical device for making and eva-luating holograms. Appl. Opt. *8*, 1235-1240 (1969)
Abramson, N.: The holo-diagram. II: A practical device for information retrieval in hologram interferometry. Appl. Opt. *9*, 97-101 (1970)
Abramson, N.: The holo-diagram. III: A practical device for predicting fringe patterns in hologram interferometry. Appl. Opt. *9*, 2311-2320 (1970)
Abramson, N.: The holo-diagram. IV: A practical device for simulating fringe patterns in hologram interferometry. Appl. Opt. *10*, 2155-2161 (1971)
Abramson, N.: The holo-diagram. V: A device for practical interpreting of hologram interference fringes. Appl. Opt. *11*, 1143-1147 (1972)
Ajovalasit, A.: Holographic photoelasticity: Influences of inaccuracies of optical retarders on isochromatics and isopachics. J. Strain Anal. *9*, 52-60 (1974)
Ajovalasit, A.: A single hologram technique for the determination of absolute retardations in holographic photoelasticity. J. Strain Anal. *10*, 148-152 (1975)
Ajovalasit, A.: Discussion of "On the effect of quarter-wave-plate errors in stress-holo-interferometry". Exp. Mech. *15*, 429 (1975)
Ajovalasit, A., Bardi, A.: Holographic photoelasticity: Determination of absolute retardations by a single hologram. Exp. Mech. *16*, 273-275 (1976)
Ajovalasit, A., Carollo, S., Tschinke, M.: "Holographic Analysis of Thermal Deformations in A Bimetallic Cylindrical Joint", Proc. 5th Int. Conf. Exp. Stress Analysis, Udine, 1974, ed. by G. Bartolozzi (Tecnoprint-Pitagora, Bologna 1974) pp.4.59-4.65
Aleksandrov, E.B., Bonch-Bruevich, A.M.: Investigation of surface strains by the hologram technique. Soviet Phys.-Techn. Phys. *12*, 258-265 (1967)
Aleksoff, C.C.: Time average holography extended. Appl. Phys. Lett. *14*, 23-24 (1969)
Aprahamian, R., Evensen, D.A.: Applications of holography to dynamics: High-frequency vibrations of beams. J. Appl. Mech. *37*, 287-291 (1970)
Aprahamian, R., Evensen, D.A.: Applications of holography to dynamics: High-frequency vibrations of plates. J. Appl. Mech. *37*, 1083-1090 (1970)

Aprahamian, R., Evensen, D.A., Mixson, J.S.: Transient-response measurements using holographic techniques. Exp. Mech. *10*, 421-426 (1970)

Aprahamian, R., Evensen, D.A., Mixson, J.S., Wright, J.E.: Application of pulsed holographic interferometry to the measurement of propagating transverse waves in beams. Exp. Mech. *11*, 309-314 (1971)

Aprahamian, R., Evensen, D.A., Mixson, J.S., Jacoby, J.L.: Holographic study of propagating transverse waves in plates. Exp. Mech. *11*, 357-362 (1971)

Archbold, E., Ennos, A.E.: Observation of surface vibration modes by stroboscopic hologram interferometry. Nature (London) *217*, 942-943 (1968)

Archbold, E., Ennos, A.E.: "Techniques of Hologram Interferometry for Engineering Inspection and Vibration Analysis", Proc. Symp. on the Eng. Uses of Holography, Univ. of Strathclyde, Glasgow, 1968, ed. by E.R. Robertson, J.M. Harvey (Cambridge University Press, Cambridge 1970)pp.381-396

Assa, A., Betser, A.A.: The application of holographic multiplexing to record separate isopachic- and isochromatic-fringe patterns. Exp. Mech. *14*, 502-504 (1974)

Balas, J., Nasch, L., Szabo, V.: "Finite Element Method Using Data Obtained by Holographic Interferometry", Proc. 5th Int. Conf. Exp. Stress Analysis, Udine, 1974, ed. by G. Bartolozzi (Tecnoprint-Pitagora, Bologna 1974) pp.1.73-1.84

Barbato, G., Barbisio, E.: "An Analysis on Vibrational Modes and Frequencies of an Aircraft Wing Panel by Holographic Methods", Proc. 5th Conf. Exp. Stress Analysis, Udine, 1974, ed. by G. Bartolozzi (Tecnoprint-Pitagora, Bologna 1974) pp.4.66-4.71

Bazelaire, E.: "Photoélasticimétrie par Holographie et Polarimétrie", Thèse de Ing. D., Université de Paris VI (1972)

Boone, P.M.: Holographic determination of in-plane deformation. Opt. Technol. *2*, 94-98 (1970)

Boone, P.M.: "Surface Measurements Using Deformation Following Holograms", Proc. Symp. on the "Eng. Appl. of Holography", Los Angeles Calif. (1972) pp.167-175

Boone, P.M.: "Measurement of Displacement, Strain and Stress by Holography", Proc. Int. Conf. Eng. Uses Coherent Light, Univ. of Strathclyde, Glasgow (1975)

Boone, P.M., De Backer, L.C.: Determination of three orthogonal displacement components from one double-exposure hologram. Optik *37*, 61-81 (1973)

Boone, P., Verbiest, R.: Application of hologram interferometry to plate deformation and translation measurements. Opt. Acta *16*, 555-567 (1969)

Brandt, G.B., Taylor, L.H.: "Holographic Strain Analysis Using Spline Functions", Proc. Symp. on the "Eng. Appl. of Holography", Los Angeles, Calif. (1972) pp.123-131

Brinson, H.F.: An interpretation of inelastic birefringence. Exp. Mech. *11*, 467-471 (1971)

Brooks, R.E.: Low-angle holographic interferometry using tri-X pan film. Appl. Opt. *6*, 1418-1419 (1967)

Brooks, R.E., Heflinger, L.O., Wuerker, R.F.: Interferometry with a holographically reconstructed comparison beam. Appl. Phys. Lett. *7*, 248-249 (1965)

Bryngdahl, O.: Polarizing holography. J. Opt. Soc. Am. *57*, 545-546 (1967)

Bryngdahl, O.: Shearing interferometry by wavefront reconstruction. J. Opt. Soc. Am. *58*, 865-871 (1968)

Bryngdahl, O., Lohmann, A.W.: Interferograms are image holograms. J. Opt. Soc. Am. *58*, 141-142 (1968)

Burch, J.M.: Holographic measurement of displacement and strain-an introduction. J. Strain Anal. *9*, 1-3 (1974)

Butters, J.N.: "Application of Holography to Instrument Diaphragm Deformations and Associated Topics", Proc. Symp. on the Eng. Appl. of Holography (Cambridge University Press, Cambridge 1968) pp.151-169

Cadoret, G.: Nouvelle méthode permettant l'obtention expérimentale d'un moiré holographique du premier ordre. Matér. Constr. 9, 33-42 (1976)

Cadoret, G.: Sur l'obtention expérimentale d'un moiré holographique du premier ordre. C.R. Acad. Sci. Paris B 282, 227-230 (1976)

Cadoret, G.: "Contribution à l'Utilisation des Methodes de l'Optique Cohérente dans l'Analyse Expérimentale des Contraintes", Thèse de Ing. Docteur, Université Pierre et Marie Curie, Paris VI (1976)

Cascarano, F.: Computer simulation of isochromatic-isopachic fringes in photo-holoelasticity. Oakland Univ. Grad. Res. Rpt. School of Eng. (1973)

Cernosek, J., McDonald, S.: On the effect of quarter-wave-plate errors in stress-holo-interferometry. Exp. Mech. 14, 403-407 (1974)

Chatelain, B.: "Photoêlasticimêtrie par Holographie: Étude Expérimentale de deux Méthodes d'Observation Simultanée ou Indépendante des Réseaux d'Iso-chromes ou d'Isopaches Relatifs a un Modèle Unique Soumis à une Seule Sollicitation", Thèse Docteur en Physique, Université de Besançon (1972)

Chatelain, B.: Holographic photo-elasticity: Independent observation of the isochromatic and isopachic fringes for a single model subjected to only one process. Opt. Lasers Technol. 5, 201-204 (1973)

Chau, H.H.M.: Holographic interferometer for isopachic stress analysis. Rev. Sci. Instrum. 39, 1789-1792 (1968)

Chau, H.H.M., Horman, M.H.: Demonstration of the application of wavefront reconstruction to interferometry. Appl. Opt. 5, 1237-1239 (1966)

Clark, J.A., Durelli, A.J.: A simple holographic interferometer for static and dynamic photomechanics. Exp. Mech. 10, 497-505 (1970)

Collins, M.C., Watterson, C.E.: Surface-strain measurements on a hemispherical shell using holographic interferometry. Exp. Mech. 15, 128-132 (1975)

Dandliker, R., Eliasson, B., Ineichen, B., Mottier, F.M.: "Quantitative Determination of Bending and Torsion Through Holographic Interferometry", Proc. Conf. on the Eng. Uses of Coherent Optics, Glasgow (1975)

Delarminat, P.M., Wei, R.P.: A fringe-compensation technique for stress analysis by reflection holographic interferometry. Exp. Mech. 16, 241-248 (1976)

Der, V.K., Holloway, D.C., Fourney, W.L.: Four exposure holographic moiré techniques. Appl. Opt. 12, 2552-2554 (1973)

Der, V.K., Holloway, D.C., Fourney, W.L.: A technique for reducing the fringe frequency in large-displacement holography. Exp. Mech. 14, 286-289 (1974)

Dhir, S.K., Peterson, H.A.: An application of holography to complete stress analysis of photoelastic models. Exp. Mech. 11, 560-564 (1971)

Dhir, S.K., Sikora, J.P.: Holographic dispacement measurements on a highly skewed propeller blade. NSRDC Rpt. 3680 (1971)

Dhir, S.K., Sikora, J.P.: An improved method for obtaining the general-displacement field from a holographic interferogram. Exp. Mech. 12, 323-327 (1972)

Dudderar, T.D.: Applications of holography to fracture mechanics. Exp. Mech. 9, 281-285 (1969)

Dudderar, T.D., Doerries, E.M.: A study of effective crack length using holographic interferometry. Exp. Mech. 16, 300-304 (1976)

Dudderar, T.D., Gorman, H.J.: The determination of mode I stress-intensity factors by holographic interferometry. Exp. Mech. 13, 145-149 (1973)

Dudderar, T.D., Koch, F.B., Doerries, E.M.: Measurement of the shapes of foil bulge-test samples. Exp. Mech. 17, 133-140 (1977)

Dudderar, T.D., O'Regan, R.: Measurement of the strain field near a crack tip in polymethylmethacrylate by holographic interferometry. Exp. Mech. *11*, 49-56 (1971)

Dudderar, T.D., O'Regan, R.: Holographic interferometry in materials research and fracture mechanics. Int. J. of Nondestr. Test. *4*, 119-147 (1972)

Ebbeni, J.: "Combinaison d'une Méthode de Moiré et d'une Méthode Holographique pour Déterminer l'Etat de Déformation d'un Object Diffusant", Proc. 5th Int. Conf. Exp. Stress Analysis, Udine, 1974, ed. by G. Bartolozzi (Tecnoprint-Pitagora, Bologna 1974) pp.4.20-4.25

Ebbeni, J., Coenen, J., Hermanne, A.: New analysis of holophotoelastic patterns and their application. J. Strain Anal. *11*, 11-17 (1976)

Ennos, A.E.: Holography and its applications. Contemporary Phys. *8*, 153-170 (1967)

Ennos, A.E.: Measurement of in-plane surface strain by hologram interferometry. J. Sci. Instrum. *1*, 731-734 (1968)

Fourney, M.E.: Application of holography to photoelasticity. Exp. Mech. *8*, 33-38 (1968)

Fourney, M.E.: "Advances in Holographic Photoelasticity", Applications of Holography in Mech., 1971, ed. by W.G. Gottenberg (1972) pp.17-38

Fourney, M.E.: Discussion of a paper entitled "The determination of mode I stress-intensity factors by holographic interferometry", ed. by T.D. Dudderar, H.J. Gorman. Exp. Mech. *13*, 145-149 (1973); *14*, 69-70 (1974)

Fourney, M.E., Mate, K.V.: Further applications of holography to photoelasticity. Exp. Mech. *10*, 177-186 (1970)

Fourney, M.E., Waggoner, A.P., Mate, K.V.: Recording polarization effects via holography. J. Opt. Soc. Am. *58*, 701-702 (1968)

Friesem, A.A., Vest, C.M.: Detection of micro-fractures by holographic interferometry. Appl. Opt. *8*, 1253-1254 (1969)

Froehly, C., Monneret, J., Pasteur, J., Viénot, J.C.: Etude des faibles déplacements d'objects opaques et de la distortion optique dans les lasers :à solide par interférométrie holographique. Opt. Acta *16*, 343-362 (1969)

Gabor, D.: A new microscopic principle. Nature *161*, 777-778 (1948)

Gabor, D.: Microscopy by reconstructed wave-fronts. Proc. R. Soc. London A *197*, 454-487 (1949)

Gabor, D.: Microscopy by reconstructed wave fronts: II. Proc. R. Soc. London B *64*, 449-469 (1951)

Gasvik, K.: Polarizing effects in holographic reconstruction. Optik *39*, 47-57 (1973)

Gasvik, K.: Holographic reconstruction of the state of polarization. Opt. Acta *22*, 189-206 (1975)

Gasvik, K.: Separation of the isochromatics-isopachics patterns by use of retarders in holographic photoelasticity. Exp. Mech. *16*, 146-150 (1976)

Gates, J.W.C.: Holographic measurement of surface distortion in three dimensions. Opt. Techn. *1*, 247-250 (1969)

Gottenberg, W.G.: Some applications of holographic interferometry. Exp. Mech. *8*, 405-410 (1968)

Haines, K.A., Hildebrand, B.P.: Contour generation by wavefront reconstruction. Phys. Lett. *19*, 10-11 (1965)

Haines, K.A., Hildebrand, B.P.: Surface-deformation measurement using the wavefront reconstruction technique. Appl. Opt. *5*, 595-602 (1966)

Haines, K.A., Hildebrand, B.P.: Interferometric measurements on diffuse surfaces by holographic techniques. IEEE Trans. *15*, 149-161 (1966)

Hazell, C.R., Liem, S.D.: Vibration analysis of plates by real-time strobo-scopic holography. Exp. Mech. *13*, 339-344 (1973)

Heflinger, L.O., Wuerker, R.F., Brooks, R.E.: Holographic interferometry.
  J. Appl. Phys. *37*, 642-649 (1966)
Hildebrand, B.P., Haines, K.A.: Multiple-wavelength and multiple-source
  holography applied to contour generation. J. Opt. Soc. Am. *57*, 155-162
  (1967)
Holloway, D.C.: Holography and its application to photoelasticity. T. and
  A.M. Rpt. No. 329, Univ. of Illinois, Urbana, Ill. (1969)
Holloway, D.C.: Simultaneous determination of the isopachic and isochromatic
  fringe patterns for dynamic loadings by holographic photoelasticity. T. and
  A.M. Rpt. No. 349, Univ. of Illinois, Urbana, Ill. (1971)
Holloway, D.C., Johnson, R.H.: Advancements in holographic photoelasticity.
  Exp. Mech. *11*, 57-63 (1971)
Holloway, D.C., Patacca, A.M., Fourney, W.L.: Application of holographic
  interferometry to a study of wave propagation in rock. Exp. Mech. *17*,
  281-289 (1977)
Holloway, D.C., Ranson, W.F., Taylor, C.E.: A neoteric interferometer for
  use in holographic photoelasticity. Exp. Mech. *12*, 461-465 (1972)
Horman, M.H.: An application of wavefront reconstruction to interferometry.
  Appl. Opt. *4*, 333-336 (1965)
Hosp, E., Wutzke, G.: Die Anwendung der Holographie in der ebenen Spannungs-
  optik. Materialprüfung *11*, 409-415 (1969)
Hosp, E., Wutzke, G.: Holographische Ermittlung der Hauptspannungen in
  ebenen Modellen. Materialprüfung *12*, 13-22 (1970)
Hovanesian, J.D.: Isochromatic and isopachic intensification in stress-holo-
  interferometry. Wayne State Univ. Research Rpt. (1967)
Hovanesian, J.D.: Interference of two and three reconstructed waves in photo-
  holoelasticity. U.S. Nav. J. Underwater Acoust. 563-570, October (1968)
Hovanesian, J.D.: "New Applications of Holography to Thermoelastic Studies",
  Proc. 4th Int. Conf. Exp. Stress Analysis, Cambridge, 1970, ed. by M.L.
  Meyer (The Institution of Mechanical Engineers, London 1971) pp.428-435
Hovanesian, J.D.: Elimination of isochromatics in photoholoelasticity.
  Strain *7*, 151-156 (1971)
Hovanesian, J.D.: Variable isochromatic/isopachic-fringe visibility in
  photoholoelasticity. Exp. Mech. *14*, 233-236 (1974)
Hovanesian, J.D.: "Recent Work in Absolute Measurement of Birefringence",
  in *Photoelastic Effect and Its Applications* (IUTAM Symposium), ed. by
  J. Kestens (Springer, Berlin, Heidelberg, New York 1975) pp.532-546
Hovanesian, J.D., Brcic, V., Powell, R.L.: A new experimental stress-optic
  method: Stress-holo-interferometry. Exp. Mech. *8*, 362-368 (1968)
Hovanesian, J.D., Haskell, R.E.: "Influence of Polarization Parameters on
  Isochromatics in Photoholoelasticity", Proc. Int. Symp. on Exp. Mech.
  in Research and Development, Univ. of Waterloo, Canada, 1972, ed. by
  Univ. of Waterloo, Vol.1 (Univ. of Waterloo, Waterloo, Ontario, Canada
  1972), pp.17/1-17/11
Hovanesian, J.D., Varner, J.: "Methods for Determining the Bending Moments
  in Normally Loaded Thin Plates by Hologram Interferometry", Proc. Symp. on
  the "Eng. Uses of Holography", Univ. of Strathclyde, Glasgow, 1968, ed. by
  E.R. Robertson and J.M. Harvey (Cambridge Univ. Press, Cambridge 1970)
  pp.173-184
Hovanesian, J.D., Zobel, E.C.: "Application of Photoholoelasticity in the
  Study of Crack Propagation", Proc. 7th Symp. "Non-Destructive Evaluation
  of Components and Materials in Aerospace, Weapons Systems and Nuclear
  Applications", San Antonio, Texas (1969)pp.1-9
Hsu, T.R.: Large-deformation measurements by real-time holographic inter-
  ferometry. Exp. Mech. *14*, 408-411 (1974)

Hsu, T.R., Lewak, R.: Measurements of thermal distortion of composite plates by holographic interferometry. Exp. Mech. *16*, 182-187 (1976)

Hsu, T.R., Moyer, R.G.: Application of holography in high-temperature displacement measurements. Exp. Mech. *12*, 431-432 (1972)

Hunter, A.R., Morton, T.M.: Application of holographic interferometry to predict long-time torsional relaxation. Exp. Mech. *15*, 153-160 (1975)

Johnson, C.D., Mayer, G.M.: Hologram interferometry as a practical vibration measurement technique. Shock Vib. Bull. *39* (2), 41-49 (1969)

Jones, R.: "The Application of Holographic Interferometry to the Measurement of the Elastic Constants of Solid Materials", Ph. D. Thesis, Univ. of Bradford (1972)

Jones, R., Bijl, D.: A holographic interferometric study of the end effects associated with the four-point bending technique for measuring poisson's ratio. J. Sci. Instrum. *7*, 357-358 (1974)

Khesin, G.L., Sakharov, V.N., Zhavoronok, I.V.: The use of holography in investigations of hydraulic equipment by the photoelastic method. Energetich Stroit *7*, 50-52 (1969)

Kihara, T., Kubo, H., Nagata, R.: Isopachics measurement using immersion method polarization holography. Appl. Opt. *15*, 3025-3028 (1976)

Kohler, H.: Interferenzliniendynamik bei der quantitativen Auswertung holographischer Interferogramme. Optik *47*, 9-24 (1977)

Kohler, H.: General formulation of the holographic-interferometric evaluation methods. Optik *47*, 469-475 (1977)

Kubo, H., Iwata, K., Nagata, R.: Photoelasticity using double-exposure polarization holography. Opt. Acta *22*, 59-70 (1975)

Kubo, H., Nagata, R.: Application of polarization holography by the Kurtz's method to photoelasticity. Jpn. J. Appl. Phys. *15*, 1095-1099 (1976)

Kubo, H., Nagata, R.: Further consideration of photoelasticity using polarization holography. A method in holographic photoelasticity. Opt. Acta *23*, 519-528 (1976)

Leith, E.N., Upatnieks, J.: Reconstructed wavefronts and communication theory. J. Opt. Soc. Am. *52*, 1123-1130 (1962)

Leith, E.N., Upatnieks, J.: Wavefront reconstruction with continuous-tone objects. J. Opt. Soc. Am. *53*, 1377-1381 (1963)

Leith, E.N., Upatnieks, J.: Wavefront reconstruction with diffused illumination and three-dimensional objects. J. Opt. Soc. Am. *54*, 1295-1301 (1964)

Lohmann, A.W.: Reconstruction of vectorial wavefronts. Appl. Opt. *4*, 1667-1668 (1965)

Loiseau, H.: "Progrés Récents Realisés en Photoélasticimétrie: Séparations des Contraintes Principales et Automatisation des Mesures", Proc. 4th Int. Conf. Exp. Stress Analysis, Cambridge, 1970, ed. by M.L. Meyer (The Institution of Mechanical Engineers, London 1971) pp.183-188

Loiseau, H., Nicolas, J.R., Robert, A.: Progrès récents en analyse des contraintes par photoélasticimétrie et interférométrie holographique. Récherche Aérospatiale *6*, 315-323 (1970)

Lurie, M., Zambuto, M.: A verification of holographic measurement of vibration. Appl. Opt. *7*, 2323-2325 (1968)

Luxmoore, A.R.: Holographic detection of cracks. J. Strain Anal. *9*, 50-51 (1974)

Luxmoore, A.R., House, C.: "In-Plane Strain Measurements by a Three Beam Holographic Method", Proc. Int. Symp. of Holography, Applications de l'Holographie, Besançon, 1970, ed. by J.Ch. Viénot et al., Paper 5.2

Manoranjan, De, Sevigny, L.: Polarization holography. J. Opt. Soc. Am. *57*, 110-111 (1967)

Mark, R., O'Regan, R.: Model interferometry with airholograms. Exp. Mech. *12*, 332-334 (1972)

Marom, E.: "Fatigue Detection Using Holographic Techniques", Proc. Symp. on the "Eng. Uses of Holography", Univ. of Strathclyde, Glasgow, 1968, ed. by E.R. Robertson, J.M. Harvey (Cambridge Univ. Press, Cambridge 1970) pp.237-247

Marom, E.: Real-time strain measurements by optical correlation. Appl. Opt. *9*, 1385-1391 (1970)

Marom, E., Mueller, R.K.: "Optical Strain Measurement", Proc. Int. Symp. of Holography, Applications de l'Holographie, Besançon, 1970, ed. by J.Ch. Viénot et al., Paper 5.5

Martin, D.J.V.: Holographic method giving stress levels and visualization of defects in thick cylinders. J. Strain Anal. *10*, 143-147 (1975)

Marwitz, H.: Die vollständige Auswertung, vor allem von räumlichen Spannungs-zuständen, mit Hilfe der holografischen Interferometrie. VDI Ber. *271*, 73-78 (1976)

Mate, K.V.: "Two New Applications of Holography to Photoelasticity", MS Thesis, Univ. of Washington (1968)

Mayer, G.M.: Vibration phase measurement by rotation-strobe holography. J. Appl. Phys. *40*, 2863-2866 (1969)

Meyer, M.D., Spetzler, H.A.: Material properties using holographic inter-ferometry. Exp. Mech. *16*, 434-438 (1976)

Molin, N.E., Stetson, K.A.: Measuring combination mode vibration patterns by hologram interferometry. J. Sci. Instrum. *2*, 609-612 (1969)

Monahan, M.A., Bromley, K.: Vibration analysis by holographic interferometry. J. Acoust. Soc. Am. *44*, 1225-1231 (1968)

Monneret, J., Spajer, M.: Photoélasticimétrie holographique: Séparation des contraintes dans des matériaux photoélastique; possibilités d'application aux modèles à déformations figées. Opt. Acta *24*, 843-857 (1977)

Murphy, C.G., Burchett, O.J., Matthews, C.W.: "Holometric Deformation Measurement on Carbon Carbon Biaxial Test Specimens", Proc. Symp. on the "Eng. Appl. of Holography", Los Angeles, Calif. (1972) pp.177-186

Neumann, D.B., Jacobson, C.F., Brown, G.M.: Holographic technique for de-termining the phase of vibrating objects. Appl. Opt. *9*, 1357-1362 (1970)

Neumann, D.B., Penn, R.C.: Off-table holography. Exp. Mech. *15*, 241-244 (1975)

Nicolas, J.: Sur la détermination de la somme des contraintes principales au moyen de l'interférométrie holographique. C.R. Acad. Sci. Paris A *267*, 371-373 (1968)

Nicolas, J.: "Contribution à la Détermination de la Somme des Contraintes Principales au Moyen de l'Holographie", Conf. au G.A.M.A.C., 1968, R.F.M. No. 28 (1969) pp.47-56

O'Regan, R., Dudderar, T.D.: A new holographic interferometer for stress analysis. Exp. Mech. *11*, 241-247 (1971)

Pastor, J., Evans, G.E., Harris, J.S.: Hologram-interferometry - a geometrical approach. Opt. Acta *17*, 81-96 (1970)

Pelzer-Bawin, G., De Lamotte, F.: Interprétation Géométrique de l'Holographie, Applications en Photoélasticimétrie. Sciences et Techniques de l'Armement, Mémorial de l'Artillerie française, No. 4 (1971)

Post, D.: Holography and interferometry in photoelasticity. Exp. Mech. *12*, 113-123 (1972)

Powell, R.L., Hovanesian, J.D., Brcic, V.: "Hologram Interferometry with Birefringent Objects", Proc. Symp. on the "Eng. Uses of Holography", Univ. of Strathclyde, Glasgow, 1968, ed. by E.R. Robertson, J.M. Harvey (Cambridge Univ. Press, Cambridge 1970) pp.201-224

Powell, R.L., Stetson, K.A.: Interferometric vibration analysis by wavefront reconstruction. J. Opt. Soc. Am. *55*, 1593-1598 (1965)

Pryor, T.R., Hageniers, O.L., North, W.P.T.: Diffractography vs. holography-
a stress analyst's comparison. Exp. Mech. *13*, 220-224 (1973)

Ranson, W.F., III: "Use of Holographic Interferometry to Determine the Surface Displacement Components of a Deformed Body", Ph. D. Thesis, Dept. of
Theoretical and Applied Mech., Univ. of Illinois, Urbana, Ill. (1971)

Robertson, E.R., King, W.: The technique of holographic interferometry applied
to the study of transient stresses. J. Strain Anal. *9*, 44-49 (1974)

Rogers, G.L.: Polarization effects in holography. J. Opt. Soc. Am. *56*, 831
(1966)

Rogers, G.L.: The equivalent interferometer in holography. Opt. Acta *17*,
527-538 (1970)

Rowe, S.H.: Projected interference fringes in holographic interferometry.
J. Opt. Soc. Am. *61*, 1599-1603 (1971)

Rowlands, R.E., Daniel, I.M.: Application of holography to anisotropic composite plates. Exp. Mech. *12*, 75-82 (1972)

Saito, H., Yamaguchi, I., Nakajima, T.: "Application of Holographic Interferometry to Mechanical Experiments", "Applications of Holography", Proc.
of the U.S.-Japan Seminar on Information Processing by Holography,
Washington, D.C. (1969) pp.105-126

Sampson, R.C.: Holographic-interferometry applications in experimental
mechanics. Exp. Mech. *10*, 313-320 (1970)

Sanford, R.J.: "A General Approach to the Analysis of Holographic Optical
Systems for Stress Analysis", Ph. D. Thesis, Catholic Univ. of America,
Washington, D.C. (1971)

Sanford, R.J.: Differential stress-holo-interferometry. Exp. Mech. *13*,
330-338 (1973)

Sanford, R.J., Durelli, A.J.: Interpretation of fringes in stress-holo-interferometry. Exp. Mech. *11*, 161-166 (1971)

Sanford, R.J., Parks, V.J.: Discussion of "An application of holography to
complete stress analysis of photoelastic models". Exp. Mech. *12*, 528-529
(1972)

Schumann, W.: Some aspects of the optical techniques for strain determination.
Exp. Mech. *13*, 225-231 (1973)

Schumann, W., Dubas, M.: On the motion of holographic images caused by movements of the reconstruction light source, with the aim of application to
deformation analysis. Optik *46*, 377-392 (1976)

Schumann, W., Dubas, M.: On the holographic interferometry used for deformation analysis, with one fixed and one movable reconstruction source.
Optik *47*, 391-404 (1977)

Sciammarella, C.A.: "Holographic Interferometry Analyzed from the Point of
View of Moiré Patterns", Proc. Int. Symp. on Exp. Mech. in Research and
Development, Univ. of Waterloo, Canada, 1972, ed. by Univ. of Waterloo,
Vol. 1 (Univ. of Waterloo, Waterloo, Ontario, Canada 1972) pp.20/1-20/18

Sciammarella, C.A., Chang, T.Y.: Optical differentiation of the displacement
patterns using shearing interferometry by wavefront reconstruction. Exp.
Mech. *11*, 97-104 (1971)

Sciammarella, C.A., Chang, T.Y.: Holographic interferometry applied to the
solution of a shell problem. Exp. Mech. *14*, 217-224 (1974)

Sciammarella, C.A., Gilbert, J.A.: A holographic-moiré technique to obtain
separate patterns for components of displacement. Exp. Mech. *16*, 215-220
(1976)

Sciammarella, C.A., Nyuko, H.: "Determination of Strains by Applying Holographic
Shearing Interferometry to Techniques that Provide Displacement Information",
Prog. Exp. Mech., Durelli Anniversary Volume, ed. by V.J. Parks (Catholic
Univ. of America, Washington, D.C. 1975) pp.1-9

Sciammarella, C.A., Quintanilla, G.: Techniques for the determination of absolute retardation in photoelasticity. Exp. Mech. 12, 57-66 (1972)

Shibayama, K., Uchiyama, H.: Measurement of three-dimensional displacements
by hologram interferometry. Appl. Opt. 10, 2150-2154 (1971)

Sikora, J.P., Mendenhall, F.T.: Holographic vibration study of a rotating
propeller blade. Exp. Mech. 14, 230-232 (1974)

Sirohi, R.S., Krishnamurthy, R.: Stress analysis with hologram interferometry using birefringent coating on object. Optik 40, 315-321 (1974)

Smigielski, P.: Holographie des objects de phase. Etude de quelques méthodes
d'exploitation de l'onde restituêe. Opt. Acta 18, 483-506 (1971)

Sollid, J.E.: Holographic interferometry applied to measurements of small
static displacements of diffusely reflecting surfaces. Appl. Opt. 8, 1587-
1595 (1969)

Sollid, J.E.: Translational displacements versus deformation displacements
in double-exposure holographic interferometry. Opt. Commun. 2, 282-288
(1970)

Stetson, K.A.: A rigorous treatment of the fringes of hologram interferometry.
Optik 29, 386-400 (1969)

Stetson, K.A.: Effects of beam modulation on fringe loci and localization
in time-average hologram interferometry. J. Opt. Soc. Am. 60, 1378-1388
(1970)

Stetson, K.A.: Moirê method for determining bending moments from hologram
interferometry. Opt. Tech. 2, 80-84 (1970)

Stetson, K.A., Powell, R.L.: Interferometric hologram evaluation and real-
time vibration analysis of diffuse objects. J. Opt. Soc. Am. 55, 1694-1695
(1965)

Stetson, K.A., Powell, R.L.: Hologram interferometry. J. Opt. Soc. Am. 55,
1570 (1965); 56, 1161-1166 (1966)

Stroke, G.W., Funkhouser, A., Leonard, C., Indebetouw, G., Zech, R.G.:
Hand-held holography. J. Opt. Soc. Am. 57, 110 (1967)

Surget, J.: Application de l'interférométrie holographique à l'étude des
déformations des corps transparents. Recherche Aérosp. 3, 167-174 (1970)

Taylor, L.H., Brandt, G.B.: An error analysis of holographic strains determined by cubic splines. Exp. Mech. 12, 543-548 (1972)

Theocaris, P.S.: On a geometric interpretation of holography. Techn. Ann.
10, 861-868 (1971) (in Greek)

Theocaris, P.S., Gdoutos, E.E.: A unified interpretation of interferometric
and holographic fringe patterns in photoelasticity. J. Strain Anal. 13
(2) 95-102 (1978)

Tselikov, A.I., Morosov, B.A., Makeev, I.M., Sergeev, A.W., Surkov, A.I.:
"Study of Metal Plastic Working Processes and of Deformations of Metall-
urgical Machine Components with the Application of a Holographic Interferometry Technique", Proc. 5th Int. Conf. Exp. Stress Analysis, Udine,
1974, ed. by G. Bartolozzi (Tecnoprint-Pitagora, Bologna 1974) pp.4.72-4.79

Tsujiuchi, J., Takeya, N., Matsuda, K.: Mesure de la déformation d'un
object par interférométrie holographique. Opt. Acta 16, 709-722 (1969)

Tsuruta, T., Shiotake, N., Tsujiuchi, J., Matsuda, K.: Holographic generation
of contour map of diffusely reflecting surface by using immersion method.
Jpn. J. Appl. Phys. 6, 661-662 (1967)

Tsuruta, T.: "Application of Classical Theory of Interferometry to Holography", Proc. of the U.S.-Japan Seminar on Information Processing by
Holography, Washington, D.C. (1969) pp.79-104

Tsuruta, T., Itoh, Y.: Holographic interferometry for rotating subject. Appl.
Phys. Lett. 17, 85-86 (1970)

Tsuruta, T., Shiotake, N., Itoh, Y.: Formation and localisation of holographically produced interference fringes. Opt. Acta 16, 723-733 (1969)

Tuschak, P.A., Allaire, R.A.: Axisymmetric vibrations of a cylindrical re-
sonator measured by holographic interferometry. Exp. Mech. *15*, 81-88 (1975)
Uozato, H., Nagata, R.: Holographic photoelasticity by using dual-hologram
method. Jpn. J. Appl. Phys. *16*, 95-100 (1977)
Viénot, J.C., Froehly, C., Monneret, J., Pasteur, J.: "Hologram Interfero-
metry: Surface Displacement Fringe Analysis as an Approach to the Study of
Mechanical Strains and other Applications to the Determination of Aniso-
tropy in Transparent Objects", Proc. Symp. on the "Eng. Uses of Holography",
University of Strathclyde, Glasgow, 1968, ed. by E.R. Robertson, J.M.
Harvey (Cambridge Univ. Press, Cambridge 1970) pp.133-150
Viénot, J.C., Monneret, J.: Interférométrie et photoélasticimétrie holo-
graphiques. Rev. Opt. *46*, 75-79 (1967)
Vikram, C.S.: Holographic interferometry of superposition of motions with
different time functions. Optik *45*, 55-64 (1976)
Wallach, J., Holeman, J.M., Passanti, F.A.: "Holographic Strain Measurement
on a Tensile Specimen", Proc. Symp. on the "Engineering Applications of
Holography", Los Angeles, Calif. (1972) pp.167-175
Walles, S.: Visibility and localization of fringes in holographic interfero-
metry of diffusely reflecting surfaces. Arkiv Fys. *40*, 299-403 (1969)
Walles, S.: On the concept of homologous rays in holographic interferometry
of diffusely reflecting surfaces. Opt. Acta *17*, 899-913 (1970)
Wetzels, W.: Holographie als Hilfsmittel zur Isopachenbestimmung. Optik *27*,
271-273 (1968)
Wilson, A.D.: In-plane displacement of a stressed membrane with a hole
measured by holographic interferometry. Appl. Opt. *10*, 908-912 (1971)
Wilson, A.D., Lee, C.H., Lominac, H.R., Strope, D.H.: Holographic and analytic study of a semiclamped rectangular plate supported by struts. Exp.
Mech. *11*, 229-234 (1971)
Yamaguchi, I., Saito, H.: Application of holographic interferometry to the
measurement of poisson's ratio. Jpn. J. Appl. Phys. *8*, 768-771 (1969)
Zaidel, A.N., Malkhasyan, L.G., Markova, G.V., Ostrovskii, Y.I.: Holographic
strobe method for studying vibrations. Soviet Phys.-Techn. Phys. *13*,
1470-1473 (1969)

12. The Method of Birefringent Coatings

Books

Dally, J.W., Riley, W.F.: *Experimental Stress Analysis* (McGraw-Hill, New
York 1965) Chap.11, pp.280-295
Durelli, A.J., Riley, W.F.: *Introduction to Photomechanics* (Prentice-Hall,
Englewood Cliffs, N.J. 1965) Chap.9, pp.247-253
Heywood, R.B.: *Photoelasticity for Designers* (Pergamon Press, Oxford
1969) Chap.10, pp.277-293
Holister, G.S.: *Experimental Stress Analysis* (Cambridge University Press,
Cambridge 1967) Chap.5, pp.210-238
Javornicky, J.: *Photoelasticity*, Pt.4 (Czechoslovak Academy Sci., Prague
1974) Chap.15-16, pp.262-280
Kuske, A., Robertson, G.: *Photoelastic Stress Analysis* (Wiley and Sons,
New York 1974) Chap.11, pp.263-274
McMaster, R.C.: *Nondestructive Testing Handbook*, Vol.2 (Ronald Press, New
York 1959)

Zandman, F., Redner, S., Dally, J.W.: *Photoelastic Coatings*, Soc. Exp. Stress Anal. (Iowa State University Press, Ames 1977)

Papers

Abeles, P.W.: Cracking and bond resistance in high strength reinforced concrete beams, illustrated by photoelastic coating. J. Am. Concr. Inst. *63*, 1265-1278 (1966)

Adams, C., Beese, J.G.: The application of birefringent coatings to metal surfaces which are to undergo large plastic strains. Exp. Mech. *14*, 202-203 (1974)

Akhmetzyanov, M.K.: Photoelastic coatings for determination of stresses and deformations in flexible plates and shells. NASA TN F-220 (1964)

Aleksandrov, A.Y., Akhmetzyanov, M.K.: Investigation of strains and stresses by the method of photoelastic coatings. Ind. Lab. (U.S.S.R.) *42*, 1768-1778 (1976)

Amba-Rao, C.L.: Study of birefringent coatings treated as thick plates. J. Franklin Inst. *290*, 409-418 (1970)

Andreyev, V.: Discussion of "The application of birefringent coatings to metal surfaces which are to undergo large plastic strains". Exp. Mech. *15*, 394-395 (1975)

Austin, A.L.: Measurements of thermally induced stress waves in a thin rod using birefringent coatings. Exp. Mech. *5*, 1-10 (1965)

Barron, K.: Glass insert stressmeters. AIME, Trans. Soc. Min. *232*, 287-299 (1965)

Bettany, C.P., Fessler, H.: Birefringence behaviour of an epoxy resin in compression. Br. J. Appl. Phys. *14*, 692-695 (1963)

Blackburn, B.R.: "An Investigation into the Reinforcing Effects of Birefringent Coating on Plane Structures", M. SC. Thesis, Dept. of Eng. Mech., Univ. of Texas, Austin, Texas (1967)

Blum, A.E.: The use and understanding of photoelastic coatings. Strain *13*, 96-101 (1977)

Bogdyl', P.T., Bonchenko, G.A.: Examination of the development of local elastic plastic strains in butt welded joints by means of the method of photoelastic coatings. Weld. Prod. U.S.S.R. *23*, 18-21 (1976)

Bohler, P., Schumann, W.: On a multiplication technique applied to very thin photoelastic coatings. Exp. Mech. *11*, 289-295 (1971)

Bonvalet, M.: Application de la microphotoélasticimètrie à l'étude de la déformation plastique. C.R. Acad. Sci. *140* (2), 157-158 (1955)

Bynum, D.J., Lemcoe, M.M.: Birefringent-coating analysis of laterally loaded perforated plates. Exp. Mech. *3*, 140-147 (1963)

Calcote, L.R., Bowman, C.E.: Experimental determination of the elastic-plastic boundary. Exp. Mech. *5*, 262-266 (1965)

Chen, W.T., Duffy, J.: Effects of the thickness of birefringent coatings for two-dimensional variations in surface strain. U.S. Army Reseach Office (Durham), DA Proj. No. 59901004, Technical Rpt. No. 2 (Feb. 1962)

Clyne, G., Fessler, H., Wilson, R.W.: Improvements of photoelastic technique for strain measurement on flat surfaces. Brit. J. Appl. Phys. *12*, 8-10 (1961)

Cole, C.A., Quinlan, J.F., Zandman, F.: "The Use of High-Speed Photography and Photoelastic Coatings for the Determination of Dynamic Strains", Proc. 5th Int. Congr. High-Speed Photography, Washington D.C., 1961, ed. by J.S. Courtney-Pratt (SMPTE, New York 1962) pp. 250-261

D'Agostino, J., Drucker, D.C., Liu, C.K., Mylonas, C.: An analysis of plastic behavior of metals with bonded birefringent plastic. Proc. Soc. Exp. Stress Anal. *12*, 115-122 (1955)

D'Agostino, J., Drucker, D.C., Liu, C.K., Mylonas, C.: Epoxy adhesives and casting resins as photoelastic plastics. Proc. Soc. Exp. Stress Anal. *12*, 123-128 (1955)

Dally, J.W., Alfirevich, I.: Application of birefringent coatings to glass-fiber-reinforced plastics. Exp. Mech. *9*, 97-102 (1969)

Dantu, P.: Etude des contraintes dans les milieux hêterogênes. Application au beton. Lab. Central des Ponts et Chaussées, Paris, Publ. No. 57-6 (1957)

Day, E.E., Kobayashi, A.S., Larson, C.N.: Fringe multiplication and thickness effects in birefringent coating. Exp. Mech. *2*, 115-121 (1962)

Dixon, J.R.: Elastic-plastic strain distribution in flat bars containing holes or notches. J. Mech. Phys. Solids *10*, 253-263 (1962)

Dixon, J.R.: Photoelastic surface coating technique. NEL Rpt. 277 (1967)

Duffy, J.: Effects of the thickness of birefringent coatings. Exp. Mech. *1*, 74-82 (1961)

Duffy, J., Lee, T.C.: Measurement of surface strain by means of bonded birefringent strips. Exp. Mech. *1*, 109-112 (1961)

Duffy, J., Mylonas, C.: "An Experimental Study on the Effects of the Thickness of Birefringent Coatings", Proc. Int. Symp. on Photoelasticity, Illinois Institute of Technology, Chicago, 1961, ed. by M.M. Frocht (Pergamon Press, London 1963) pp.27-42

Duffy, J., Mylonas, C.: Preliminary tests on aluminum specimens with birefringent coating permanently deformed by impact. Tech. Rpt. No. 22, Contr ct DA-020-ORD-798, Div. of Appl. Math., Brown Univ. (1955)

Fessler, H., Haines, D.J.: The photoelastic technique for strain measurements on flat aluminium alloy surfaces. Br. J. Appl. Phys. *9*, 278-279, 282-287 (1958)

Fessler, H., Haines, D.J.: Plasto-elastic stress distributions in lugs. Aeronaut. Q. *10*, 230-246 (1959)

Fessler, H., Newton, E.A.A., Wilson, R.W.: On the effect of layer thickness in photoelastic reflection techniques. Br. J. Appl. Phys. *14*, 889-893 (1963)

Fessler, H., Wilson, R.W.: Plastic-elastic strains observed near loaded fillets in stepped bars. J. Mech. Eng. Sci. *6*, 9-12 (1964)

Fleury, R., Zandman, F.: Jauge d'efforts photoêlastique. C.R. Acad. Sci. *238*, 1559-1561 (1954)

Franz, G.: Unmittelbare Spannungsmessung in Beton und Baugrund. Bauingenieur *33*, 190-195 (1958)

Furue, H., Shimamura, S.: A study on the photoelastic coating method (I) - on the ringlight polariscope. J. Mech. Eng. Lab. *30*, 299-305 (1976)

Galster, D.: Spannungsoptische Bestimmung von Dehnungen mit profilierten Oberflächenschichten. Beitr. zur Spannungs- und Dehnungsanalyse *4*, 47-88 (1967)

Gerberich, W.W.: Stress distribution about a slowly growing crack determined by photoelastic-coating method. Exp. Mech. *2*, 359-365 (1962)

Gerberich, W.W.: Plastic strains and energy density in cracked plates, Pt. I. Experimental technique and results. Exp. Mech. *4*, 335-344 (1964)

Gibbs, H.G., Hooke, C.J., Stagg, J.J.: An application of photoelastic gauges to the measurement of residual stress. VDI-Ber. Ver. Dtsch. Ing. *102*, 113-117 (1966)

Golubovic, G.: Different applications of photoelastic layers in the structural field. RAE Trans. No. 838 (1959) pp. 10

Hawkes, I.: Stress evaluation in low-modulus and viscoelastic materials using photoelastic glass inclusions. Exp. Mech. *9*, 58-66 (1969)

Hawkes, I.: Theory of the determination of the greatest principal stress in a biaxial stress field using photoelastic hollow cylinder inclusions. Int. J. Rock Mech. Min. Sci. *6*, 143-158 (1969)

Hawkes, I., Foley, J.H.: Elastic-plastic strain distributions in short-link chain under load. J. Strain Anal. *3*, 11-16 (1968)

Hizamatsu, J., Niwa, J., Oka, J.: Measurement of stress in field by application of photoelasticity. Techn. Rpts., Eng. Res. Inst., Kyoto Univ. *37* (1957)

Hoek, E.: "Experimental Study of Rock-Stress Problems in Deep Level Mining", Proc. 1st Int. Congr. Exp. Mech., New York 1963, ed. by B.E. Rossi (Pergamon Press, New York 1963) pp.177-194

Hoek, E., Bieniawski, Z.T.: Application of the photoelastic coating technique to the study of the stress redistribution associated with plastic flow around notches. South Afr. Mech. Eng. *12*, 275-282 (1963)

Holister, G.S.: Recent developments in photoelastic coating techniques. J. R. Aeronaut. Soc. *65*, 661-669 (1961)

Holister, G.S.: Photoelastic strain gauges. Appl. Mater. Res. *2*, 20-30 (1963)

Hooke, C.J., Stagg, J.J.: An approximate solution for the effect of "Shear Lag" in the measurements of residual stresses using a photoelastic gauge. Int. J. Solids Struc. *4*, 139-157 (1968)

Hovanesian, J.D.: Sign determination in oblique incidence for photoelastic coatings. Exp. Mech. *5*, 128 (1965)

Hovanesian, J.D.: New method for the determination of oblique-incidence coefficients for birefringent coatings and models. Exp. Mech. *6*, 500-501 (1966)

Hovanesian, J.D., Eggenberger, G., Hung, Y.Y.: Full-field isopachic generation in photoelastic coatings. Exp. Mech. *12*, 196-200 (1972)

Kaplan, M.F.: Strains and stresses of concrete at initiation of cracking and near failure. J. Am. Concr. Inst. *60*, 853-879 (1963)

Kawata, K.: Analysis of elasto-plastic behaviour of metals by means of photo-elastic coating method. J. Sci. Res. Inst., Tokyo *52*, 17-40 (1958)

Kawata, K.: "Studies on Photoplastic Analysis of Elasto-Plastic Behaviour of Metals with Birefringent Coating", Proc. 7th Japan Nat. Congr. Appl. Mech., Tokyo (1958) pp.148-152

Kawata, K.: "Elastoplastic Stress Analysis and Determination of Flow Limit by Means of Photoelastic Coating Method", Proc. Int. Symp. on Photoelasticity, Illinois Institute of Technology, Chicago, 1961, ed. by M.M. Frocht (Pergamon Press, London 1963) pp.219-230

Kedward, K.T., Hindle, G.R.: Analysis of strain in fibre-reinforced materials. J. Strain Anal. *5*, 309-315 (1970)

Keeton, J.R.: Strain distribution in compressively loaded concrete specimens. Proc. Am. Soc. Test. Mater. *61*, 1197-1220 (1961)

Khesin, G.L., Sakharov, V.N.: "Methods of Strain Measurement on the Surface of Concrete and Reinforced Concrete Constructions by Means of Photoelastic Coatings", Proc. 4th Int. Conf. Exp. Stress Analysis, Cambridge, 1970, ed. by M.L. Meyer (The Institution of Mechanical Engineers, London 1971) pp.47-57

Kumazawa, T.: Photoelastic technique to investigate the effect of temperature and strain rate on plastic flow in mild steel. Exp. Mech. *17*, 271-275 (1977)

Lambert, T.H., Snell, C.: Effect of yield on the interference between a pin and a plate. J. Mech. Eng. Sci. *6*, 38-43 (1964)

Lee, T.C., Mylonas, C., Duffy, J.: Thickness effects in birefringent coatings with radial symmetry. Exp. Mech. *1*, 134-142 (1961)

Linge, J.R.: Photoelastic measurement of surface strain. Aircraft Eng. *32* (378), 216-221; (379), 261-270; (380), 295-298 (1960)

Mabboux, G.: Applications de la photoélasticimétrie à l'étude des ouvrages en beton. Rev. Opt. Théor. Instrum. *11*, 501-507 (1932)

McIver, R.W.: Structural-test applications utilizing large continuous photo-elastic coatings. Exp. Mech. *5*, A19-25, A19-26 (1965)

Merskon, J.L.: Preliminary evaluation of PVRC photoelastic test data on reinforced openings vessels. Weld. Res. Counc. Bull. Eng. Foundation *113*, 53-70 (1966)

Mesnager, A.: Sur la détermination optique des tensions intérieures dans les solides à trois dimensions. C. R. Acad. Sci. *190*, 1249-1250 (1930)

Mönch, E.: Die vollständige Bestimmung des Dehnungszustandes auf Oberflächen durch photoelastische Streifenschichten. Schweiz. Bauz. *84*, 840-842 (1966)

Navaratnarajah, V.: A photoelastic method of detecting microcracking in concrete. J. Sci. Instrum. *42*, 276-277 (1965)

Nickola, W.E.: Photoelastic coatings on flat rotating axisymmetrical parts. Exp. Mech. *4*, 99-109 (1964)

Nisida, M., Takabayashi, H.: Thickness effects in "Hole Method" and applications of the method to residual stress measurement. Sci. Pap. Inst. Phys. Chem. Res. Jpn. *59*, 78-86 (1965)

Oppel, G.U.: Das polarisationsoptische Schichtverfahren zur Messung der Oberflächenspannung am beanspruchten Bauteil ohne Modell. Z. VDI *81*, 803-804 (1937)

Oppel, G.U.: Unmittelbar anzeigende polarisationsoptische Messelemente für die Dehnungs- und Spannungsanalyse an Bauteilen. Z. VDI *101*, 809-816 (1959)

Oppel, G.U.: Photoelastic strain gages. Exp. Mech. *1*, 65-73 (1961)

Oppel, G.U., Hill, P.W.: Strain measurements at the root of cracks and notches. Exp. Mech. *4*, 206-211 (1964)

O'Regan, R.: New method for determining strain on the surface of a body with photoelastic coatings. Exp. Mech. *5*, 241-246 (1965)

Palmer, R.S.J., Jarvis, J.L.: Bonding of birefringent coatings to composite flat-belt-drive materials. Exp. Mech. *17*, 198-200 (1977)

Papirno, R., Becker, H.: A multilayer-reflection technique for three-dimensional photoelastic studies of perforated plates in bending. Exp. Mech. *7*, 361-371 (1967)

Peng, S.S.: A photoelastic coating technique for rock fracture analysis. Int. J. Rock Mech. Min. Sci. *13*, 173-176 (1976)

Pih, H., Knight, C.E.: Photoelastic analysis of anisotropic fiber reinforced composites. J. Comp. Mater. *3*, 94-107 (1969)

Pipes, R.B., Dalley, J.W.: On the birefringent-coating method of stress analysis for fiber-reinforced laminated composites. Exp. Mech. *12*, 272-277 (1972)

Post, D., Zandman, F.: Accuracy of birefringent-coating method for coatings of arbitrary thickness. Exp. Mech. *1*, 21-32 (1961)

Redner, S.S.: New oblique-incidence method for direct photoelastic measurement of principal strains. Exp. Mech. *3*, 67-72 (1963)

Reitblat, Z.V., Baikov, V.V.: The photoelastic coating method and its application in investigations of steel-reinforced concrete structures. Sb. Tr. Nauchno-Issled. Inst. Poligr. Machinost. (1964)

Roberts, A., Emery, C.L., Chakravarty, P.K.: Photoelastic coating technique applied to research in rock mechanics. Trans. Am. Inst. Mining Metall. Pet. Eng. *71*, 581-617 (1961-1962)

Roberts, A., Hawkes, I., Williams, F.T.: Some field applications of the photoelastic stressmeter. Int. J. Rock Mech. Min. Sci. *2*, 93-103 (1965)

Roberts, A., Hawkes, I., Williams, F.T., Dhir, S.K.: A laboratory study of the photoelastic stressmeter. Int. J. Rock Mech. Min. Sci. *1*, 441-457 (1964)

Rowlands, R.E., Daniel, I.M., Whiteside, J.B.: Stress and failure analysis of a glass-epoxy composite plate with a circular hole. Exp. Mech. *13*, 31-37 (1973)

Schumann, W., Mylonas, C., Bucci, R.: The separation of membrane and bending
shears in shell with two birefringent coatings. J. Appl. Mech. *38*, 217-226
(1971)

Schwaighofer, J.: Applications of photoelastic strain gages. Exp. Mech. *1*,
198-202 (1961)

Schwaighofer, J.: Extended frozen stress method. Proc. Am. Soc. Civ. Eng. *88*,
(EM6), 1-12 (1962)

Sirohi, R.S., Krishnamurthy, R.: Stress analysis with hologram interferometry
using birefringent coating on object. Optik *40*, 315-321 (1974)

Stephen, R.M., Pirtz, D.: Application of birefringent coating to the study
of strains around circular inclusions in mortar prisms. Exp. Mech. *3*,
91-97 (1963)

Swamy, R.N.: "Application of Photoelastic Coating Techniques to the Determination of Internal Strain Distribution in Cementitious Materials", Proc.
4th Int. Conf. Exp. Stress Analysis, Cambridge, 1970, ed. by M.L. Meyer
(The Institution of Mechanical Engineers, London 1971) pp.58-67

Swedlow, J.L., Gerberich, W.W.: Plastic strains and energy density in cracked
plates, Pt. II. Comparison with elastic theory. Exp. Mech. *4*, 345-351
(1964)

Theocaris, P.S.: Experimental solution of elastic-plastic plane-stress problems. J. Appl. Mech. *29*, 735-743 (1962); *30*, 631-634 (1963)

Theocaris, P.S.: Combined photoelastic and electrical-analog method for
solution of plane-stress plasticity problems. Exp. Mech. *3*, 207-214 (1963)

Theocaris, P.S.: Discussion of "Extended frozen stress method". Proc. Am.
Soc. Civ. Eng. *89*, (EM4), 73-77 (1963)

Theocaris, P.S.: The effect of plasticity on the stress-distribution of thin
notched plates in tension. J. Franklin Inst. *279*, 22-38 (1965)

Theocaris, P.S.: Discussion of "Methods of Strain Measurement on the Surface
of Concrete and Reinforced Concrete Constructions by Means of Photoelastic
Coatings", Proc. 4th Int. Conf. Exp. Stress Analysis, Cambridge, 1970, ed.
by M.L. Meyer (The Institution of Mechanical Engineers, London 1971)
pp.71-73

Theocaris, P.S.: Discussion of "Application of Photoelastic Coating Techniques
to the Determination of Internal Strain Distribution in Cementitious
Materials", Proc. 4th Int. Conf. Exp. Stress Analysis, Cambridge, 1970,
ed. by M.L. Meyer (The Institution of Mechanical Engineers, London 1971)
pp.73-75

Theocaris, P.S., Dafermos, K.: The elastic strip under mixed boundary
conditions. J. Appl. Mech. *31*, 714-716 (1964)

Theocaris, P.S., Dafermos, K.: A critical review of the thickness effect of
birefringent coatings. Exp. Mech. *4*, 271-276 (1964)

Theocaris, P.S., Marketos, E.: The elastic-plastic strain and stress distribution in notched plates under of plane-stress. J. Mech. Phys. Solids
11, 411-428 (1963)

Theocaris, P.S., Marketos, E.: Elastic-plastic analysis of perforated thin
strips of a strain-hardening material. J. Mech. Phys. Solids *12*, 377-390
(1964)

Theocaris, P.S., Mylonas, C.: Viscoelastic effects in birefringent coatings.
J. Appl. Mech. *28*, 601-607 (1961); *29*, 599-603 (1962)

Timby, E.K., Hedrick, I.G.: Photoelastic analysis broadened. Eng. News-Rec.
121, 179-181 (1938)

Underwood, J.H.: Residual-stress measurement using surface displacements
around an indentation. Exp. Mech. *13*, 373-380 (1973)

Visser, W.: Theoretical analysis of the error due to layer thickness in the
photoelastic coating technique when measuring strain gradients. NEL Rpt.
PM 309, AB Div.2/61 (1961)

Watkins, D.J.: Feasibility study of the photoelastic coating technique for the direct measurement of principal strains for application in the deep-drawing of sheet metal. MEEP Rpt. UWIST Cardiff, England (1970)

Wilson, I.H., White, D.J.: Cruciform specimens for biaxial fatigue tests: An investigation using finite-element analysis and photoelastic-coating techniques. J. Strain Anal. *6*, 27-37 (1971)

Yew, C.H., Blackburn, B.R.: On Reinforcing effect of birefringent coatings on plate structures. Exp. Mech. *8*, 91-93 (1968)

Zandman, F.: Mesures photoélastiques des déformations élastiques et plastiques et des fragmentations cristallines dans les métaux. Rev. Métall. *53*, 638-644 (1956)

Zandman, F.: Analyse des contraintes par vernis photoélastiques. Groupement pour l'Avancement des Méthodes d'Analyse des Contraintes *2* (6), 3-14 (1956)

Zandman, F.: Photoelastic-coating technique for determining stress distribution in welded structures. Weld. Journal *39*, 191-198 (1960)

Zandman, F.: Stress analysis of a guided missile tail section with the photoelastic coating technique. Proc. Soc. Exp. Stress Anal. *17*, 135-150 (1960)

Zandman, F.: Maximum shear strain measurements and determination of initial yielding by use of photoelastic coating technique. ASTM STP 289 (1960)

Zandman, F.: Stress analysis with a photoelastic coating. Metal Prog. *78*, 111-117 (1960)

Zandman, F.: Discussion of "Thickness effects in birefringent coatings with radial symmetry". Exp. Mech. *1*, 141-142 (1961)

Zandman, F.: Concepts of the photoelastic stress gage. Exp. Mech. *2*, 225-233 (1962)

Zandman, F., Maier, H.N.: Six new techniques for photoelastic coatings. Prod. Eng. London *42* (1961)

Zandman, F., Redner, S.S., Riegner, E.I.: Reinforcing effect to birefringent coatings. Exp. Mech. *2*, 55-64 (1962)

Zandman, F., Redner, S.S., Post, D.: Photoelastic-coating analysis in thermal fields. Exp. Mech. *3*, 215-221 (1963)

Zandman, F., Watter, M., Redner, S.S.: Stress analysis of rocket-motor case by birefringent-coating method. Exp. Mech. *2*, 215-221 (1962)

Zandman, F., Wood, M.R.: Photostress. A new technique for photoelastic stress analysis for observing and measuring surface strains on actual structures and parts. Prod. Eng. London *27*, 167-178 (1956)

13. Graphical and Numerical Methods in Polarization Optics, Based on the Poincaré Sphere and the Jones Calculus

Books

Aben, H.K.: *Integrated Photoelasticity* (Valgus Tallin, USSR 1975)

Aubouin, J., Brousse, R., Lehman, J.-P.: *Précis de Géologie* (Dunod, Paris 1968) Chap.3, pp.74-83

Badgley, P.: *Structural Methods for the Exploration Geologist* (Peter C. Badgley, New York 1959) Chap.8, pp.187-242

Kuske, A., Robertson, G.: *Photoelastic Stress Analysis* (Wiley and Sons, New York 1974) Chap.13, pp.304-328

Papers

Cernosek, J.: Simple geometrical method for analysis of elliptical polarization. J. Opt. Soc. Am. *61*, 324-327 (1971)

Cernosek, J.: Towards the achromatic quarterwave plate. Exp. Mech. *13*, 83-85 (1973)

Cernosek, J.: On the effect of rotating secondary principal stresses in scattered-light photoelasticity. Exp. Mech. *13*, 273-279 (1973)

Cernosek, J.: On photoelastic response of composites. Exp. Mech. *15*, 354-357 (1975)

Cernosek, J.: New compensation method in photoelasticity. Exp. Mech. *16*, 263-266 (1976)

Kuske, A.: "Beiträge zur spannungsoptischen Untersuchung von Flächentragwerken", Int. Spannungsopt. Symp., Berlin, 1961 (Akademie Verlag, Berlin 1962) pp.115-126

Kuske, A.: Die Gesetzmässigkeiten der Doppelbrechung. Optik *19*, 261-272 (1962)

Kuske, A.: L'analyse des phénonènes optiques en photoélasticité a trois dimensions par la méthode du cercle de "J". Rev. Fr. Méc. *9*, 49-58 (1964)

Kuske, A.: The J-circle method. Exp. Mech. *6*, 218-224 (1966)

Kuske, A.: "Photoelastic Effect Analysed by Means of the J-Circle Method", in *Photoelastic Effect and Its Applications* (IUTAM Symposium), ed. by J. Kestens (Springer, Berlin, Heidelberg, New York 1975) pp.507-524

Menges, H.J.: Die experimentelle Ermittlung räumlicher Spannungszustände an durchsichtigen Modellen mit Hilfe des Tyndalleffektes. Z. Angew. Math. Mech. *20*, 210-217 (1940)

Richartz, M., Hsü, H.Y.: Analysis of elliptical polarization. J. Opt. Soc. Am. *39*, 136-157 (1949)

Schwieger, H.: Graphical methods for determining the resulting photoelastic effect of compound states of stress. Exp. Mech. *9*, 67-74 (1969)

Whitney, C.: Pauli-algebraic operators in polarization optics. J. Opt. Soc. Am. *61*, 1207-1213 (1971)

Wright, F.E.: A spherical projection chart for use in the study of elliptically polarized light. J. Opt. Soc. Am. *20*, 529-564 (1930)

Author Index

Page numbers in *italic* refer to the Bibliography

Subject Index

Holographic Recording Materials

Editor: **H.M. Smith**

1977. 96 figures, 17 tables. XIII, 252 pages
(Topic in Applied Physics, Volume 20)
ISBN 3-540-08293-X

Contents:
H.M. Smith: Basic Holographic Princip-
les. – **K. Biedermann:** Silver Halide Pho-
tographic Materials. – **D. Meyerhofer:**
Dichromated Gelatin. – **D.L. Staebler:**
Ferroelectric Crystals. – **R. Duncan,
D.L. Staebler:** Inorganic Photochromic
Materials. – **J.C. Urbach:** Thermoplastic
Hologram Recording. – **R.A. Bartolini:**
Photoresists. – **J. Bordogna,
S.A. Keneman:** Other Materials and
Devices.

Holographic Recording Materials is a technical treatise detailing the fundamentals
and state-of-the-art knowledge of virtually all known recording materials that have
been used for recording holograms. The
book contains authoritative dissertations
by recognized experts in the fields of
materials research and holography. Each
chapter discusses not only basic theory
and mechanisms, but methods of preparation and processing techniques. The
book is intended to be an up-to-date
compendium of virtually all available information on every recording material
suitable for holography.

Integrated Optics

Editor: **T. Tamir**

1975. 99 figures. XIII. 315 pages
(Topics in Applied Physics, Volume 7)
ISBN 3-540-07297-7

Contents:
T. Tamir: Introduction. – **H. Kogelnik:**
Theory of Dielectric Waveguides. –
T. Tamir: Beam and Waveguide Coup-
lers. – **J.M. Hammer:** Modulation and
Switching of Light in Dielectric Wave-
guides. – **F. Zernike:** Fabrication and
Measurement of Passive Components. –
E. Garmire: Semiconductor Compo-
nents for Monolithic Applications.

"The busy reader will want to get to the
'bottom line' quickly, so here it is: this is
an excellent book, a research monograph
rather than a textbook, certainly the most
valuable single volume presently available on its subject. The discussions and
references are eminently up-to-date..."
*J.J. Burke in: Journal of the Optical
Society of America*

"...The excellent presentation and the
very clear figures permit a quick first grasp
of the problems considered. The book is
well composed and presents an excellent
synthesis of the different aspects of Integrated Optics..."
C. Imbert in: Applied Physics

Springer-Verlag
Berlin
Heidelberg
New York

Optical Data Processing

Applications

Editor: **D. Casasent**

1978. 170 figures, 2 tables. XIII, 286 pages
(Topics in Applied Physics. Volume 23)
ISBN 3-540-08453-3

Contents:
D. Casasent, H. J. Caulfield: Basic Concepts. – **B. J. Thompson:** Optical Transforms and Coherent Processing Systems with Insights from Crystallography. – **P. S. Considine, R. A. Gonsalves:** Optical Image Enhancement and Image Restoration. – **E. N. Leith:** Synthetic Aperture Radar. – **N. Balasabramanian:** Optical Processing in Photogrammetry. – **N. Abramson:** Nondestructive Testing and Metrology. – **H. J. Caulfield:** Biomedical Applications of Coherent Optics. – **D. Casasent:** Optical Signal Processing.

This is an updated summary of the present status of the rapidly advancing field of optical data processing. It is intended for those researchers presently engaged in various aspects of optical processing, or for those in any type of data processing, who may be contemplating the use of optical processing techniques. Anyone who desires to know what has recently been achieved in this area - and why, will find this book indispensable.

Specific application areas in which optical processing has made significant contributions are discussed, such as crystallography, image restoration and enhancement, synthetic aperture radar, photogrammetry, metrology and non-destructive testing, biomedical processing and signal processing. An understanding is provided as to why optical processing is a viable candidate technique for each application, how fundamental optical concepts are used to define candidate solutions, and how these are subsequently refined and new ones developed for each application. The potential user is thereby shown the applicability of optical processing to his area, and methods for modifying existing optical systems to his needs are discussed.

X-Ray Optics

Applications of Solids

Editor: **H.-J. Queisser**

1977. 133 figures, 14 tables. XI, 227 pages
(Topics in Applied Physics, Volume 22)
ISBN 3-540-08462-2

Contents:
H.-J. Queisser: Introduction. – Structure and Structuring of Solids. – **M. Yoshimatsu, S. Kozaki:** High Brilliance X-Ray Sources. – **E. Spiller, R. Feder:** X-Ray Lithography. – **U. Bonse, W. Graeff:** X-Ray and Neutron Interferometry. – **A. Authier:** Section Topography. **W. Hartmann:** Live Topography.

This volume is concerned with novel developments in those fields of X-ray optics which influence research and development in the solid-state sciences. X-ray optics serves to identify defects in single crystals; topography and interferometry are treated in detail. X-rays may replace visible light in the lithography for modern semiconductor microelectronics with improved spatial resolution; one chapter covers this rapidly developing field of X-ray lithography in depth. A detailed description of the latest state-of-the art of high-brilliance X-ray sources is included.

Springer-Verlag
Berlin
Heidelberg
New York